Lecture Notes in Physics

Editorial Board

R. Beig, Wien, Austria
W. Beiglböck, Heidelberg, Germany
W. Domcke, Garching, Germany
B.-G. Englert, Singapore
U. Frisch, Nice, France
P. Hänggi, Augsburg, Germany
G. Hasinger, Garching, Germany
K. Hepp, Zürich, Switzerland
W. Hillebrandt, Garching, Germany
D. Imboden, Zürich, Switzerland
R. L. Jaffe, Cambridge, MA, USA
R. Lipowsky, Golm, Germany
H. v. Löhneysen, Karlsruhe, Germany
I. Ojima, Kyoto, Japan
D. Sornette, Nice, France, and Los Angeles, CA, USA
S. Theisen, Golm, Germany
W. Weise, Garching, Germany
J. Wess, München, Germany
J. Zittartz, Köln, Germany

The Lecture Notes in Physics

The series Lecture Notes in Physics (LNP), founded in 1969, reports new developments in physics research and teaching – quickly and informally, but with a high quality and the explicit aim to summarize and communicate current knowledge in an accessible way. Books published in this series are conceived as bridging material between advanced graduate textbooks and the forefront of research to serve the following purposes:

- to be a compact and modern up-to-date source of reference on a well-defined topic;
- to serve as an accessible introduction to the field to postgraduate students and nonspecialist researchers from related areas;
- to be a source of advanced teaching material for specialized seminars, courses and schools.

Both monographs and multi-author volumes will be considered for publication. Edited volumes should, however, consist of a very limited number of contributions only. Proceedings will not be considered for LNP.

Volumes published in LNP are disseminated both in print and in electronic formats, the electronic archive is available at springerlink.com. The series content is indexed, abstracted and referenced by many abstracting and information services, bibliographic networks, subscription agencies, library networks, and consortia.

Proposals should be sent to a member of the Editorial Board, or directly to the managing editor at Springer:

Dr. Christian Caron
Springer Heidelberg
Physics Editorial Department I
Tiergartenstrasse 17
69121 Heidelberg/Germany
christian.caron@springer-sbm.com

Radyadour Kh. Zeytounian

Topics in Hyposonic Flow Theory

Springer

Author

Professor Radyadour Kh. Zeytounian
12 Rue St Fiacre
75002 Paris
France
E-mail: zeytounian@aol.com

Radyadour Kh. Zeytounian, *Topics in Hyposonic Flow Theory*,
Lect. Notes Phys. 672 (Springer, Berlin Heidelberg 2006), DOI 10.1007/b135495

Library of Congress Control Number: 2005930476

ISSN 0075-8450
ISBN-10 3-540-25549-4 Springer Berlin Heidelberg New York
ISBN-13 978-3-540-25549-9 Springer Berlin Heidelberg New York

Springer is a part of Springer Science+Business Media
springer.com

Printed in The Netherlands

Typesetting: by the author and TechBooks using a Springer LaTeX macro package

Printed on acid-free paper SPIN: 11414346 54/TechBooks 5 4 3 2 1 0

...after subsonic, transonic, supersonic and
hypersonic flows, now the fluid dynamicists
must cope with a new challenge:
the theory of 'hyposonic fluid flows'

Preface

It is well known that the characteristic Mach number (M), the ratio of the reference constant velocity to the constant magnitude of the speed of sound, is the main dimensionless parameter which characterizes the effects of compressibility in the problem of the flow of a Newtonian fluid. Indeed, in the investigation of the flow of a compressible fluid, much of the analysis is considered in terms of the Mach number, and four cases are usually studied, namely subsonic ($M < 1$), transonic ($M \approx 1$), supersonic ($M > 1$) and hypersonic ($M \gg 1$). The main purpose of this book is to present various facets of the unsteady, very slow flow of a fluid at low Mach number which are strongly related to a fifth category of fluid flow problems, called 'hyposonic' here ($M \ll 1$). It seems that the word 'hyposonic', in relation to fluid mechanics, was mentioned for the first time in a short paper of mine ('Aerodynamics of hyposonic speeds') written at the end of 1981 and published in the Russian journal *Applied Mechanics and Technological Physics* (vol. 2, pp. 53–61, 1983), published by the Siberian branch of the Russian Academy of Sciences. That paper was, in fact, a 'syllabus' for a future book, which is the present book.

Low-Mach-number, unsteady, very slow hyposonic motion plays a dominant role in various situations, even if this has not actually been established in such a firm way as is the case for flows with a high or low Reynolds number. If hypersonic fluid flow problems are closely linked to space fluid dynamics, hyposonic fluid flow problems, in contrast, are mainly linked to the fluid dynamics of the terrestrial environmental. More precisely, we shall emphasize in the seven chapters of this book that at least five areas of Newtonian hyposonic fluid dynamics are concerned. Namely, slightly compressible unsteady external and internal aerodynamics, combustion, nonlinear acoustics, slow atmospheric motions and thermal convection are important parts of the general subject area related to hyposonic flow theory. The first three chapters of this book play the role of an 'enlarged introduction'. The next three Chaps. 4, 5 and 6, are devoted to three main areas where the theory of low-Mach-number flows has various applications, namely external and internal hyposonic aerodynamics, and slow atmospheric motions. The last Chap. 7, is a miscellany of low-Mach-number fluid problems and motions (related, for instance, to nonlinear acoustics and thermal convection). At the end of each chapter there is a list of references.

To tell the truth, the smallness of the Mach number in hyposonic flow theory is particularly significant when it occurs simultaneously with the smallness of one or several other basic dimensionless parameters in a problem of the flow of a Newtonian fluid governed by the unsteady Navier–Stokes–Fourier (NSF) equations for a viscous, compressible, heat-conducting fluid. In such cases, a fundamental role is played by similarity rules between $M \ll 1$ and the other small dimensionless parameters (for instance, the inverses of the Reynolds and Strouhal numbers, the Rossby/Kibel number, and the Froude/Boussinesq number). In particular, via these similarity rules, we also obtain various criteria for the validity and consistency of the hyposonic models derived.

A very difficult problem in hyposonic flow theory is closely linked to the filtering of fast acoustic waves in low-Mach-number unsteady flows. This filtering of 'parasite waves' is very well justified from a physical point of view for very slow motions, but changes strongly the mathematical nature of the unsteady evolution problem considered (in particular, the well-posedness is lost), and for this reason one encounters two basic problems. First, for an *external* flow, we must ask what initial conditions one may prescribe for the derived model (for instance, this concerns the 'Navier–Fourier' model for a divergenceless, temperature-dependent flow). Second, for an *internal* flow with combustion in a time-dependent domain, we must elucidate the role of the interplay with fast acoustics, since the averaged functions in the model equations are coupled to high-frequency oscillations produced by the motion of the boundary through the linear acoustic system and the averaged equations. Matching is possible in the case of an external flow, between the main Navier incompressible region and the two singular far-field and initial-time acoustic regions; unfortunately, this is not the case for an internal flow with combustion, where a family of fast times nonlinearly related to the slow evolution time of the boundary velocity must be introduced in the framework of a multiple-timescale analysis.

It is clear that the main practical goal in hyposonic flow theory is to derive various simplified, approximate but significant model problems which can be solved with less numerical and financial effort than the 'exact' full problem governed by the NSF or Euler equations for a slightly compressible fluid flow – often, time-dependent numerical compressible-flow schemes are ineffective at low Mach numbers. Concerning the rigorous mathematical theory (initiated in France by P.-L. Lions in 1993), the reader can find, in Sect. 2.6, a short comment with some recent references. In that section, we have also mentioned various references concerning numerical computations and simulations. Many problems remain unsolved (some of these problems are mentioned in the three first chapters of this book), in spite of the fact that their applications are vital to the understanding of our environment. I hope that the present book will provide a motivation for fluid dynamicists to investigate these unsolved hyposonic flow problems in depth later on. It would be a good (but difficult!)

exercise for postgraduate students to work through the various parts of this book and attempt to understand it for the purpose of applying the theory, and also to refer to the many original papers quoted. This book gives essentially an analytical view of hyposonic flow problems, but I think (and I hope) that, on the basis of the analytical results derived, it will be possible to consider various practical problems in combustion theory, in road and rail traffic aerodynamics, and in meteorology.

Many friends and colleagues have provided stimulation and encouragement that have made this book possible, but I wish particularly to thank Professor Jean-Pierre Guiraud, who has played an important role, over a period of twenty years, in the formulation and asymptotic modelling of the various low-Mach-number flow problems presented in this book. The reader should take into account that the present book is the first book devoted to hyposonic flow theory, and it is the author's hope that the various unavoidable 'deficiencies' in this book will encourage others to write new and improved books on a similar theme. After subsonic, transonic, supersonic and hypersonic flows, now the fluid dynamicists must cope with a new challenge, the theory of 'hyposonic fluid flows'. Finally, I thank Dr Christian Caron, Physics Editor, and the members of the Springer Physics Editorial Department, where the camera-ready manuscript was produced in LaTeX and reread by a native English speaker for publication in the Lecture Notes in Physics Series with the agreement of the LNP Board.

Yport/Paris
January 2005

R. Kh. Zeytounian

Contents

1

Aerothermodynamics, Nonlinear Acoustics and the Meteorological Equations

The study of low-Mach-number flows occurs in the general context of asymptotic modelling of fluid flows, and in a recent book (Zeytounian 2004, Chap. 6) an argument was given for the value of asymptotic modelling of flows at low Mach number as a effective way of building consistent models amenable to numerical simulation, because it is well known that time-dependent compressible-flow schemes often become ineffective at low Mach numbers.

Unfortunately, most fluid dynamicists believe that low-Mach-number expansions are straightforward and think that an asymptotic expansion for large or small Mach number is assuredly a regular perturbation – the latter case being the 'Janzen–Rayleigh' or 'M^2' expansion. As a matter of fact, some subtleties are involved when we consider an unsteady weakly compressible flow at $M \ll 1$, and they concern both unsteady external aerodynamics, which leads to sound generation by low-speed flows, and internal aerodynamics, which raises a number of interesting questions relating to the long-time persistence of acoustic oscillations in a cavity with a deformable wall. Many problems in internal aerodynamics have scarcely been investigated, but resonance phenomena and the formation of shock waves in closed tubes, which are mainly related to low-Mach-number asymptotics, have been investigated for various applications. However, the most intriguing features are found when one is dealing with low-Mach-number flows in the atmosphere. It is somewhat discouraging to realize that the low-Mach-number approximation looks very good from the outset, but that very few useful results have been obtained by using it. As a matter of fact, the pioneering work (see Chap. 6) on quasi-non-divergent flows involves some intriguing features tied to the phenomenon of blocking, well known in stratified flows at low Froude number, and this might explain why such flows have not been used in weather prediction. However, blocking is avoided whenever the low-Mach-number approximation is coupled with another one. When both the Rossby and the Mach numbers are low one obtains the useful quasi-geostrophic approximation, which has been systematically investigated from the point of view of asymptotic modelling. When both the Boussinesq number and the Mach number are simultaneously

R. Kh. Zeytounian: *Topics in Hyposonic Flow Theory*, Lect. Notes Phys. **672**, 1–26 (2006)
www.springerlink.com

low, via an asymptotic expansion one obtains a useful model, the Boussinesq equations. An obvious interpretation, for an Eulerian flow, is that in this way we recover Long's model for lee waves in the troposphere generated by relief, with a length scale $O(M)$ in comparison with the height of the troposphere. It is important to note also the singular features of low-Mach-number expansions in the vicinity of the initial time. As a consequence, it is necessary to derive appropriate initial-layer expansions, which may deal with adjustment processes. Furthermore, it is necessary also to investigate the singular nature of the far field in the case of external inviscid aerodynamics – the process of going to infinity cannot be exchanged, without caution, with that of letting the Mach number go to zero.

In fact, the acoustic model may be extracted from the equations for the initial and far fields. An important problem is the discussion of the matching between the various expansions and some issues related to this, namely the powers of the Mach number which occur in these expansions. A very difficult problem arises in the multiple-scaling approach to the persistence of initial acoustic oscillations in a deformable cavity, and the 'long life' of these acoustic oscillations, generated within the cavity by the initial process; how one can predict their evolution when the shape of the cavity has been deformed a great deal from the original shape?

Naturally, the ineffectiveness of time-dependent weakly compressible flow schemes occurs because a wide disparity exists (this is a stiff problem) between the timescales associated with slow convection and the fast-propagating acoustic waves. In this respect, one should notice that the filtering of acoustic waves in the compressible-fluid-flow equations changes drastically the type of the equations, and the initially well-posed Navier–Stokes–Fourier (NSF) equations for compressible, viscous flow with heat conduction often become 'ill-posed'. However, the experienced reader will be aware of the fact that the partial differential equations of fluid dynamics are not sufficient for discussing flow problems, and this seemingly anodyne remark has far-reaching consequences.

In this chapter, we present briefly the Navier–Stokes–Fourier unsteady equations governing the flow of a viscous, compressible and heat-conducting fluid. We work mainly with non-dimensional quantities, and corresponding dimensionless equations and initial and boundary conditions. Our main assumption is the smallness of the reference (constant) Mach number (M), which is the ratio of the characteristic velocity (U_{c}) for the fluid flow under consideration, to the constant speed of sound (a_{c}) at a constant reference temperature T_{c}. In low-Mach-number theory, there are difficult problems linked with the consideration of the full unsteady NSF equations for various problems in hyposonic aerodynamics, meteorology and nonlinear acoustics, and also with the formulation of realistic, physically consistent initial and boundary conditions.

An another question is related to the *validity* of the NSF equations as the Mach number tends to zero. It may be objected that the NSF equations are no longer valid in the limit considered in this book, especially not in the limit where M^2/Re tends to infinity – Re tends to zero faster than M^2. This question has beeen discussed by Lagerstrom (1964), and here we give only a brief comment. As the Mach number tends to zero, at constant free-stream velocity U_c, the speed of sound and hence the temperature T_c tend to infinity. At sufficiently high temperatures, any real gas certainly has properties which are not accounted for in the NSF equations. In fact, the asymptotic approach based on the consideration of expansions of NSF solutions for small values of M is physically significant only if there are values of M which are sufficiently large for the NSF equations to be adequate, and at the same time sufficiently small for the asymptotic method used to be approximately valid. As yet, it has not been investigated carefully whether or not such values of the parameter M exist in the cases considered in this book – but consideration of the various similarity rules, in the cases when, together with M, we have an another small or large parameter, give us the possibility to obtain some validity criteria for the approximate models derived.

Various forms of the fluid dynamic equations are presented below in this chapter to prepare us for their application in the framework of low-Mach-number flows, which is described mainly in Chaps. 4–7. Chapters 2 and 3 give an overview of the low-Mach-number theory with many references and also mention various open problems – for instance, a very challenging area in hyposonic flows is that of the aerodynamics of high-speed trains and of road vehicles.

1.1 The Dynamics of a Newtonian Fluid

Fluids may be defined as materials which continue to deform in the presence of a shear stress. In fluid mechanics, fluids are considered to be continuous, although, like any substance, they consist of discrete molecules. This approach is taken not only because of the resultant simplicity in the analysis, but also because the behaviour of the individual molecules is not usually of primary interest in technology. The average properties of the molecules in a small parcel of fluid are used as the properties of the continuous material. For example, the mass of all the molecules per unit volume of the parcel is called the density of the fluid. However, for this approach to be successful, the size of the flow system must be much larger than the mean free path of the molecules.

Classical fluids may be classified into two groups, namely *liquids* and *gases.* The density of a fluid is related to the temperature and the pressure exerted on an element of the fluid. The density of a gas varies much more readily than that of a liquid. In many cases, however, the effects of the variation in density are very small, and the density of a fluid element can be assumed to remain constant as an approximation – under this assumed condition, the flow is called (usually) an *incompressible flow* and the velocity vector $\boldsymbol{u}$ of the fluid flow is divergence-free. The forces acting on an infinitesimal fluid element may be classified into two groups: *body* forces and *surface* forces. A *body force* is one whose magnitude is proportional to the volume $\mathrm{d}V$ of the fluid element, for example a gravitational force, and the main body force to be considered in this book is the weigth due to gravity ρg, where the density ρ of a material is defined as the mass per unit volume of the material, and g is the magnitude of the gravitational acceleration (equal to 981 cm/s^2 at sea level). A *surface force* is a force acting on the surface bounding the element, exerted by the adjacent fluid elements or by solid boundaries. The force on a portion $\mathrm{d}S$ of the surface may be resolved into two components, one *normal* and the other *tangential* to the area $\mathrm{d}S$. The normal force per unit area is called the *normal stress* – this stress is compressive and is often called the *pressure*. The tangential force per unit area is called the *shear stress*, and, according to the definition of a fluid, a fluid element will continue to deform when there are shear stresses acting on its surface. The relationship between the *rate of deformation* and the shear stress defines a mechanical property of the fluid. Most common fluids deform at a rate U/d proportional to the shear stress, i.e. $\tau = \mu(U/d)$, where the coefficient μ is independent of the speed U, and d is the (macroscopic) size of the flow system. Such fluids, including air and water, are called *Newtonian* fluids. The coefficient μ is called the *dynamic viscosity* and, under ordinary pressure, μ for a Newtonian fluid varies only with temperature (at least when the density is constant).

In any case, all real fluids offer some resistance to a finite rate of deformation; however, in many cases, the shear stresses in most parts of the flow are unimportant and can be neglected in an approximate analysis. Such an idealized fluid is called a *frictionless* (i.e. inviscid, or non-viscous) *Eulerian fluid*. The ratio μ/ρ will be encountered very often and is therefore given the name *kinematic viscosity*, ν. Since U/d is the rate of change of speed with depth, we can also write $\tau = \mu\,\mathrm{d}u/\mathrm{d}y$. However, we have written the above relation for a particular flow system usually referred to as *plane Couette flow* (laminar flow where the velocity $u(y)$ varies only in the direction y (> 0), perpendicular to the direction of flow).

In a more general theory, we observe that, for the specification of a Newtonian fluid (also called a *Navier–Stokes fluid*), which is a particular case of a Stokesian fluid (see, for instance, pp. 230–239 in the pioneer review paper Serrin, 1959), it is necessary to introduce the following Cauchy–Poisson stress tensor:

$$\mathbf{T} = (-p + \lambda\nabla\,.\,\boldsymbol{v})\mathbf{I} + \mu\mathbf{D}(\boldsymbol{v})\,. \tag{1.1}$$

This is a continuous linear function of the deformation tensor $\boldsymbol{D}(\boldsymbol{v})$, defined as the symmetric tensor

$$\mathbf{D}(\boldsymbol{v}) = \nabla\boldsymbol{v} + \nabla(\boldsymbol{v})^*\,, \tag{1.2}$$

where $(\nabla\boldsymbol{v})^*$ is the transpose of the tensor $\nabla\boldsymbol{v}$. In (1.1), $\mathbf{I}$ is the unit matrix, p the pressure, $\boldsymbol{v}$ the vector velocity and ∇ the gradient operator. The fully general expression was given by Poisson (1831), and Saint-Venant (1843) proposed a version of (1.1) in the special case when the two viscosity coefficients

μ and λ are both constant:

$$\mu \equiv \mu_0 = \text{const}, \quad \text{such that } \lambda_0 + \frac{2}{3}\mu_0 \equiv 0 \,. \tag{1.3}$$

The relation

$$\lambda = -\frac{2}{3}\mu \,, \tag{1.4}$$

used for a thermally perfect monatomic gas (in the general case for non-constant μ and λ), was proposed by Stokes (1845) and is called the *Stokes relation.* However, in general (for polyatomic gases), μ and λ are both functions of the temperature T and the density ρ. In fact, rigorously, from a microscopic point of view, for the determination of μ and λ it is necessary to consider the kinetic theory of gases (see, for instance, the book by Cercignani, Illner and Pulvirenti 1994 for a modern exposition). For a *thermally perfect gas*, we can write two simple *equations of state*, namely

$$p = R\rho T \quad \text{and } e = C_v T \,, \tag{1.5}$$

where $R = C_p - C_v$ is the gas constant, $e(\equiv e(T))$ is the internal energy, and C_p and C_v $(\equiv \mathrm{d}e/\mathrm{d}T)$ are the specific heat capacities at constant pressure and constant volume. The ratio C_p/C_v is the (constant) adiabatic exponent and is denoted by γ. The *equation of motion* for the velocity vector $\boldsymbol{v}(\boldsymbol{x}, t)$, where t is the time and $\boldsymbol{x}$ is the vector position in 3D space, is a direct consequence of the Cauchy (1828) stress principle, which can be written simply in the following elegant form:

$$\rho \frac{\mathrm{D}\boldsymbol{v}}{\mathrm{D}t} = \nabla \cdot \mathbf{T} \,, \tag{1.6}$$

where $\mathrm{D}/\mathrm{D}t \equiv \partial/\partial t + \boldsymbol{v} \cdot \nabla$. The evolution equation (1.6) for $\boldsymbol{v}$ can be written if we neglect the body force per unit mass and the Coriolis force, which are driving forces in the meteorological equations governing atmospheric slow motions (see Sect. 7).

In the general case of a viscous, compressible, non-barotropic/baroclinic motion, the coefficients of the viscosity λ and μ are functions of the thermodynamic state (for instance, of ρ and T) and it is assumed that the following thermodynamic relations hold good for the (specific) entropy S and pressure p introduced in (1.1):

$$S = -\frac{\partial \Psi}{\partial T} \quad \text{and} \quad p = -\frac{\partial \Psi}{\partial (1/\rho)} \,, \tag{1.7a}$$

where

$$\Psi(\rho, T) = e(\rho, T) - TS(\rho, T) \,, \tag{1.7b}$$

$e(\rho, T)$ being the (specific) internal energy such that

$$\mathrm{d}e = T\,\mathrm{d}S - p\,\mathrm{d}\left(\frac{1}{\rho}\right) \,. \tag{1.8}$$

As a first companion equation associated with the equation of motion (1.6) for $\boldsymbol{v}$, we have the following compressible ($0 < \rho \neq$ const) *equation of continuity*:

$$\frac{\partial \rho}{\partial t} + \boldsymbol{v} \,.\, \nabla \rho + \rho \nabla \,.\, \boldsymbol{v} = 0 \,, \tag{1.9}$$

which is an evolution equation for ρ.

For the internal energy $e(\rho, T)$, we have the following *energy evolution equation*:

$$\rho \left[\frac{\partial e}{\partial t} + \boldsymbol{v} \,.\, \nabla e \right] = \mathbf{T} : \mathbf{D}(\boldsymbol{v}) - \nabla \,.\, \boldsymbol{q} \,, \tag{1.10}$$

where $\boldsymbol{q}$ is the heat conduction vector, assumed to satisfy Fourier's law (Fourier 1833):

$$\boldsymbol{q} = -k \nabla T \,, \tag{1.11}$$

where k is a (positive, since $\boldsymbol{q} \,.\, \nabla T \leq 0$) scalar function of $|\nabla T|$ and of the thermodynamic variables. However, in place of (1.10), it is possible also to write the energy equation with the help of the specific entropy S. Namely,

$$\rho T \left[\frac{\partial S}{\partial t} + \boldsymbol{v} \,.\, \nabla S \right] = -\nabla \,.\, \boldsymbol{q} + \Phi(\boldsymbol{v}) \,, \tag{1.12a}$$

where

$$\Phi(\boldsymbol{v}) = \lambda (\nabla \,.\, \boldsymbol{v})^2 + \frac{\mu}{2} [\mathbf{D}(\boldsymbol{v}) : \mathbf{D}(\boldsymbol{v})] \,, \tag{1.12b}$$

denotes the rate at which work is done by the viscous stress per unit volume (the dissipation function) and is a positive definite form in D_{ij}, the components of $\mathbf{D}(\boldsymbol{v})$ defined by (1.2). We observe that the components of $\mathbf{T} : \mathbf{D}(\boldsymbol{v})$ are $T_{ij} D_{ij}$, where T_{ij} are the components of the tensor $\mathbf{T}$. The three equations (1.6), (1.9) and (1.10), for v, ρ and e, together with (1.1), (1.2), (1.7) and (1.11), are called the Navier–Stokes–Fourier aerothermodynamic equations below, when we write these equations in dimensionless form.

We shall make a final observation concerning the NSF equations derived above. In accordance with Lagerstrom (1964), the concepts of density, internal energy and velocity in continuous media are defined phenomenologically without reference to the molecular structure. Two concepts from equilibrium thermodynamics, namely pressure and temperature, are assumed to be defined for non-equilibrium states by their dependence on density and internal energy as observed in the equilibrium state. The stress tensor and heat conduction vector are assumed to be highly specialized linear isotropic functions of the velocity gradient and temperature gradient, respectively. These assumptions are all of a physical nature, and underlie any mathematical statements about the behaviour of the hypothetical (weakly compressible) fluids treated in this book. In other words, these assumptions are always part of any theoretical analysis based on the above NSF equations, whether or not the assumptions in question are valid for the real fluids encountered in nature – it may be possible that the NSF equations are inadequate under certain extreme conditions.

1.2 The Dimensionless NSF Equations

In this book, we work mainly with dimensionless equations, and for this purpose we take into account that in aerothermodynamic applications, we usually have reference characteristic values for length (L_c), time (t_c), velocity (U_c) and thermodynamic functions (p_c, T_c, ρ_c), such that $p_c = R\rho_c T_c$, for a thermally perfect gas, and reference values for the dissipative coefficients (μ_c, λ_c, k_c) where $\nu_c \equiv \mu_c/\rho_c$. If we choose $t_c \equiv L_c/U_c$ then the Strouhal number $St = (L_c/t_c U_c)$ to be equal to 1, and, in the dimensionless NSF aerothermodynamic equations (see the system of equations (1.29) in Sect. 1.5), the following three main reduced dimensionless parameters appear:

$$\gamma M^2 = \frac{U_c^2}{RT_c}, \quad Re = \frac{L_c U_c}{\nu_c}, \quad Pr = \frac{C_p \mu_c}{k_c} . \tag{1.13a}$$

Here we assume that in the equation of motion for $\boldsymbol{v}$ (1.6), gravity and the Coriolis forces are absent, as is the case in the usual aerothermodynamic applications. In (1.13a), M is the Mach number, Re the Reynolds number and Pr the Prandtl number.

In nonlinear acoustics (see, for instance, Coulouvrat, 1992), in place of the Reynolds number $Re = L_c U_c/\nu_c$, it is usually more judicious to introduce the *acoustic Reynolds number* Re_{ac}, where, in place of L_c, we use the inverse of $k = \omega/c_0$, the reference wave number, where ω is the reference frequency and $c_0 = \sqrt{\gamma R T_0}$ is the speed of sound for a thermally perfect gas, T_0 being a reference constant temperature. As the reference velocity U_c, we choose the velocity amplitude at the (acoustic) source, and $M = U_c/c_0$ plays the role of an *acoustic Mach number*. In this case we obtain the following relation between Re_{ac}, Re and M:

$$Re_{ac} = \frac{\rho_c c_0^2}{\mu_c \omega} = \frac{L_c c_0}{\nu_c} = \frac{U_c L_c}{M \nu_c} = \frac{Re}{M} . \tag{1.13b}$$

In Sect. 1.6 (see the dimensionless system of equations (1.37)), we introduce the NSF acoustic equations for small acoustic Mach number, keeping only linear (acoustic) terms and terms proportional to M, and assuming $Re_{ac} = O(1)$.

Meteorological applications require another approach, and in Sect. 1.7 we consider the dimensionless meteorological (see the dimensionless systems of equations (1.40), (1.54) and (1.57)), where the main driving forces are gravity and the Coriolis acceleration. In Zeytounian (1990, 1991), the reader can find a detailed derivation of these three systems of equations. The dimensionless meteorological equations can be derived if we introduce two length scales, horizontal (L_c) and vertical (H_c), and consider also the mean radius of the rotating Earth a_0. As a consequence, we have the possibility to form two dimensionless parameters from these three length scales, the *sphericity parameter* (δ) and the *hydrostatic parameter* (ε):

$$\delta = \frac{L_c}{a_0}, \quad \varepsilon = \frac{H_c}{L_c} . \tag{1.13c}$$

On the other hand, the vector of rotation of the Earth $\boldsymbol{\Omega} = \Omega_0 \boldsymbol{e}$ is directed from south to north parallel to the axis joining the poles, such that $\boldsymbol{e} = \boldsymbol{k} \sin\phi + \boldsymbol{j} \cos\phi$, where ϕ is the algebraic latitude, $\boldsymbol{k}$ is the unit vector directed towards the zenith (in the direction opposite to $\boldsymbol{g} = -g\boldsymbol{k}$, the force of gravity) and $\boldsymbol{j}$ is the unit vector directed towards the north. If $f^0 = 2\Omega_0 \sin\phi^0$ is the Coriolis parameter, then the *Rossby number* is

$$Ro = \frac{Uc}{f^0 L_c} , \tag{1.13d}$$

where ϕ^0 is a reference latitude. In place of Ro, we often use instead the *Kibel number*,

$$Ki = St\, Ro = \frac{1}{t_c f^0} . \tag{1.13e}$$

The effect of gravity, represented by its magnitude g, is governed by the *Froude number*,

$$Fr = \frac{U_c}{\sqrt{gH_c}} . \tag{1.13f}$$

Finally, we observe that we have the following relation between Fr, M and γ:

$$\frac{\gamma M^2}{Fr^2} \equiv Bo = \frac{H_c}{RT_c/g} . \tag{1.13g}$$

Here, Bo is the *Boussinesq number*, the ratio of the vertical atmospheric length H_c to $H_s \equiv RT_c/g$, where H_s is the altitude of a homogeneous atmosphere. In this book, the main formalism used is asymptotic modelling, and our main small parameter is the characteristic (constant) Mach number $M \ll 1$. As a consequence, we investigate very slow motions where $U_c \ll \sqrt{gRT_c} \approx 300\,\text{m/sec}$, which we term 'hyposonic'. However, in various situations we also take low or high values of the other dimensionless parameters into account, and write associated similarity rules (see, for instance, Chaps. 2 and 3, where various cases are considered).

1.3 From the Euler Incompressible Equations to the Navier Equations

1.3.1 The Euler Incompressible, Non-Viscous Equations

The first mathematical description of the motion of an 'ideal' – non-viscous and incompressible – fluid was formulated by Euler (1755) as a statement of Newton's second law of motion (discovered and published by Newton (1687) in his *Philosophiae Naturalis Principia Mathematica*) applied to a fluid moving under the influence of an internal force known as the pressure gradient (without any infuence of viscosity). The *Euler equations* governing the time evolution of the velocity vector field $\boldsymbol{v}(\boldsymbol{x},t)$ and the (scalar) pressure $p(\boldsymbol{x},t)$ of an incompressible, non-viscous (ideal) fluid have the following very simple dimensionless form (where $\rho \equiv 1$):

$$\frac{\partial \boldsymbol{v}}{\partial t} + (\boldsymbol{v}\,.\,\nabla)\boldsymbol{v} = -\nabla p \quad (\boldsymbol{x} \in \Re^n, t > 0)\,, \tag{1.14}$$

$$\nabla\,.\,\boldsymbol{v} = 0\,, \tag{1.15}$$

$$\boldsymbol{v}(\boldsymbol{x},0) = \boldsymbol{v}^0(\boldsymbol{x})\,, \tag{1.16}$$

where $\boldsymbol{v}^0(\boldsymbol{x})$ is a given divergence-free vector field. In the above, we have restricted our attention to incompressible, non-viscous fluids (where $\rho \equiv 1$, $Re \equiv \infty$ and thermodynamics is absent), filling all of $\Re^n$, where n is the spatial dimension, which we take to be 2 or 3. However, it is necessary to observe that in the Euler equation (1.14), the term ∇p on the right-hand side is not an unknown quantity in the initial-value problem (1.14)–(1.16). In fact, ∇p is the force term acting on the particles of fluid, allowing them to move as freely as possible but in a way compatible with the constraint of a divergence-free velocity vector $\boldsymbol{v}$ (1.15). The truncated equation (which is a Lagrangian invariant for $\boldsymbol{v}$)

$$\frac{\mathrm{D}\boldsymbol{v}}{\mathrm{D}t} \equiv \frac{\partial \boldsymbol{v}}{\partial t} + (\boldsymbol{v}\,.\,\nabla)\boldsymbol{v} = 0\,, \tag{1.17}$$

where $\mathrm{D}/\mathrm{D}t$ is the material derivative (along the trajectories of the fluid flow), admits solutions violating the condition $\nabla\,.\,\boldsymbol{v} = 0$ at $t > 0$, even if the divergence of $\boldsymbol{v}$ vanishes at time $t = 0$. In fact, the pressure can be determined (via a Poisson elliptic equation) after we have found the velocity field $\boldsymbol{v}$ (the solution of the above Euler incompressible, non-viscous equations (1.14)). First, we take the divergence of (1.14)

$$\Delta p = -\nabla\,.\,[(\boldsymbol{v}\,.\,\nabla)\boldsymbol{v}]\,. \tag{1.18}$$

Knowing $\boldsymbol{v}$, we can then find p by solving (1.18) with the Neumann boundary condition

$$\frac{\partial p}{\partial n} = -(\nabla\,.\,[(\boldsymbol{v}\,.\,\nabla)\boldsymbol{v}])\,.\,\boldsymbol{n} \tag{1.19}$$

in a domain with a boundary, after the Euler equation (1.14) has been projected onto the outward unit normal $\boldsymbol{n}$, to the boundary. As a consequence of the above, it is sufficient to consider the Euler incompressible equation (1.14) in terms of the vorticity ω:

$$\omega = \nabla \wedge \boldsymbol{v} \Rightarrow \frac{\partial \omega}{\partial t} + (\omega\,.\,\nabla)\omega = (\omega\,.\,\nabla)\boldsymbol{v}, \quad \text{with } \nabla\,.\,\boldsymbol{v} = 0\,. \tag{1.20}$$

In the recent book Zeytounian (2002a), the reader can find a theory of non-viscous Eulerian, incompressible and compressible fluid flows with various applications.

1.3.2 The Navier Equation

The above Euler equation (1.14), while important theoretically, omits the effect of friction (since in reality a fluid is viscous) and brings about, as pointed out by d'Alembert (1752), 'a singular paradox which I leave to geometricians to explain'. According to Stewartson (1981), the mathematical conjectures associated with d'Alembert's paradox can be stated most clearly in terms of the Navier equations for an incompressible but viscous fluid.

To incorporate friction, the French mathematician–engineer Navier published a paper (Navier, 1822) containing a derivation of the equation of motion for a viscous fluid in which he included the effects of attraction and repulsion between neighbouring molecules. From purely theoretical considerations, he derived the following modification of the Euler equation (1.14) – the *Navier equation*. In dimensionless form, this Navier equation is (again $\rho \equiv 1$)

$$\frac{\partial \boldsymbol{v}}{\partial t} + (\boldsymbol{v} \cdot \nabla)\boldsymbol{v} = \frac{1}{Re}\Delta \boldsymbol{v} - \nabla p \quad (\boldsymbol{x} \in \Re^n, t > 0) \ . \tag{1.21a}$$

The Navier equation, together with (1.15) and (1.16), repeated below,

$$\nabla \cdot \boldsymbol{v} = 0 \ , \tag{1.21b}$$

$$\boldsymbol{v}(\boldsymbol{x}, 0) = \boldsymbol{v}^0(\boldsymbol{x}) \ , \tag{1.21c}$$

forms the *Navier initial-value problem.*

We observe that, for Navier, the coefficient in front of $\Delta \boldsymbol{v}$ (the Laplacian of $\boldsymbol{v}$) in (1.21a), written for the dimensional velocity $\boldsymbol{v}$ and pressure p, was simply a function of the molecular spacing, to which he attached no particular physical significance.

Rigorously speaking, in the above Navier model (1.21), the pressure p and the initial velocity $\boldsymbol{v}^0(\boldsymbol{x})$ are different from what they are in the Euler model (1.14)–(1.16), since the Euler model is obtained when Re tends to infinity (vanishing-viscosity limit) with t and $\boldsymbol{x}$ fixed, and is singular close to a boundary: the no-slip condition for the velocity vector is replaced by a slip condition on the boundary for the Euler model (see Sect. 1.5.2).

Navier's seminal paper (Navier 1822) was presented at the Académie des Sciences in Paris and was well received. He was elected a member of the Académie, in the mechanics section, in January 1824. In a recent paper by Cannone and Friedlander (2003), the reader can find various pieces of information concerning the scientific activities of Navier and a few observations about the (open!) challenges that the Navier initial-value problem (1.21) poses to mathematicians (see, for instance, Constantin (1995)).

In a recent book (Zeytounian 2004, Chap. 8), the reader can find a fluid-dynamic point of view of some aspects of the mathematically rigorous theory of the viscous (mainly Navier) equations, with various recent references.

In view of the above short discussion, it seems very judicious to associate the name of Navier with the viscous equation (1.21a) and the equation for an incompressible/divergence-free velocity vector (1.21b), which, in the case of vanishing viscosity ($(1/Re) \downarrow 0$), lead again to the Euler incompressible equations (1.14) and (1.15) for a non-viscous incompressible (ideal) fluid.

1.4 From the Euler Compressible Equations to the Navier–Stokes Equations

1.4.1 The Euler Compressible Equations

In the compressible case (where the density ρ is a function of $(t, \boldsymbol{x})$), we take into account the compressible continuity equation (1.9) for the density $\rho(t, \boldsymbol{x})$, in place of $\nabla \cdot \boldsymbol{v} = 0$, and we write the *Euler compressible equation*

$$\rho \left[\frac{\partial \boldsymbol{v}}{\partial t} + (\boldsymbol{v} \cdot \nabla) \boldsymbol{v} \right] = -\nabla p \,. \tag{1.22}$$

According to Euler (1755), if we add to (1.9) and (1.22) the specifying equation

$$p = g(\rho) \,, \tag{1.23}$$

which gives the relation between the pressure and the density, we have three equations (a closed system) which include all of the theory of the motion of ideal fluids.

With this formulation, Euler believed that he had reduced fluid dynamics, in principle, to a mathematical–physical science. However, it is important to note that, in reality, and unfortunately, (1.23) is not an equation of state, but specifies only the particular type of motion (called 'barotropic') under consideration, and in this case the fluid is called an 'elastic' fluid. In general, it is necessary to write a relation between three thermodynamic functions, characterizing the fluid in baroclinic motion (a *trivariate* fluid). An example of such an equation of state is (if, together with p and ρ, we choose the absolute temperature T), (1.5), or in a more general case

$$p = P(\rho, T) \,, \tag{1.24}$$

where T must be a solution of the energy equation (1.10) if we take into account (1.7), (1.8) and (1.11).

1.4.2 The Navier–Stokes Equations

Now, by analogy with the Navier incompressible, viscous case, we consider the viscous, compressible barotropic case with (1.23). Here, in place of (1.21b), we write the compressible continuity equation (1.9), and from the Euler compressible equation (1.22) we pass to the following *Navier–Stokes* viscous, compressible, barotropic dimensionless equation, using (1.6) and (1.1):

$$\rho\left[\frac{\partial \boldsymbol{v}}{\partial t}+(\boldsymbol{v}\,.\,\nabla)\boldsymbol{v}\right]=-\nabla p+\frac{1}{Re}\{\nabla\,.\,[\mu\mathbf{D}(\boldsymbol{v})]+\nabla(\lambda\nabla\,.\,\boldsymbol{v})\}\,. \tag{1.25a}$$

Here μ and λ are the shear/dynamic viscosity and the second coefficient of viscosity, respectively, and $\mathbf{D}(\boldsymbol{v})$ is the deformation tensor given by (1.2). The closed system of three equations formed by (1.25a) together with

$$\frac{\partial \rho}{\partial t}+\boldsymbol{v}\,.\,\nabla\rho+\rho\nabla\,.\,\boldsymbol{v}=0\,, \tag{1.25b}$$

$$p=g(\rho)\,, \tag{1.25c}$$

where μ and λ are assumed to be functions only of the density ρ, is referred to here as the system of Navier–Stokes equations for the three fields $\boldsymbol{v}$, p and ρ, which are strongly related to a *barotropic, compressible, viscous* motion. These Navier–Stokes (NS) equations are mainly considered by applied mathematicians in their rigorous mathematical analyses (see, for instance the book by Lions 1998). Unfortunately, a barotropic *viscous* motion does not have any physical reality, because viscosity always generates entropy (baroclinity!) and, in particular, the rigorous mathematical results concerning the 'incompressible limit' related to the limiting process $M \downarrow 0$ seem very questionable (see also the short paper by Leray 1994).

1.5 The Navier–Stokes–Fourier Equations

With (1.11), we can derive from (1.12a) the following equation for the entropy $S(t, \boldsymbol{x})$, written for dimensional variables:

$$T\rho\left[\frac{\partial S}{\partial t}+(\boldsymbol{v}\,.\,\nabla)S\right]=\nabla\,.\,(k\,\mathrm{grad}\,T)+\Phi(\boldsymbol{v})\,, \tag{1.26}$$

where $\Phi(\boldsymbol{v})$ is given as a function of the velocity vector $\boldsymbol{v}$, according to (1.12b). We see that even with $k \equiv 0$ (no heat conduction) and $\lambda \equiv 0$ (zero second coefficient of viscosity), we have $T\rho\ DS/Dt = (\mu/2)[\mathbf{D}(\boldsymbol{v}):\mathbf{D}(\boldsymbol{v})] > 0$, and the NSF fluid flow is not isentropic (or barotropic); obviously, we do not have the possibility to recover the NS equations (1.25).

In technological applications, it is often assumed that the fluid is a thermally perfect viscous, heat-conducting gas, such that (using dimensional quantities, see also (1.5))

$$S=C_v\log\left(\frac{p}{\rho^{\gamma}}\right),\quad \gamma=\frac{C_p}{C_v},\quad \mathrm{d}e=C_v\,\mathrm{d}T,\quad \frac{p}{\rho}=RT\,, \tag{1.27}$$

and, according to the kinetic theory of gases, $\gamma = 5/3$ for a monoatomic gas, for which the Stokes relation (1.4) is obeyed. With (1.27), we obtain for $\boldsymbol{v}$, p, ρ and T the following usual NSF closed system (with dimensional quantities) of equations, for a thermally perfect gas:

$$
\begin{aligned}
\frac{\mathrm{D}\rho}{\mathrm{D}t} + \rho \nabla \cdot \boldsymbol{v} &= 0\,, \\
\rho \frac{\mathrm{D}\boldsymbol{v}}{\mathrm{D}t} &= -\nabla p + \nabla \cdot [2\mu \mathbf{D}(\boldsymbol{v})] + \nabla(\lambda \nabla \cdot \boldsymbol{v})]\,, \\
\rho C_v \frac{\mathrm{D}T}{\mathrm{D}t} + p\nabla \cdot \boldsymbol{v} &= \nabla \cdot (k\,\nabla T) + \Phi(\boldsymbol{v})\,, \\
p &= R\rho T
\end{aligned}
\tag{1.28}
$$

For our purposes in this book, we write the following *dimensionless* NSF aerothermodynamic equations for the dimensionless velocity vector $\boldsymbol{v}$ and the dimensionless thermodynamic functions p, ρ and T, where we use the same notation for dimensionless quantities as for dimensional quantities:

$$
\begin{aligned}
\frac{\mathrm{D}\rho}{\mathrm{D}t} + \rho \nabla \cdot \boldsymbol{v} &= 0\,, \\
\rho \frac{\mathrm{D}\boldsymbol{v}}{\mathrm{D}t} + \frac{1}{\gamma M^2}\nabla p &= \frac{1}{Re}\{\nabla \cdot [2\mu \mathbf{D}(\boldsymbol{v})] + \nabla(\lambda \nabla \cdot \boldsymbol{v})\}\,, \\
\rho \frac{\mathrm{D}T}{\mathrm{D}t} + (\gamma - 1)p\nabla \cdot \boldsymbol{v} &= \frac{\gamma}{Pr\,Re}\nabla \cdot (k\,\nabla T) + \frac{\gamma - 1}{Re}\gamma M^2 \Phi(\boldsymbol{v})\,, \\
p &= \rho T\,.
\end{aligned}
\tag{1.29}
$$

The dimensionless parameters γM^2, Re and Pr in the system (1.29) are given by (1.13a).

1.5.1 The Eulerian Case: $Re \equiv \infty$

In the case of the motion of a non-viscous, adiabatic, but compressible (Eulerian) fluid, we obtain from the NSF equations (1.29), when $Re \equiv \infty$, the dimensionless *Euler compressible, baroclinic, adiabatic equations:*

$$
\begin{aligned}
\frac{\mathrm{D}\rho}{\mathrm{D}t} + \rho \nabla \cdot \boldsymbol{v} &= 0\,, \\
\rho \frac{\mathrm{D}\boldsymbol{v}}{\mathrm{D}t} + \frac{1}{\gamma M^2}\nabla p &= 0\,, \\
\frac{\mathrm{D}T}{\mathrm{D}t} + (\gamma - 1)T\,\nabla \cdot \boldsymbol{v} &= 0\,,
\end{aligned}
\tag{1.30a}
$$

and

$$
p = \rho T\,. \tag{1.30b}
$$

These Euler equations (1.30) apply to a non-viscous, thermally perfect gas.

A Simple Model for Combustion

We note that the system of Euler inviscid, compressible equations (1.30) is often used in low-Mach-number combustion modelling (see, for instance, Majda 1984, Majda and Sethian 1985, and Dwyer 1990), where the third equation in (1.30a), for the temperature T, is replaced (see below) by (1.32b). In an idealized special case, where we assume that there are only two species present, the unburnt gas and the burnt gas, and that the unburnt gas is converted to burnt gas through a one-step, irreversible, exothermic chemical reaction with Arrhenius kinetics mechanism, we have the following two *continuum chemistry equations* (Majda 1984, pp. 5–7):

$$\frac{\partial(\rho Z)}{\partial t} + \nabla \cdot (\rho \boldsymbol{v} Z) = -\rho W(Z, T) \tag{1.31a}$$

and

$$W(Z, T) = K \exp\left(-\frac{A_0}{T}\right) Z . \tag{1.31b}$$

These are used with the first two Euler non-viscous, compressible equations in (1.30a). In (1.31), Z is the mass fraction of unburnt gas, so that $1 - Z$ is the mass fraction of burnt gas. As regards the classical Euler equations (1.30), a crucial modification is the definition of the internal energy of the mixture, written as (to within a constant)

$$e(p, \rho, Z) = Z e_{\mathrm{u}}(p, \rho) + (1 - Z) e_{\mathrm{b}}(p, \rho) , \tag{1.31c}$$

where, for simplicity, we assume that $e_{\mathrm{u}}(p, \rho) = C_v(T - T_0) + q_0$, where q_0 is the normalized energy of formation at the reference temperature T_0 for the unburnt gas, and $q_0 > 0$ for an exothermic reaction. Since $e_{\mathrm{b}}(p, \rho) = e_{\mathrm{u}}(p, \rho) - q_0$, then

$$e(p, \rho, Z) = C_v T + q_0 Z , \tag{1.31d}$$

if the inessential constant shift is ignored in e. Finally, the system of non-viscous adiabatic (Euler) equations of the simplified theory of combustion, where the diffusion of species is neglected, consists of the first two Euler dimensionless equations in (1.30a), for ρ and $\boldsymbol{v}$, and the equation of state (1.30b), with two dimensionless equations:

$$\frac{\mathrm{D}(\rho Z)}{\mathrm{D}t} = -K \rho \exp\left(-\frac{A_0}{T}\right) Z , \tag{1.32a}$$

$$\rho \frac{\mathrm{D}T}{\mathrm{D}t} + (\gamma - 1) p \nabla \cdot \boldsymbol{v} = (\gamma - 1) K q_0 \rho \exp\left(-\frac{A_0}{T}\right) Z . \tag{1.32b}$$

Here, the chemical activity has been modelled by a one-step, irreversible, overall chemical reaction of Arrhenius type with an activation energy A_0 and a pre-exponential factor K (for a particular definition of the parameter K, see

Harten, Kapila and Matkowsky 1984). In a recent paper by Matalon and Metzener (1997), a nonlinear evolution equation that describes the propagation of a premixed flame in a closed tube has been derived from conservation equations expressing conservation of the mixture's mass, momentum and energy and a mass balance for the deficient reactant, very similar to our equations (1.32), but those authors take into account the thermal conductivity and the molecular diffusivity associated with the deficient reactant in the mixture (which is, in general, temperature-dependent). In Majda (1984, p. 7), the reader can find the full system of dissipative equations of combustion theory, where the viscosity, heat conduction and species diffusion are taken into account (if these coefficients are non-dimensionalized with macroscopic length and time scales, they are typically of the order of 10^{-3} or 10^{-4}).

We observe that flame fronts are significant waves, which do not move at the characteristic speeds associated with the inviscid combustion equations defined above. These waves are generally slow-moving fronts with velocities of 2 or $3\,\mathrm{m/sec}$, and the wave speeds are governed by a subtle balance between the reaction term $W(Z,T)$ defined by (1.31b) and the diffusion coefficients. In fact, even for detonation fronts, the fast-moving supersonic wave fronts in combustion theory, which are analogous to shock fronts, the inviscid shock-layer approximation (known as the Chapman–Jouget theory in this context – see, for instance, Courant and Friedrichs 1949), which works so well for ideal gas dynamics, can break down completely in describing the actual solutions of our equations (see Gardner 1981 and Majda 1981). In fact, for fluid dynamics, with very general equations of state, the standard conclusions of shock-layer analysis which allow one to ignore the detailed effects of diffusion can be completely wrong when heat conduction dominates over viscosity. The above remarks (due to A. Majda) indicate the subtlety involved in ignoring dissipative mechanisms in the theory of conservation laws, and, in fact, in more complex models of physical systems such as the dissipative equations of combustion theory, a more detailed assessment of their effect on macroscopic length scales is always needed. The simpler inviscid combustion equations described above might be an excellent approximation in a given regime (usually a detonation regime) in practice, but one always needs to assess the effects of these diffusion mechanisms in that regime through careful analysis of simpler problems.

Finally, we note that a consistent mathematical derivation, by asymptotic methods, of a diffusional–thermal-type model from the general equations of combustion was presented by Matkowsky and Sivashinsky (1979), and, in a sequel, Harten and Matkowsky (1982) included the effects of a weak coupling. For more recent references concerning flames and detonation theories, the reader can consult the book edited by Godreche and Manneville (1991).

1.5.2 Initial and Boundary Conditions

Obviously, the dimensionless systems of the NSF equations (1.29) or the Euler equations (1.30) are not sufficient for the analysis of aerodynamic problems, and both these systems must be solved under convenient initial and boundary conditions. The detailed discussion of these conditions is postponed mainly to Chaps. 4, 5 and 6, devoted to external and internal low-Mach-number aerodynamics and to slow atmospheric motions respectively.

Here, for the moment, we observe only that it is necessary to assume that at the initial time $t \leq 0$, the fluid functions $\boldsymbol{v}$, ρ and T, which are solutions of the NSF evolution equations (1.29) or the Euler equations (1.30), are known, namely

$$t \leq 0: \quad \boldsymbol{v} = \boldsymbol{v}^0(\mathbf{x}), \quad \rho = \rho^0(\boldsymbol{x}), \quad T = T^0(\boldsymbol{x}) , \tag{1.33}$$

where $\boldsymbol{v}^0(\boldsymbol{x}), \rho^0(\boldsymbol{x})$ and $T^0(\boldsymbol{x})$ are given data (however, in general, these data can be different for the NSF and Euler systems).

As *boundary conditions*, for the full NSF equations (1.29), it is necessary to assume a *no-slip condition* for the velocity vector $\boldsymbol{v}$ on the wall Σ in contact with the viscous fluid, and also to write a condition for the temperature T, since the viscous fluid is also a heat conductor. If $\boldsymbol{U}_\Sigma(t, \boldsymbol{P}), \boldsymbol{P} \in \Sigma$, is the velocity of the point $\boldsymbol{P}$ of a solid body Ω, in motion, so that $\partial\Omega \equiv \Sigma$, then we write the following as the no-slip condition for a moving solid, impermeable wall:

$$\boldsymbol{v} - \boldsymbol{U}_\Sigma = 0 \quad \text{on } \Sigma . \tag{1.34a}$$

For the Euler system (1.30), in place of (1.34a) we write only the usual *slip condition*

$$\boldsymbol{n} \,.\, (\boldsymbol{v} - \boldsymbol{U}_\Sigma) = 0 \quad \text{on } \Sigma , \tag{1.34b}$$

where $\boldsymbol{n}$ denotes the unit outward normal vector to the boundary $\partial\Omega$. The second boundary condition for the NSF system (1.29), for the dimensionless temperature $T(t, \boldsymbol{x})$, is often written as

$$k\frac{\partial T}{\partial n} + h(T - T_\Sigma) = \Theta \quad \text{on } \Sigma , \tag{1.35}$$

where $T_\Sigma > 0$ and Θ are known functions and $h > 0$ is a given constant.

In the case of external aerodynamics in an unbounded fluid domain, when we consider the flow of a fluid around a bounded solid body we assume also that we have suitable *behaviour conditions* for the velocity vector $\boldsymbol{v}$ and the thermodynamic functions p, ρ, T at infinity, when $|\boldsymbol{x}| \uparrow \infty$. But it is necessary to note that the region close to the initial time and the far-field region, as a consequence of the filtering of acoustic waves by the limiting process $M \downarrow 0$ when t and $\boldsymbol{x}$ are fixed, are both singular regions in the framework of an external-aerodynamics problem (see Chap. 4). On the other hand, in the case of an internal-aerodynamics problem (in a bounded domain), the process of filtering of acoustic waves (generated, for instance, at the initial time in a time-dependent container with a slow boundary velocity) is a very difficult problem, which is considered in Chap. 5. Finally, we note that in a paper by Oliger and Sundström (1978), initial-boundary-value problems for several systems of partial differential equations in fluid dynamics are discussed.

1.6 The Equations of Nonlinear Acoustics

The order of magnitude of the acoustic perturbations is measured by the acoustic Mach number M, which is the ratio of the velocity U_c at the source (the characteristic velocity chosen by the observer) to the sound velocity c_0. This Mach number is very small compared with unity, i.e. $U_c \ll c_0$. Indeed, the values of the Mach number collected from data from various experiments in water and biological media, listed in Appendix 1 of Coulouvrat (1992), show that M is actually extremely small, not much more than 10^{-4} even for the most strongly nonlinear situation. If the source signal is not too wide-band then the timescale is determined by the main frequency $f = \omega/2\pi$ of the source signal, i.e. $t_c = 1/\omega$. From the wave number k associated with the reference frequency $\omega, k = \omega/c_0$, the reference length scale is deduced, i.e. $L_c = 1/k$. The acoustic Reynolds number Re_{ac} defined by (1.13b) compares the orders of magnitude of the propagation terms (the left hand-side of (1.37)) and the viscous dissipative terms. For the frequencies and media commonly used in nonlinear acoustics, Re_{ac} is always very large compared with unity (see, again, Appendix 1 of Coulouvrat 1992). This means that the sound attenuation is small over a few wavelengths, or that the medium is weakly dissipative (at the chosen frequency). Consequently, (1.37) turns out to be a perturbation problem with two small parameters, M and $1/Re_{ac}$. In Sect. 7.1, a similarity rule

$$\frac{1}{Re_{ac}} \approx M \tag{1.36a}$$

is assumed to be obeyed in the framework of the asymptotic derivation of the KZK model equation. On the other hand, in problems of nonlinear acoustics, it is necessary to take into account the following similarity rule, between $M \ll 1$ and $St \gg 1$:

$$M\,St = 1\,. \tag{1.36b}$$

In such a case, the reference acoustic time is just

$$t_c = \frac{L_c}{c_0} \equiv t_{ac} = \frac{1}{\omega}\,. \tag{1.36c}$$

In this case, in place of the dimensionless thermodynamic functions p, ρ, T, we introduce the corresponding perturbations π, ω, θ, such that

$$p = 1 + M\pi, \quad \rho = 1 + M\omega, \quad T = 1 + M\theta\,. \tag{1.36d}$$

If we now assume that the dissipative coefficients are constant, then, with the help of the relation (1.13b) and assuming a constant bulk viscosity $\mu_{vc} = \lambda_c + (\frac{2}{3})\mu_c$, we can write the following 3D system of unsteady dimensionless NSF acoustic equations for $\boldsymbol{v}$ and π, ω, θ:

$$\frac{\partial \omega}{\partial t} + \nabla \cdot \boldsymbol{v} = -M \nabla \cdot (\omega \boldsymbol{v}) ,$$

$$\frac{\partial \boldsymbol{v}}{\partial t} + \frac{1}{\gamma} \nabla \pi = \frac{1}{Re_{\mathrm{ac}}} \left\{ \nabla^2 \boldsymbol{v} + \left[\frac{1}{3} + \frac{\mu_{\mathrm{vc}}}{\mu_{\mathrm{c}}} \right] \nabla (\nabla \cdot \boldsymbol{v}) \right\}$$
$$-M \left[\omega \frac{\partial \boldsymbol{v}}{\partial t} + (\boldsymbol{v} \cdot \nabla) \boldsymbol{v} \right] + O(M^2) , \tag{1.37a}$$

$$\frac{\partial \theta}{\partial t} + (\gamma - 1) \nabla \cdot \boldsymbol{v} = \frac{\gamma}{Pr\, Re_{\mathrm{ac}}} \nabla^2 \theta \tag{1.37b}$$
$$+M \left\{ \frac{\gamma(\gamma-1)}{Re_{\mathrm{ac}}} \left[\frac{1}{2} [\boldsymbol{D}(\boldsymbol{v}) \cdot \boldsymbol{D}(\boldsymbol{v})] + \left[\frac{1}{3} + \frac{\mu_{\mathrm{vc}}}{\mu_{\mathrm{c}}} \right] (\nabla \cdot \boldsymbol{v})^2 \right] \right.$$
$$\left. - \left[\omega \frac{\partial \theta}{\partial t} + \boldsymbol{v} \cdot \nabla \theta + (\gamma - 1) \pi \nabla \cdot \boldsymbol{v} \right] \right\} + O(M^2) , \tag{1.37c}$$

$$\pi - (\theta + \omega) = M \theta \omega . \tag{1.37d}$$

These 'dominant' acoustic equations (1.37) are the main starting point for the derivation of various model (approximate low-Mach-number) equations in nonlinear acoustics (see, for instance, Chap. 8 in the book Zeytounian 2002b). In particular, in Chap. 7 of this book, which presents a miscellany of low-Mach-number flows, we derive from this system of equations (1.37), (see Sect. 7.1) the parabolic KZK-model single equation for small acoustic Mach number M, first derived by Kuznetsov (1970) and Zabolotskaya and Khokhlov (1969). In the one-dimensional case, we recover from this parabolic KZK equation the well-known equation obtained by Burgers (1948) illustrating the theory of turbulence (see Sect. 4.2).

1.7 The Meteorological Equations

Chapter 6 of this book is devoted to slow atmospheric motions, the Mach number for these 'slow motions' being a very natural small parameter. Unfortunately, the limiting process $M \downarrow 0$ for the meteorological NSF equations including the Coriolis and gravitational forces is strongly singular, and a number of unsolved problems arise. Here, for the moment, we formulate first a system of atmospheric Eulerian adiabatic equations, with a term proportional to Fr^{-2} driving the effect of gravity, for the velocity vector and the thermodynamic perturbations; these equations are very judicious for an asymptotic derivation of the Boussinesq non-viscous, adiabatic model equations. Then we consider the Kibel primitive equations, which are derived in the framework of the long-wave approximation ($\varepsilon \downarrow 0$) when dissipative effects (proportional to Re^{-1}) are neglected. However, for large Reynolds number ($Re \uparrow \infty$) and small $\varepsilon \downarrow 0$, with a judiciously chosen similarity rule (see (1.56)), it is possible to take into account the dominant viscous and conduction terms in the long-wave approximate equations and derive a hydrostatic dissipative system of equations.

As the subject is quite extensive, attention is focused in Chap. 6 first, on the asymptotic justification of the Boussinesq approximation for the inviscid problem of lee waves and also for the free-circulation problem, when the dissipative and Coriolis effects are taken into account. A second atmospheric low-Mach-number problem, related to the Rossby/Kibel number Ki, is considered when $Ki = O(1)$ and when $Ki \ll 1$ such that $Ki \approx M$.

1.7.1 The Eulerian Inviscid, Adiabatic Meteorological Equations

For further analysis of atmospheric motions, it is very useful to postulate the existence of a 'standard atmosphere' (which is assumed to exist on a day-to-day basis) in the form of a basic thermodynamic reference situation (designated by p_s, ρ_s, T_s); these latter quantities are functions solely of the standard altitude, denoted by z_s, which has $H_s = RT_s(0)/g$ as a characteristic length scale. In fact, if the relative velocities are small, then the 'true' atmospheric pressure will be only slightly disturbed from the basic static value $p_s(z_s)$, defined by the dimensionless relations (written with the same notation)

$$\frac{\mathrm{d}p_s}{\mathrm{d}z_s} + Bo\ \rho_s = 0 \tag{1.38a}$$

and

$$\rho_s(z_s) = \frac{p_s(z_s)}{T_s(z_s)} , \tag{1.38b}$$

where $T_s(z_s)$ is assumed to be known a priori (in the adiabatic case considered). The basic standard state is assumed to be known, although its determination from first principles of thermodynamics requires the consideration of the mechanism of radiative transfer in the atmosphere (see, for instance, Sect. 1.4 of the book Kibel 1963 for a pertinent discussion of this problem). To describe the true atmospheric motions, which represent departures from the static standard state, we introduce the perturbation of the pressure π, a perturbation of the density ω and a perturbation of the temperature θ, defined by the dimensionless relations

$$p = p_s(z_s)(1+\pi), \quad \rho = \rho_s(z_s)(1+\omega), \quad T = T_s(z_s)(1+\theta) , \tag{1.39}$$

where $z_s = Bo\, z$, in dimensionless form. In this case, using (1.39) we can write, in place of the Euler equations (1.30), the following equations for the dimensionless velocity components (denoted by u, v, w) and the perturbations θ, π, ω (we use the same notation, for simplicity):

$$\frac{\mathrm{D}\omega}{\mathrm{D}t} + (1+\omega)\left(\frac{\partial u}{\partial x} + \frac{\partial v}{\partial y} + \frac{\partial w}{\partial z}\right) = (1+\omega)\frac{Bo}{T_s(z_s)}[1 - \Gamma_s(z_s)]w , \tag{1.40a}$$

$$(1+\omega)\frac{\mathrm{D}u}{\mathrm{D}t} + \frac{T_s(z_s)}{\gamma M^2}\frac{\partial \pi}{\partial x} = 0 , \tag{1.40b}$$

$$(1+\omega)\frac{\mathrm{D}v}{\mathrm{D}t}+\frac{T_{\mathrm{s}}(z_{\mathrm{s}})}{\gamma M^2}\frac{\partial\pi}{\partial y}=0\,, \tag{1.40c}$$

$$(1+\omega)\frac{\mathrm{D}w}{\mathrm{D}t}+\frac{T_{\mathrm{s}}(z_{\mathrm{s}})}{\gamma M^2}\frac{\partial\pi}{\partial z}-(1+\omega)\frac{Bo}{\gamma M^2}\theta=0\,, \tag{1.40d}$$

$$(1+\omega)\frac{\mathrm{D}\theta}{\mathrm{D}t}-\frac{\gamma-1}{\gamma}\frac{\mathrm{D}\pi}{\mathrm{D}t}+(1+\pi)\frac{Bo}{T_{\mathrm{s}}(z_{\mathrm{s}})}\left[\frac{\gamma-1}{\gamma}-\Gamma_{\mathrm{s}}(z_{\mathrm{s}})\right]w=0\,, \tag{1.40e}$$

$$\pi=\omega+(1+\omega)\theta\,. \tag{1.40f}$$

In the second dimensionless equation in (1.30a), for $\boldsymbol{v}$, we take into account the hydrostatic dimensionless equation $\mathrm{d}p_{\mathrm{s}}/\mathrm{d}z_{\mathrm{s}}+\rho_{\mathrm{s}}=0$, and also the term proportional to $+Fr^{-2}\rho$, and we neglect the influence of the Coriolis acceleration. We observe that we have as dimensional reference thermodynamic functions

$$p_{\mathrm{c}}=p_{\mathrm{s}}(0),\quad \rho_{\mathrm{c}}=\rho_{\mathrm{s}}(0),\quad T_{\mathrm{c}}=T_{\mathrm{s}}(0),\quad \varepsilon\equiv 1\quad (L_{\mathrm{c}}\equiv H_{\mathrm{c}})\,. \tag{1.41}$$

In (1.40a) and (1.40e), the dimensionless standard temperature gradient (in the adiabatic case),

$$\Gamma_{\mathrm{s}}(z_{\mathrm{s}})=-\frac{\mathrm{d}T_{\mathrm{s}}}{\mathrm{d}z_{\mathrm{s}}}\,, \tag{1.42}$$

which characterizes, together with (1.38), the standard atmosphere, is assumed to be known. We observe that, if in the above equations (1.40a–e) we have (since we are working with dimensionless quantities) $T_s(0)\equiv 1$, then $\Gamma_s(0)$ is different from zero. We stress again that the above dimensionless Euler meteorological inviscid adiabatic equations (1.40a–f), for u,v,w,π,ω and θ, are *exact* equations, and this remark is important for a consistent asymptotic derivation of the inviscid Boussinesq-model equations for slow atmospheric motions (see Chap. 6) when Bo and M are both small parameters.

1.7.2 The f^0-Plane Approximation Equations for Atmospheric Inviscid Motion

For atmospheric motions, it is helpful to employ spherical coordinates λ,ϕ,r, and to let u,v,w denote the corresponding relative velocity components in these directions, i.e. with increasing azimuth (λ), latitude (ϕ) and radius (r). However, for our purposes here, it is more judicious to introduce the following transformations:

$$x=a_0\cos\phi^0\,\lambda,\quad y=a_0(\phi-\phi^0),\quad z=r-a_0\,, \tag{1.43a}$$

where ϕ^0 is the reference latitude. From (1.43a), it follows immediately that

$$\frac{\partial}{\partial\lambda}=a_0\cos\phi^0\frac{\partial}{\partial x},\quad \frac{\partial}{\partial\phi}=a_0\frac{\partial}{\partial y},\quad \frac{\partial}{\partial r}=\frac{\partial}{\partial z}\,. \tag{1.43b}$$

The origin of the above right-handed curvilinear coordinate system lies on the surface of the Earth (for flat ground, where $r = a_0$) at latitude ϕ^0 and longitude $\lambda = 0$. We assume therefore that the atmospheric motion occurs in a mid-latitude region, distant from the equator, around some central latitude ϕ^0, and therefore that $\sin\phi^0$, $\cos\phi^0$ and $\tan\phi^0$ are all of order unity.

Although x and y are, in principle, new longitude and latitude coordinates in terms of which the basic fluid-dynamic equations (here the Euler equations) may be rewritten without approximation, they have obviously been introduced in the expectation that for small $\delta(= L_c/a_0)$ they will be the Cartesian coordinates of the 'f^0-plane approximation'. However, when $\delta \ll 1$, the quantity $\delta y'$ in $\phi = \phi^0 + \delta y$ will be small (we assume that $y = O(1)$ when δ tends to zero) compared with unity, and $\cos\phi$ $\sin\phi$ and $\tan\phi$ can be expanded in a convergent Taylor series about the reference latitude ϕ^0:

$$\cos\phi = \cos\phi^0[1 - \delta\tan\phi^0\, y + O(\delta^2)]\,,$$
$$\sin\phi = \sin\phi^0\left[1 + \delta\left(\frac{1}{\tan\phi^0}\right) y + O(\delta^2)\right]\,,$$
$$\tan\phi = \tan\phi^0\left[1 + \delta\left(\frac{1}{\cos\phi^0\sin\phi^0}\right) y + O(\delta^2)\right]\,. \tag{1.44}$$

As a consequence, when $\delta \ll 1$ and for any value of $\varepsilon = O(1)$, the Coriolis acceleration becomes

$$2\Omega \wedge \boldsymbol{u} = f^0 U_c\left\{u\boldsymbol{j} - v\boldsymbol{i} + \frac{1}{\tan\phi^0}(\varepsilon w\boldsymbol{i} - u\boldsymbol{k})\right\} + O(\delta)\,, \tag{1.45}$$

where $f^0 = 2\Omega_0\sin\phi^0$ is the constant *Coriolis parameter*. In the f^0-plane approximation, in place of the Euler vectorial dimensionless equation:

$$\rho\left\{\frac{\mathrm{D}\boldsymbol{v}}{\mathrm{D}t} + 2\Omega \wedge \boldsymbol{v}\right\} + \frac{1}{\gamma M^2}\nabla p + \frac{1}{Fr^2}\rho = 0 \tag{1.46}$$

for atmospheric inviscid motion, when we take into account gravitational and Coriolis forces, we can write three scalar dimensionless f^0-plane approximation equations for the dimensionless components, u, v, w of the velocity vector in a moving frame which is characterized by the vector of rotation of the Earth Ω. Namely,

$$\rho\left\{\frac{\mathrm{D}u}{\mathrm{D}t} - \frac{1}{Ro}\left[v - \frac{\varepsilon}{\tan\phi^0}w\right]\right\} + \frac{1}{\gamma M^2}\frac{\partial p}{\partial x} = 0\,, \tag{1.47a}$$
$$\rho\left\{\frac{\mathrm{D}v}{\mathrm{D}t} + \frac{1}{Ro}u\right\} + \frac{1}{\gamma M^2}\frac{\partial p}{\partial y} = 0\,, \tag{1.47b}$$
$$\rho\left\{\varepsilon^2\frac{\mathrm{D}w}{\mathrm{D}t} - \frac{1}{Ro}\frac{\varepsilon}{\tan\phi^0}u\right\} + \frac{1}{\gamma M^2}\frac{\partial p}{\partial z} + \frac{1}{Fr^2}\rho = 0\,, \tag{1.47c}$$

where $\mathrm{D}/\mathrm{D}t \equiv \partial/\partial t + u\,\partial/\partial x + v\,\partial/\partial y + w\,\partial/\partial z$, and Fr^2 is the square of the Froude number formed from the vertical length scale H_c (see (1.13f)).

Together with the three Euler f^0-plane approximation adiabatic equations (1.47a–c), we have the following two dimensionless evolution equations for the density (continuity) and temperature (adiabaticity):

$$\frac{\mathrm{D}\rho}{\mathrm{D}t} + \rho\left(\frac{\partial u}{\partial x} + \frac{\partial v}{\partial y} + \frac{\partial w}{\partial z}\right) = 0\,, \tag{1.47d}$$

$$\gamma\rho\frac{\mathrm{D}T}{\mathrm{D}t} - (\gamma - 1)\frac{\mathrm{D}p}{\mathrm{D}t} = 0\,. \tag{1.47e}$$

We also have the dimensionless equation of state (we assume that dry atmospheric air is a thermally perfect gas)

$$p = \rho T\,. \tag{1.47f}$$

1.7.3 The Kibel Primitive Equations in the Pressure Coordinates System

When we consider the Kibel limiting process

$$\varepsilon \text{ tends to zero with } t, x, y, z, Ro, \gamma M^2 \text{ and } Fr^2 \text{ fixed}\,, \tag{1.48}$$

then, in place of (1.47c) for w, we obtain (see (1.13g)) the hydrostatic balance

$$\frac{\partial p}{\partial z} + Bo\,\rho = 0 \Rightarrow \frac{\partial}{\partial z} = -Bo\frac{p}{T}\frac{\partial}{\partial p}\,, \tag{1.49}$$

and we have the possibility to use the pressure coordinates system. In the system of pressure coordinates, the usual horizontal coordinates (x, y) denote a point's position projected onto a horizontal plane, but the pressure p denotes its location on the vertical axis. The 'horizontal' derivatives of a variable are its differences between one point and another on the same isobaric (p = constant) surface divided by the corresponding differences of position projected onto a horizontal plane. The 'vertical' derivative of a variable is its derivative with respect to pressure, but is directed along the vertical axis. The dependent variables are unaffected by the coordinate transformation except that p (pressure) becomes one of the independent variables, the height z of any particular isobaric surface becomes a dependent variable, and the role of the vertical speed of the air $\mathrm{d}z/\mathrm{d}t$ is taken over by ϖ, the total derivative of the pressure:

$$\varpi = \frac{\mathrm{D}p}{\mathrm{D}t}\,. \tag{1.50a}$$

In this case, in place of the continuity equation (1.47d), we can write

$$\frac{\partial u}{\partial x} + \frac{\partial v}{\partial y} + \frac{\partial \varpi}{\partial p} = 0\,, \tag{1.50b}$$

and note that the primitive equations of the tangential (f^0-plane) atmospheric motions written in the isobaric system do not involve the density.

Naturally, the main advantage of the pressure coordinates stems from the fact that, according to the hydrostatic approximation (1.49), the atmosphere, for the synoptic-scale tangential motions, is in hydrostatic equilibrium to leading order, and this guarantees that the pressure is a monotonic function of the altitude z for x and y fixed.

Consequently, the change is mathematically sound. The procedure for transforming derivatives in the coordinates (x, y, z) into derivatives in the coordinates (x, y, p) is very simple. In dimensionless form, we have

$$\begin{aligned} \frac{\partial}{\partial x} &= \frac{\partial}{\partial x} + Bo\frac{p}{T}\frac{\partial H}{\partial x}\frac{\partial}{\partial p} \, , \\ \frac{\partial}{\partial y} &= \frac{\partial}{\partial y} + Bo\frac{p}{T}\frac{\partial H}{\partial y}\frac{\partial}{\partial p} \, , \\ \frac{\partial}{\partial z} &= -Bo\frac{p}{T}\frac{\partial}{\partial p} \end{aligned} \tag{1.51}$$

and

$$\frac{\mathrm{D}}{\mathrm{D}t} = \frac{\partial}{\partial t} + \boldsymbol{v}_{\mathrm{T}} \, . \, \boldsymbol{D} + \varpi\frac{\partial}{\partial p} \, , \tag{1.52}$$

where $\boldsymbol{v}_T = (u, v)$ is the horizontal velocity vector $(\boldsymbol{v} = \boldsymbol{v}_T + w\boldsymbol{k}), \boldsymbol{D} = (\partial/\partial x, \partial/\partial y)$ is the horizontal gradient operator, and

$$z = H(t, x, y, p) \tag{1.53a}$$

is a dependent variable, the local height of an isobaric surface above flat ground. We note that in place of the vertical component of the velocity w, we have the pseudo-vertical component

$$\varpi = Bo\, p\left[\frac{\partial H}{\partial t} + \boldsymbol{v}_{\mathrm{T}} \, . \, \boldsymbol{D}H - w\right] \, . \tag{1.53b}$$

Summarizing the simplifications attached to the use of pressure coordinates, we can rewrite the complete set of the resulting tangential primitive Kibel equations in the following form, in accordance with (1.47) and (1.51)–(1.53):

$$\begin{gathered} \boldsymbol{D} \, . \, \boldsymbol{v}_{\mathrm{T}} + \frac{\partial \varpi}{\partial p} = 0 \, , \\ \frac{\partial \boldsymbol{v}_{\mathrm{T}}}{\partial t} + (\boldsymbol{v}_{\mathrm{T}} \, . \, \boldsymbol{D})\boldsymbol{v}_{\mathrm{T}} + \varpi\frac{\partial \boldsymbol{v}_{\mathrm{T}}}{\partial p} + \frac{1}{Ro}(\boldsymbol{k} \wedge \boldsymbol{v}_{\mathrm{T}}) + \frac{Bo}{\gamma M^2}\boldsymbol{D}H = 0 \, , \\ T = -Bo\, p\frac{\partial H}{\partial p} , \\ \frac{\partial T}{\partial t} + \boldsymbol{v}_{\mathrm{T}} \, . \, \boldsymbol{D}T + \left[\frac{\partial T}{\partial p} - \frac{\gamma - 1}{\gamma}\frac{T}{p}\right]\varpi = 0 \, . \end{gathered} \tag{1.54}$$

In Chap. 6, devoted to slow (low-Mach-number) atmospheric motions, the Kibel equations (1.54) are used mainly in the derivation of the equations of the quasi-non-divergent model. The quasi-geostrophic model (when Ki is also small) is derived from a *dissipative system* of equations, derived below when we take (1.56) into account.For the above non-viscous adiabatic system (1.54), we have, according to (1.53b), the following slip condition on flat ground:

$$\varpi = \rho\, Bo \left[\frac{\partial H}{\partial t} + \boldsymbol{v}_{\mathrm{T}} \,.\, \boldsymbol{D} H \right] , \quad \text{on } H = 0 \,. \tag{1.55}$$

1.7.4 The Hydrostatic Dissipative System of Equations

For large Reynolds number and small ε, such that

$$\varepsilon^2\, Re \equiv Re_{\perp} = O(1) \,, \tag{1.56}$$

we can derive a *hydrostatic dissipative* system of equations. In this case, in place of the above system of adiabatic non-viscous equations (1.47), we obtain a system of hydrostatic dissipative dimensionless equations which take into account, in the framework of the hydrostatic long-wave approximation, the viscous and non-adiabatic effects. These equations are

$$\rho \left\{ \frac{\mathrm{D}u}{\mathrm{D}t} - \frac{1}{Ro} v \right\} + \frac{1}{\gamma M^2} \frac{\partial p}{\partial x} = \frac{1}{Re_{\perp}} \frac{\partial^2 u}{\partial z^2} ,$$

$$\rho \left\{ \frac{\mathrm{D}v}{\mathrm{D}t} + \frac{1}{Ro} u \right\} + \frac{1}{\gamma M^2} \frac{\partial p}{\partial y} = \frac{1}{Re_{\perp}} \frac{\partial^2 v}{\partial z^2} ,$$

$$\frac{\partial p}{\partial z} + Bo\, \rho = 0 \,,$$

$$\frac{\mathrm{D}\rho}{\mathrm{D}t} + \rho \left(\frac{\partial u}{\partial x} + \frac{\partial v}{\partial y} + \frac{\partial w}{\partial z} \right) = 0 \,,$$

$$\rho \frac{\mathrm{D}T}{\mathrm{D}t} - \frac{\gamma - 1}{\gamma} \frac{\mathrm{D}p}{\mathrm{D}t} = \frac{1}{Pr\, Re_{\perp}} \frac{\partial^2 T}{\partial z^2} + (\gamma - 1) \frac{M^2}{Re_{\perp}} \left[\left(\frac{\partial u}{\partial z} \right)^2 + \left(\frac{\partial v}{\partial z} \right)^2 \right] ,$$

$$p = \rho T \,, \tag{1.57}$$

where $\mathrm{D}/\mathrm{D}t \equiv \partial/\partial t + u\, \partial/\partial x + v\, \partial/\partial y + w\, \partial/\partial z$. The above equations (1.57) are used as the starting equations when, for instance, we consider the influence of a localized thermal non-homogeneity on a flat ground surface $z = 0$. In this case, we write as the thermal dimensionless boundary condition

$$T = 1 + \tau_0 \Theta(t, x, y) \quad \text{on } z = 0 \,, \tag{1.58}$$

where Θ is a known function decribing the temperature field on the ground in a localized region characterized by a reference horizontal length scale l_{c}, in the proximity of some origin (0, 0). In (1.58), the thermal parameter

$$\tau_0 = \frac{\Delta T_0}{T_s(0)} \tag{1.59}$$

is obviously, for the usual atmospheric thermal problem, a small parameter. More precisely, in the above dimensionless system of equations (1.57), the reference scales for x and y are just l_c, while for z the 'natural' vertical reference scale is

$$h_c = R\frac{\Delta T_0}{g} \ll H_s\ , \tag{1.60}$$

where ΔT_0 is the rate of change of temperature associated with Θ. With the scales l_c and h_c, the parameter $\varepsilon = h_c/l_c$ is always a small parameter, since $l_c \approx 10^5$ m (because, in such a case $Ro \approx 1$) and $\Delta T_0 \approx 10°$ C.

In Chap. 6, the above equations (1.57), with (1.58), are used as starting (exact) equations for the derivation of a low-Mach-number model problem for breezes, in the framework of a hydrostatic dissipative Boussinesq system of equations.

References

J. M. Burgers, A mathematical model illustrating the theory of turbulence. Adv. Appl. Mech., **1** (1948), 171–199.

M. Cannone and S. Friedlander, Navier: blow-up and collapse. Notices Amer. Math. Soc., **50** (2003), 7–13.

A. Cauchy, Ex. de Math. 3 = Oeuvres (2), **8** (1828), 195–226, 227–252, 253–277.

C. Cercignani, R. Illner and M. Pulvirenti, *The Mathematical Theory of Dilute Gas.* AMS 106, Springer, New York, 1994.

P. Constantin, A few results and open problems regarding incompressible fluids. Notices Amer. Math. Soc., **42** (1995), 658–663.

F. Coulouvrat, On the equations of nonlinear acoustics. J. Acoustique, **5** (1992), 321–359.

R. Courant and K. O. Friedrichs, *Supersonic Flow and Shock Waves.* Springer, New York, 1949.

J. Le R. d'Alembert, Essai d'une nouvelle théorie de la résistance des fluides 1752.

H. A. Dwyer, Calculation of low Mach number flows. AIAA J. **28** (1990), 98–105.

L. Euler, Principes généraux du mouvement des fluides. Mém. Acad. Sci. Berlin, **11** (1755), 274–315.

J. Fourier, Mémoire d'analyse sur le mouvement de la chaleur dans les fluides. Mém. Acad. Roy. Sci. Inst. France (2), **12** (1833), 507–530.

R. A. Gardner, On the detonation of a combustible gas. Preprint, 1981.

C. Godreche and P. Manneville, *Hydrodynamics and Nonlinear Instabilities.* Cambridge University Press, Cambridge, 1991.

A. van Harten and B. J. Matkowsky, A new model in flame theory. SIAM J. Appl. Math., (1982).

A. van Harten, A. K. Kapila and B. J. Matkowsky, Acoustic coupling of flames. SIAM J. Appl. Math., **44**(5) (1984), 982–995.

I. A. Kibel, *An Introduction to the Hydrodynamical Methods of Short Period Weather Forecasting.* Macmillan, London, 1963 (English translation from Russian edition, Moscow, 1957).

V. P. Kuznetsov, Equations of nonlinear acoustics. Sov. Phys. Acoust., **16** (1970), 467–470.

P. A. Lagerstrom, Laminar flow theory. In: *Theory of Laminar Flows*, edited by F. K. Moore. Princeton University Press, Princeton, NJ, 1964, pp. 61–83.

J. Leray, Aspects de la mécanique théorique des fluides. La Vie des Sciences, C. R. Acad. Sci., Série. Gén., **11** (1994), No. 4, 287–290.

P.-L. Lions, *Mathematical Topics in Fluid Mechanics*, Vol. 2: *Compressible Models.* Oxford Lecture Series in Mathematics and its Applications, No. 10, Clarendon Press, Oxford, 1998.

A. Majda, A qualitative model for dynamic combustion. SIAM J. Appl. Math., **41** (1981), 70–93.

A. Majda, *Compressible Fluid Flow and Systems of Conservation Laws in Several Space Variables.* Springer, New York, 1984.

A. Majda and J. Sethian, The derivation and numerical solution of the equations for zero Mach number combustion. Combust. Sci. Technol., **42** (1985), 185–205.

M. Matalon and P. Metzener, The propagation of premixed flames in closed tubes. J. Fluid Mech., **336** (1997), 331–350.

B. J. Matkowsky and C. I. Sivashinsky, An asymptotic derivation of two models in flame theory associated with the constant density approximation. SIAM J. Appl. Math., **37** (1979), 686.

C. L. M. H. Navier, Mémoire sur les lois du mouvement des fluides. Mém. Acad. Sci. Inst. France, **6** (1822), 389–440.

I. Newton, *Philosophia Naturalis Principia Mathematica.* London, 1687.

J. Oliger and A. Sundström, Theoretical and practical aspects of some initial boundary value problems in fluid dynamics. SIAM J. Appl. Math., **35**(1) (November 1978), 419–446.

S. D. Poisson. J. Ecole Polytech., **13**(20) (1831), 1–174.

A. J. B. de Saint-Venant, C. R. Acad. Sci. **17** (1843), 277–292.

J. Serrin, *Mathematical Principles of Classical Fluid Mechanics.* Handbuch der Physik,Vol.VIII/1, Springer, Berlin, 1959, pp. 125–263.

K. Stewartson, D'Alembert paradox. SIAM Rev., **23** (1981), 308–345.

G. G. Stokes, On the theories of the internal friction of fluids in motion. Trans. Cambridge Philos. Sci., **8** (1845).

E. A. Zabolotskaya and R. V. Khokhlov, Quasi-plane waves in the nonlinear acoustics of confined beams. Sov. Phys. Acoust., **15** (1969), 35–40.

R. Kh. Zeytounian, *Asymptotic Modeling of Atmospheric Flows.* Springer, Heidelberg, 1990.

R. Kh. Zeytounian, *Meteorological Fluid Dynamics.* Lecture Notes in Physics, Vol. m5, Springer, Heidelberg, 1991.

R. Kh. Zeytounian, *Theory and Applications of Non-viscous Fluid Flows.* Springer, Heidelberg, 2002a.

R. Kh. Zeytounian, *Asymptotic Modelling of Fluid Flow Phenomena.* Fluid Mechanics and Its Applications, Vol. 64, Series Ed. R. Moreau, Kluwer Academic, Dordrecht, 2002b.

R. Kh. Zeytounian, *Theory and Applications of Viscous Fluid Flows.* Springer, Heidelberg, 2004.

2

The Many Faces of the Asymptotics of Low-Mach-Number Flows

It has long been known (Janzen 1913 and Rayleigh 1916) that asymptotic methods provide a sound way of describing incompressible aerodynamics (hydrodynamics) in terms of a low-Mach-number ($M \ll 1$) flow, and a comprehensive discussion of this topic was given by Imai (1957), as far as steady flows of an inviscid fluid are concerned. A rigorous mathematical treatment has been provided by Klainerman and Majda (1982) and Ebin (1982), although the physical situation dealt with was rather simple, the physical space being devoid of material boundaries, and this, unfortunately, is often the case in rigorous formal mathematical investigations by applied mathematicians (see, for instance, Lions 1998). Obviously, the case of a steady inviscid potential external flow, linked to the 'Janzen–Rayleigh' ('M^2') expansion, is rather straightforward conceptually and provides a high degree of achievement, and this is well demonstrated by Van Dyke (1998). But this result is also somewhat misleading, and might give the erroneous impression that the low-Mach-number expansion is very straightforward. As a matter of fact, we see, on the contrary, that the low-Mach-number asymptotics of fluid-dynamic equations are often very singular, and some subtleties are involved when we consider unsteady (slow) motions of a weakly compressible fluid. These subtleties concern, according to Zeytounian and Guiraud (1984), mainly unsteady external and internal aerodynamics (see Chaps. 4 and 5) and slow atmospheric motions (analysed in Chap. 6), but also nonlinear acoustics, combustion theory and thermal convection (see Chap. 7, where some problems are investigated). Indeed, understanding the role of the interplay with fast acoustics has an enormous relevance in the case of combustion in confined domains (considered in Chap. 5) for the development of a consistent low-Mach-number model (using averaged equations) – in particular, the fast acoustics have an effect on the pressure which would be felt by a gauge, and this pressure is not related to the mean, averaged, motion only (see, for instance, Sect. 5.3).

R. Kh. Zeytounian: *Topics in Hyposonic Flow Theory*, Lect. Notes Phys. **672**, 27–43 (2006)
www.springerlink.com

2.1 External Aerodynamics

In external aerodynamics, we are confronted first with sound generation by low-speed flows (initiated by Lighthill 1952); see, for example, Lauvstad (1968), Crow (1970), Viviand (1970) and Obermeier (1976) for asymptotic approaches.

2.1.1 Far Field

More precisely, the process of going to infinity (to the 'far field', $|\boldsymbol{x}| \uparrow \infty$) cannot be exchanged without caution with the process of letting the Mach number go to zero. For this purpose, it is necessary to consider the following *outer-far-field* limiting process:

$$\lim{}^{\text{Off}} = \{M \downarrow 0, \text{ with time and space variables } t \text{ and } \boldsymbol{\xi} = M\boldsymbol{x} \text{ fixed}\} . \tag{2.1}$$

The consistency of the matched-asymptotic-expansion (MAE) method, in relation to the singular far-field outer region, is based on the matching of outer-far-field limiting process (2.1) with the main Navier limiting process

$$\lim{}^{\text{Nav}} = \{M \downarrow 0, \text{ with space–time variables } (\boldsymbol{x}, t) \text{ fixed}\} , \tag{2.2}$$

via

$$\lim_{|\xi|\downarrow 0}(\text{Lim}^{\text{Off}}) = \lim_{|\boldsymbol{x}|\uparrow\infty}(\text{Lim}^{\text{Nav}}) , \tag{2.3}$$

and was extensively investigated by Tracey (1988) and Leppington and Levine (1987). It was pushed up to $O(M^6)$ in the main incompressible (Navier) region, checking consistency with the outer region, in Sery Baye's thesis (Sery Baye 1994).

2.1.2 Initial-Time Region and Adjustment Problem

On the other hand, close to the initial time ($t = 0$), as a consequence of the filtering of acoustic waves by the main (Navier) limiting process (2.2), it is necessary to consider an *adjustment* acoustic problem, linked to the inner acoustic limiting process in time

$$\lim{}^{\text{Ac}} = \{\text{M} \downarrow 0, \text{ with time and space variables } \tau = \frac{t}{M} \text{ and } \boldsymbol{x} \text{ fixed}\} , \tag{2.4}$$

to find a consistent initial condition for the Navier divergenceless velocity vector, by matching of (2.4) with (2.2):

$$\lim_{\tau\uparrow\infty}(\text{Lim}^{\text{Ac}}) = \lim_{t\downarrow 0}(\text{Lim}^{\text{Nav}}) . \tag{2.5}$$

In Zeytounian (2000), with the help of the scattering theory of Wilcox (1975) for the d'Alembert (acoustic wave) equation in the exterior domain, a Neumann problem for the Laplace equation has been derived from an unsteady adjustment acoustic problem (significant close to the initial time), which gives the possibility to determine the initial condition (at $t = 0$) for the Navier divergence-free velocity vector $\boldsymbol{v}_{\mathrm{N}}(t, \boldsymbol{x}) = \lim^{\mathrm{Nav}} \boldsymbol{v}(t, \boldsymbol{x}; M)$.

In such a case, using the pseudo-pressure

$$p_{\mathrm{N}}(t, \boldsymbol{x}) = \lim^{\mathrm{Nav}} \left[\left(1 - \frac{p(t, \boldsymbol{x}; M)}{M^2} \right) \right] , \tag{2.6}$$

we derive the *Navier viscous model* for $\boldsymbol{v}_{\mathrm{N}}(t, \boldsymbol{x})$ and $p_{\mathrm{N}}(t, \boldsymbol{x})$, with a consistent asymptotically well-balanced divergenceless initial condition at time $t = 0$. In fact, a more complete approximate model can be derived from the full NSF unsteady-equations model if, together with the main limiting process (2.2), the above two local limiting process (2.1) and (2.4) are considered.

2.1.3 Navier–Fourier Model Problem

More precisely, from a starting initial-boundary-value problem for the full 3D NSF equations governing the unsteady motion of a viscous, compressible, heat-conducting fluid (see the dimensionless equations (1.29)), we are able to derive an approximate model in a rational asymptotic way, via the above three limiting processes (2.1), (2.2) and (2.4) and two matching relations (2.3) and (2.5), in the framework of the MAE method. This model is called the Navier–Fourier model in Sect. 4.1.6, and gives $\boldsymbol{v}_{\mathrm{N}}(t, \boldsymbol{x}), p_{\mathrm{N}}(t, \boldsymbol{x})$ and the perturbation of the temperature θ_{N}. The model consistently replaces the full NSF problem for a low-Mach-number limiting flow. This opens the gate to obtaining a consistent second-order asymptotic low-Mach-number model with dominant effects of compressibility and thermal effects.

2.1.4 Burgers' Equation

In external aerodynamics, the long-time behaviour of small-amplitude acoustic waves experiencing weak dissipation in a homogeneous, unbounded fluid at rest is also relevant in the framework of a low-Mach-number asymptotic approach. For this problem, it is very useful to assume that the Reynolds and Strouhal numbers Re and St are both large parameters, such that

$$M\, St = 1 \quad \text{and} \quad M\, Re = R^* = O(1) . \tag{2.7}$$

However, in the NSF evolution equations (1.29) for ρ, p, $\boldsymbol{v}$ and T, it is necessary to replace the operator $\mathrm{D}/\mathrm{D}t$ by $St\, \mathrm{D}/\mathrm{D}t \equiv St\, \partial/\partial t + \boldsymbol{v} \, . \, \nabla$. As a result, Burgers' well-known dissipative model equation can be asymptotically derived rationally from the (unsteady, one-dimensional) NSF equations; this equation (discovered in 1948; see Burgers 1948) appears as a compatibility condition. However, Burgers' equation (in a slightly generalized form) is, in fact, also a consistent model equation for an acoustic wave train trapped inside a thin region close to an acoustic wavefront; outside this thin region (characterized by a thickness of order $O(M)$), in the purely acoustic region, the classical linear acoustic equations are valid at the leading order (see, for instance, Sect. 4.2.2).

2.1.5 Vanishing-Viscosity Case

Finally, we observe that the above acoustic limiting process (2.4) is *singular* in the framework of the NSF equations (1.29), since close to the initial time $t = 0$ we derive (see Sect. 3.3.4) the classical (non-viscous) system of equations of acoustics. On the wall bounding the fluid flow, as the boundary condition for the acoustic velocity, we have only the possibility to choose a *slip condition*. As a consequence, in the vicinity of the wall and near the initial time, it may be necessary to consider a new limiting process in place of (2.4), and to derive a new set of consistent ('Rayleigh') dissipative layer equations in place of the acoustic equations. In such a case, the slip condition (see, for instance, (1.34b)) appears as a matching condition (relating to the coordinate normal to the wall of the body considered). This open problem (at the present time), which has theoretical interest in itself, seems to be consistent only for large Reynolds numbers if a similarity rule is assumed between the two small parameters M and $1/Re$. Indeed, it is interesting to consider, in the framework of the full 3D unsteady NSF equations (1.29), an asymptotic theory for a slightly compressible fluid with a vanishing viscosity, when simultaneously both the Mach number and the inverse of the Reynolds number tend to zero. In the *steady* case, in the framework of the classical (but compressible) problem of Blasius (1908) (see Sect. 4.3), of steady 2D flow past a solid flat plate (in the half-plane $y > 0$, with $0 < x < \infty$) placed in a uniform stream parallel to the plate and normal to its edge, this problem has been considered by Godts and Zeytounian (1990), and in this case a more consistent similarity rule between the two small parameters M and $1/Re$ is

$$M^2 = \frac{1}{\sqrt{Re}} \,. \tag{2.8}$$

2.2 Internal Aerodynamics

2.2.1 Time-Dependent Domains

In the case of internal aerodynamics, we consider in particular the unsteady evolution of a gas confined in a closed, time-dependent container, where the (slow) time variation of the boundary (the 'boundary velocity') is small compared with the characteristic acoustic speed of the gas. This problem raises a number of interesting questions, relating first to the derivation of consistent model equations for the long time persistence of acoustic oscillations, and second to the obtaining of averaged equations (excluding acoustics) for the nearly incompressible motion. It seems that this complicated internal problem was first considered, at the end of the 1970s, by Zeytounian and Guiraud (1980a,b), via a multiple-time-scale asymptotic analysis with a family of fast times functionally related to the slow time. Müller (1996, 1998) and, more recently, Ali (2003) considered the case of two time scales with only one fast time variable, $\tau = t/M$, where t is a slow time linked to the boundary velocity. Unfortunately, this approach is obviously not sufficient to describe the (infinite but 'enumerable') sequence of free modes produced by a generic motion of the boundary. As a consequence, the unsteady motion of the gas within the time-dependent closed container must be considered as a superposition of acoustic fast oscillations (depending on a family of fast times functionally related to the slow time) and an averaged, pseudo-incompressible flow depending on the slow time t. Our multiple-time-scale analysis is carried out via the introduction of two operators, $\langle . \rangle$ and $D(.)$, applied to the unsteady compressible, viscous, heat-conducting field $U = [\boldsymbol{v}, p, \rho, T]$, governed by the unsteady NSF equations (1.29) and considered as a function of ($t, \boldsymbol{x}; M$ and the infinity of fast times) such that

$$U = \langle U^* \rangle + U^{*\prime} \,, \tag{2.9a}$$

$$\frac{\partial U}{\partial t} = \frac{\partial U^*}{\partial t} + \frac{1}{M} D U^* \tag{2.9b}$$

and

$$D\langle U^* \rangle \equiv 0 \,. \tag{2.9c}$$

Here $\langle U^* \rangle$ is an 'averaged' solution, a function only of the slow time t and of the position vector $\boldsymbol{x}$ when the Mach number tends to zero, and $U^{*\prime}$ is the 'average-free' part, a function of ($t, \boldsymbol{x}; M$ and the infinity of fast times). The operator D, by definition, cancels out the averaged functions, because it plays the role of a time derivative relative to the set of fast times. Concerning dissipative motions, where the Reynolds number Re is present in the initial full, unsteady NSF equations (1.29), we have the possibility to consider formally two simple cases,

$$Re \equiv \infty, \quad \text{and} \quad Re = O(1), \quad \text{when} \quad M \downarrow 0 \,, \tag{2.10}$$

and a general case

$$Re \gg 1 \quad \text{and } M \ll 1 \quad \text{such that } (Re)^{\alpha} M = O(1), \quad \alpha > 1 \,. \tag{2.11}$$

In relation to the viscous damping of acoustics oscillations, when we consider a slightly viscous flow (large Reynolds number, i.e. $Re \gg 1$), an open problem arises in the precise asymptotic analysis with two small parameters M and $1/Re$. The analysis of this damping phenomenon appears to be a difficult task and raises many questions. Nevertheless, in Chap. 5 (see in Sect. 5.4.2) a particular case is considered that is simpler and is solvable via a boundary-layer analysis; it was considered briefly in Zeytounian and Guiraud (1980a). In this case, over a much longer time period, the oscillations are damped by viscosity and heat conduction within a Stokes-like boundary layer. As an example of a classical asymptotic model of low-Mach-number flows in a bounded domain (via matching?), we mention that Rhadawan and Kassoy (1984) investigated the acoustic response due to boundary heating in a confined inert gas (between two infinite, parallel, planar walls), to get a better understanding of the low-Mach-number limit of the (compressible) one-dimensional (t, x) Navier–Stokes equations and to derive simplified equations which account for the net effect of periodic acoustic waves on slow flow over a long time. The result provides an explicit expression for the piston analogy of boundary heat addition.

2.2.2 Application to Combustion

In Müller's habilitation thesis (Müller 1996) and in papers by Klein and Peters (1988), Klein et al. (2001) and Meister (2003a,b), the reader can find numerous references concerning the application of low-Mach-number asymptotics to combustion problems. In particular, in Klein and Peters (1988) it is shown that the accumulation of small-amplitude gas-dynamic perturbations is able to accelerate the process of self-ignition of a homogeneous explosive mixture. We observe also that Schneider (1978), in his book, provided a 'tentative elucidating' description of the non-reacting, inviscid piston–cylinder problem and suggested a first asymptotic solution ansatz, which, however, did not lead to a uniformly valid long-time solution at that time. Nevertheless, Schneider's approach greatly inspired the work of Klein and Peters (1988).

In addition, the linear acoustic system (derived in Chap. 5 in the framework of internal aerodynamics) for the average-free leading-order velocity and the first-order average-free pressure deserves special attention, and its solution plays a fundamental role in the modelling problem for bounded time-dependent domains.

Thus far, the problem of high-frequency acoustics has been addressed only for unbounded domains, by Majda and Rosales (1983), DiPerna and Majda (1985), Hunter, Majda and Rosales (1986), Pego (1988), Joly, Métivier and Rauch (1993), and Schochet (1994a,b), while the moving-boundary probem has been considered satisfactorily only in a single space dimension (the 'piston problem') by Klein and Peters (1988), Schneider (1978), and Tiberi Timperi (1997).

In Ali (2003), the problem of high-frequency acoustics is considered in more than one space dimension and in a closed, variable-boundary domain; it is shown that it is possible to resolve completely the average-free functions for a relevant class of motion of the time-dependent domain. For the one-dimensional piston problem, this approach (Ali 2003, Sect. 6) leads to results equivalent to those obtained by Klein and Peters (1988). Unfortunately, the analysis (see Sect. 5.3.3) is not conclusive, since Ali's 'theory' is not capable of providing a full resolution of high-frequency acoustics.

Obviously, the key point is that one fast time variable is not sufficient to describe the sequence of modes produced by a generic motion of the boundary.

Finally, we observe that flame–acoustic interaction is an important aspect of combustion in an enclosed volume. The interaction process is governed by the full equations of reactive gas dynamics, which are quite complex, even for the relatively elementary one-step, first-order Arrhenius kinetics (see, for instance, the two equations (1.32)). As a consequence, it is necessary to derive a simpler set of model equations which is appropriate for studying acoustic interaction with 'slender' flames. For this purpose, we can consider the limits of large activation energy and small Mach number $M \downarrow 0, A_0 \uparrow \infty$. The first limit is in fact implied by the second, since M is exponentially small in A_0 (see, for instance, Buckmaster and Ludford 1982). However, in various analyses, these limits are frequently applied in the order $M \downarrow 0, A_0 \uparrow \infty$, the first limit eliminating exponentially small terms, and the second providing the basis for expansions in powers of A_0^{-1}.

2.3 Atmospheric Motions

The most intriguing features related to slow motions characterized by $M \downarrow 0$ are found when one deals with atmospheric flows with $Bo = 1$ (see, for instance, the Chap. 12 in Zeytounian 1990). The case where $Bo \ll 1$ is another particularly interesting case and, as is explained consistently in Chap. 6 of the present book, the Boussinesq limiting process

$$\lim{}^{Bo} = \left\{ M \downarrow 0 \text{ and } Bo \downarrow 0 \text{ such that } \frac{Bo}{M} = B^*, t \text{ and } \boldsymbol{x} \text{ fixed} \right\} , \quad (2.12)$$

leads, for $B* = O(1)$, from the equations (1.40), to the inviscid adiabatic Boussinesq equations, applicable to the motion of lee waves downstream of a mountain. These equations are the *least simplified* limiting system when $M \downarrow 0$, with t and $\boldsymbol{x}$ fixed under the similarity rule

$$\frac{Bo}{M} = B^* \quad (2.13)$$

between M and Bo (Zeytounian 1974). We observe that when $\varepsilon \equiv 1 (L_c \equiv H_c)$ and $\gamma M^2 \equiv Fr^2$, we also have $Bo \equiv 1$, and as a consequence $L_c \equiv H_c = RT_c/g$ – this is the case considered by Klein et al. (2001, Sect. 2.3), which uses new small-scale horizontal and time (dimensionless) coordinates:

$$(\xi, \eta) = \frac{1}{M}(x, y) \quad \text{and } \tau = \frac{t}{M} . \quad (2.14)$$

$(X, Y) = M(x, y)$ represents a new large-scale horizontal coordinates. In particular, in Klein et al. (2001, Sect. 2.3), the space–time coordinates (ξ, η, z, τ)

are considered for the analysis of deep convection on very small horizontal scales; for details, see the report by Botta, Klein and Almgren (1979). In Zeytounian (1979), the reader can find various 2D steady, inviscid model equations for lee waves, derived, when $M \downarrow 0$, from an exact, rather awkward-looking quasi-linear second-order elliptic partial differential equation; in Chap. 6, we consider again the various low-Mach-number limiting forms of this equation.

However, it is somewhat discouraging to realize that the $M \downarrow 0$ limit looks so much good from the outset and that very few useful results have been obtained by using it. It is true that the application of the Boussinesq equations to the local atmospheric lee waves motions (over and downstream of a mountain) and to free circulation (to the phenomenon of breezes) is actually well justified (Zeytounian 2003b). As a matter of fact, the pioneering works of Monin (1961), Drazin (1961) and Charney (1962), on quasi-non-divergent flows, involves some intriguing features tied to the phenomenon of blocking, well known (Drazin 1961) in stratified flows at low Froude number, and this might explain why these models have not been used in weather prediction (see, for instance, Phillips 1970 and Monin 1972). However, fortunately, blocking is avoided whenever the main low-Mach-number approximation is coupled (i.e. matched in the framework of the MAE) with another one, for instance, to the inner approximation in time close to the initial time, and to the outer approximation valid in the far field, as has been shown by Guiraud and Zeytounian (see, Zeytounian 1990, Chap. 12, and Chap. 6 of the present book). On the other hand, the case of low Rossby/Kibel number in the framework of the Kibel primitive equations (1.54), when the sphericity parameter δ and the hydrostatic, parameter ε are both small and the f_0-plane approximation is used, is also strongly related to low-Mach-number asymptotics. Namely, for a consistent derivation of the Kibel quasi-geostrophic (single) equation, it is necessary to consider the following Kibel geostrophic limiting process (Guiraud and Zeytounian 1980):

$$\lim{}^{Ki} = \{M \downarrow 0 \text{ and } Ki \downarrow 0 \text{ such that } Ki = F(M),\ t \text{ and } \boldsymbol{x} \text{ fixed}\}\,, \quad (2.15)$$

where the function $F(M)$ is chosen such that the estimates of the horizontal length scale L_c and of the reference velocity U_c are in agreement with a short-range weather prediction (see Chap. 6). However, on the other hand, the present author believes that by using a judiciously chosen similarity rule between $\varepsilon \ll 1$ and $M \ll 1$,

$$\varepsilon \sim M^{\alpha}, \quad \alpha > 0\,, \quad (2.16)$$

it may be possible to derive a consistent low-Mach-number atmospheric model.

2.4 Nonlinear Acoustics

We observe that nonlinear aeroacoustics is a privileged area for low-Mach-number asymptotics, and the curious reader can consult, in particular, the papers by Crighton (1993), Coulouvrat (1992), and Lesser and Crighton (1975), and the book edited by Hardin and Hussaini (1993) relating to computational aeroacoustics. The system of equations (1.37) is a starting system for the derivation of various low-Mach-number model equations. For instance, in Coulouvrat (1992), as basic assumptions of nonlinear acoustics, we have the following (see (1.13b) and Sect. 1.5):

$$\begin{aligned}
\text{small perturbations:} \quad & M \ll 1 \,; \\
\text{length scale given by acoustics:} \quad & L_c = \tfrac{1}{k} = \tfrac{\omega}{c_0} \,; \\
\text{weakly dissipative fluid:} \quad & Re_{ac} \gg 1 \,; \\
\text{consistency of the equations:} \quad & M^2 \ll \tfrac{1}{Re_{ac}} \,; \\
\text{irrotational modes:} \quad & \nabla \wedge \boldsymbol{v} = 0 \,.
\end{aligned} \tag{2.17}$$

In particular, Coulouvrat (1992) derived, from the system of equations (1.37), the Kuznetsov equation for the dimensionless acoustic potential Φ, such that $\boldsymbol{v} = -\nabla\Phi$; this is a nonlinear dissipative scalar wave equation. Namely, Coulouvrat (1992) obtained, in a rather ad hoc manner, the following approximate Kuznetsov (1970) equation:

$$\frac{\partial^2 \Phi}{\partial t^2} - \Delta\Phi - 2S\frac{\partial(\Delta\Phi)}{\partial t} = M\frac{\partial}{\partial t}\left\{\frac{\gamma-1}{\gamma}\left(\frac{\partial\Phi}{\partial t}\right)^2 + (\nabla\Phi)^2\right\} , \tag{2.18}$$

where

$$S = \frac{1}{2\,Re_{ac}}\left[\frac{4}{3} + \left(\frac{\mu_v}{\mu_c}\right) + \frac{\gamma-1}{Pr}\right] , \tag{2.19}$$

is called the *Stokes number* and measures the combined influence of all dissipative effects due to viscosity or thermal conduction (see the discussion of some problems of nonlinear acoustics in Sect. 7.1). In particular, an interesting problem is the derivation (inspired by Coulouvrat 1992, pp. 334–338), in a consistent asymptotic way, of the KZK equation, which generalizes the unsteady, dissipative Burgers' model equation to an equation with a transverse diffraction effect (absent in Burgers' one-dimensional equation). This shows that we can *really* consider acoustics as a branch of fluid mechanics (see, for instance, Crighton 1981).

2.5 Some Special Problems

First, we mention that in an extensive report by Klein et al. (2001), the issue of the 'asymptotic adaptivity' of a numerical scheme with respect to the (low) Mach number is addressed; this requires a close interplay between application-oriented asymptotics and numerical analysis, and such an approach plays an important role in various computational applications. For instance, it is used in relation to combustion, in both confined and unbounded domains (Majda 1984, Majda and Sethian 1985), and in relation to astrophysics, in the framework of magnetohydrodynamics, in order to justify the treatment of small density fluctuations, called *pseudo-sound*, in an incompressible model of the solar wind (see Matthaeus and Brown 1988, which considers nearly incompressible magnetohydrodynamics at low Mach number, and Zank and Matthaeus 1991 and 1993, which consider hydrodynamics, magnetohydrodynamics, turbulence and waves in the framework of a nearly incompressible fluid).

Such 'practical asymptotics' seem efficient for numerical simulations, but bear little relation to rational and consistent asymptotic modelling (in the spirit of Zeytounian 2002). We observe also that in Rehm and Baum (1978), the equations of motion for a thermally driven buoyant, inviscid flow with heat release are derived as a low-Mach-number limit of the compressible Euler equations, and in Fedorchenko (1997), a model of unsteady subsonic flow *with acoustics excluded* is analysed and a number of exact solutions are provided.

The properties of a relatively uncommon compressible, low-Mach-number regime of fluid flow in a channel with a very large value (10^6) of the aspect ratio Λ = length/diameter are investigated in Shajii and Freidberg (1996), and Matalon and Metzener (1997) derived a system of approximate equations for the propagation of a premixed flame in a closed tube. The paper by Dwyer (1990) is devoted to the calculation of low-Mach-number reacting flows, and Majda and Sethian (1985) derived model equations for zero-Mach-number combustion.

We observe that the theory developed in Chap. 5 is very efficient for combustion problems when we want to take into account quantitatively the influence of acoustics, and in Chap. 7 (devoted to miscellaneous topics) some of the above-mentioned problems are discussed.

2.5.1 Shallow-Convection Equations

In the recent concise review paper Zeytounian (2003a), the Boussinesq equations for atmospheric motions (where dry air is considered as a perfect gas) and also for a weakly dilatable liquid (as in the classical Bénard thermal-instability problem) were derived, as low-Mach-number and low-Froude-number asymptotics of the Euler and NSF equations, respectively. Here we observe only that for the derivation of the limiting shallow-convection ('à la Boussinesq') equations for the Rayleigh–Bénard (RB) instability problem, the Froude number can be interpreted as a 'pseudo-Mach' number for a weakly dilatable liquid (see, for instance, for the application to 'thin films', the book edited by Velarde and Zeytounian 2002). Namely, in the case of the Bénard thermal problem (heated from below), an infinite horizontal layer of a viscous, thermally conducting, weakly expansible liquid of temperature T and density ρ, with an equation of state $\rho = \rho(T)$, is in contact with a solid wall ($z = 0$) at a constant temperature T_w. At the level $z = d$ there is a free surface in contact with an atmosphere, and $T_f (\equiv T_s(d))$ is the constant temperature of the free surface in the purely static basic state, where

$$T_{\mathrm{s}}(z) = T_{\mathrm{w}} - \beta z, \quad \beta = -\frac{\mathrm{d}T_{\mathrm{s}}(z)}{\mathrm{d}z} > 0\,, \tag{2.20}$$

and as a consequence, $\Delta T_0 \equiv T_{\mathrm{w}} - T_{\mathrm{f}} = \beta d$. If ν_{f} and C_{f} are the kinematic viscosity and specific heat of the liquid at $T = T_{\mathrm{f}}$ then, in the dimensionless dominant NSF equations for the weakly expansible liquid, we have the following three parameters:

$$Fr_d^2 = \frac{(\nu_{\mathrm{f}}/d)^2}{gd}, \quad \varepsilon = \beta d\alpha(T_{\mathrm{f}}), \quad Bq = \frac{g}{\beta C_{\mathrm{f}}}, \quad \text{where} \quad \alpha(T) = -\frac{\mathrm{d}\log\rho(T)}{\mathrm{d}T}\,. \tag{2.21}$$

A consistent asymptotic derivation of the shallow-convection equations for the RB instability problem can be performed when

$$Fr_d^2 \text{ and } \varepsilon \text{ are both} \ll 1, \quad \text{such that } \frac{\varepsilon}{Fr_d^2} = Gr = O(1)\,, \tag{2.22}$$

where $Gr = \beta d^4\alpha(T_{\mathrm{f}})/(\nu_{\mathrm{f}})^2$ is the Grashof number. In this case, we have the following constraint for the thickness d of the liquid layer:

$$d \gg \left[\frac{(\nu_{\mathrm{f}})^2}{g}\right]^{1/3} \approx 1\,\mathrm{mm}\,. \tag{2.23}$$

However, on the other hand, a detailed dimensionless, asymptotic analysis shows that in the shallow convection equations, the dissipation function is absent because

$$Bq = \frac{g}{\beta C_{\mathrm{f}}} \approx 1 \quad \Rightarrow d \approx \frac{C_{\mathrm{f}}\Delta T_0}{g}\,, \tag{2.24}$$

which is a *validity criterion*. Using (2.24), we can write

$$Fr_d^2 = \frac{(\nu_{\mathrm{f}}/d)^2}{gd} \approx \frac{(\nu_{\mathrm{f}}/d)^2}{C_{\mathrm{f}}\,\Delta T_0} = M_1^2 \ll 1\,, \tag{2.25}$$

and M_1 is indeed a small Mach number for the liquid, based on the reference velocity $U_1 = \nu_{\mathrm{f}}/d$ and the pseudo-sound speed $c_1 = \sqrt{C_{\mathrm{f}}\Delta T_0}$. This Bénard thermal-instability problem is considered in Chap. 7, and an asymptotic 'low-Mach/Froude-number' derivation (from the exact formulation of the Bénard thermal-instability problem) of the model equations for the RB instability problem is given, with a discussion of the validity of these model shallow-convection equations.

2.5.2 Compressible Stokes–Oseen Exterior Problem

Finally, for a weakly compressible ($M \to 0$) flow, when $Re \to 0$, it is necessary to specify also the role of the Mach number M. In fact, it is necessary to pose the problem in terms of the behaviour of the solutions of the full dimensionless NSF equations (1.29) when simultaneously $Re \to 0$ and $M \to 0$. Naturally, for the NSF equations to be valid, it is obvious that it is assumed that the limiting weakly compressible, viscous fluid flow at low Reynolds and Mach numbers remains a continuous medium, and this implies that the Knudsen (dimensionless) number is a small parameter, i.e. $Kn = M/Re \ll 1$, according to the 'fluid dynamic limit' of the Boltzmann equation. As a consequence, the limiting process $Re \to 0$ and $M \to 0$ must be performed with the following similarity relation:

$$Re = \beta M^{1-a}, \quad \text{with } \beta = O(1) \text{ and } a > 0 \text{ when } M \to 0\,. \tag{2.26}$$

In Sect. 4.4, the equations for the weakly compressible, viscous model are derived from the similarity rule (2.26) in the framework of the classical Stokes–Oseen exterior problem.

2.6 Rigorous Mathematical Results and Numerical Simulations

Concerning mathematically rigorous results linked to the 'incompressible limit' (initiated in France by Lions 1993), we should mention a recent, very representative paper by Masmoudi (2000), where the reader can find various pertinent, recent references, and also a paper by Métivier and Schochet (2001) devoted to the incompressible limit of the *non-isentropic* Euler equations. For example, in Masmoudi (2000, Sect. 4) the author of that paper studies in detail the incompressible limit when the Mach number goes to 0 and gives a complete rigorous justification for the formal derivation: one passes to the limit in the global weak solutions, for a compressible isentropic/barotropic flows governed by the Navier–Stokes equations, which were recently proven to exist by Lions (1998). We observe that the general setup for such an asymptotic problem is a straightforward adaptation of that introduced by Klainerman and Majda (1981, 1982) in the inviscid case for local strong solutions. In fact, the heuristic approach which leads to the incompressible 'Navier model' is basically correct even globally in time, for global weak solutions; but the limiting process for the pressure is much more involved and may, depending on the initial conditions, incorporate additional terms coming from oscillations in the convective quasi-linear term $\rho(\boldsymbol{v}\,.\nabla)\boldsymbol{v}$. Concerning the rigorous results, Masmoudi (2000, pp. 127–128) writes:

> In the first subsection, we state the convergence result in the periodic case. In the second part, we proof some uniform estimates and give a formal (but wrong!) proof. In the third subsection, we present a first rigorous method of proof using the group method introduced by Schochet (1994b) and see also Grenier (1997). This method will be called a global method since it is based on some type of Fourier decomposition and depends highly on the boundary conditions. In the fourth subsection, we present another method based on some spectral

properties of the wave equations. Indeed, as will be seen later on one of the major difficulties in the passage to the limit is the presence of acoustic waves. This latter method has the advantage of being a local one and holds for any type of boundary conditions. In the fifth subsection, the case of Dirichlet boundary conditions will be studied and more precise results will be stated. In the sixth subsection, we shall consider the passage to the limit towards solutions of (incompressible) equations using an energy method.

Section 5 of Masmoudi (2000, pp. 145–156) is devoted to the limit where γ (the adiabatic constant) goes to ∞, and, depending on the total mass, the author of that work recovers in the limit either a mixed model, which behaves as a compressible one if $\rho < 1$ and as an incompressible one if $\rho = 1$, or the classical incompressible Navier system.

Unfortunately, in these rigorous papers, the mathematical analysis is carried out almost entirely by consideration of the equations, without boundary conditions corresponding to a physically realistic problem. Indeed, the experienced reader will be aware of the fact that the fluid dynamic equations are not sufficient for discussing flow problems, and this seemingly anodyne remark has far-reaching consequences, as is shown in the various applications mentioned in Chap. 3. In conclusion, the present author's impression is that, in recently published papers by applied mathematicians, the formulation of the compressible fluid flow problems considered is 'too well adapted' to the use of only a precise, rigorous, but abstract functional-analysis method.

From a mathematical point of view, it is important to observe that the filtering of acoustic waves, in the equations of compressible fluid flow during the limiting process $M \downarrow 0$, changes drastically the type of these equations, and often a fluid flow problem that is well-posed at the start becomes 'ill-posed' (see Oliger and Sundström 1978 and the book by Kreiss and Lorenz 1989). In the book by Majda (1984) and in the paper by Meister (1999), the reader can find a low-Mach-number approach within a systematic mathematical framework. We note also that it seems that the authors of mathematically rigorous results do not have a clear understanding of the important role played by the acoustic region close to the initial time for the limiting incompressible equations, even if the purely acoustic flow has died out while radiating to infinity. For instance, in external aerodynamics, when matching is possible, we obtain consistent initial condition for the Navier velocity $\boldsymbol{v}_N(t, x)$, as discussed in Chap. 4.

Finally, in Müller (1998), in recent papers by Müller (1999), Klein et al. (2001) and Meister (2003a,b), and the theses of Müller (1996) and Viozat (1998), the reader who is interested in the various results of numerical simulations of low-Mach-number flows can find many very pertinent references. Here we note that in his conclusion Meister (2003b) writes 'Numerical experiments for a wide range of viscous and inviscid flow fields demonstrate the high accuracy and robutstness of the scheme from the low Mach number regime up to hypersonic flow fields' – but this seems to the present author too optimistic. On the other hand, in Meister (2003a), a finite-volume method for reliable simulations of inviscid fluid flows at high as well as low Mach numbers, based on a preconditioning technique proposed by Guillard and Viozat(1999), is presented. Whereas in Meister (2003a) the asymptotic analysis of the scheme is focused on the behaviour of continuous and discrete pressure distributions for inviscid low-speed simulations, Meister (2003b) obtains both physically sensible discrete pressure fields for viscous, low-Mach-number flows, and a divergence-free condition for the discrete velocity field in the limit of a vanishing Mach number with respect to the simulation of inviscid fluid flow. In this book, we shall not discuss papers devoted to rigorous mathematical results and numerical simulations any further.

References

G. Ali. Low Mach number flows in time-dependent domain. SIAM J. Appl. Math., **63**(6) (2003), 2020–2041.

H. Blasius. Z. Math. Phys., **56** (1908), 1–37.

N. Botta, R. Klein and A. S. Almgren. *Dry Atmosphere Asymptotics.* PIK Potsdam Institute for Climate Impact Research, 55 Potsdam, 1999.

J. Buckmaster and G. S. S. Ludford. *Theory of Laminar Flames.* Cambridge University. Press, London, 1982.

J. M. Burgers. A mathematical model illustrating the theory of turbulence. Adv. Appl. Mech., **1** (1948), 171–199.

J. G. Charney. In: *Proceedings of the International Symposium on Numerical Weather Prediction*, Tokyo, edited by Meteorological. Society of Japan, Tokyo, 1962, pp. 82–111.

F. Coulouvrat. On the equations of nonlinear acoustics. J. Acoustique, **5** (1992), 321–359.

S. C. Crow. Aerodynamic sound emission as a singular perturbation problem. Stud. Appl. Math., **49** (1970), 21–44.

D. G. Crighton, J. Fluid Mech., **106** (1981), 261–298.

D. G. Crighton. Computational aeroacoustics for low Mach number flows. In: *Computational Aeroacoustics*, edited by J. C. Hardin and M. Y. Hussaini. Springer, New York, 1993, pp. 50–68.

R. DiPerna and A. Majda. The validity of nonlinear geometric optics for weak solutions of conservation laws. Commun. Math. Phys., **98** (1985), 313–347.

P. G. Drazin. Tellus, **13** (1961), 239–251.

H. A. Dwyer. AIAA J., **28** (1990), 98–105.

D. G. Ebin. Motion of slightly compressible fluids in a bounded domain, I. Commun. Pure Appl. Math., **35** (1982), 451–487.

A. T. Fedorchenko. J. Fluid Mech., **334** (1997), 135–155.

S. Godts and R. Kh. Zeytounian. ZAMM, **70**(1) (1990), 67–69.

E. Grenier. Oscillatory perturbations of the Navier–Stokes equations. J. Math. Pures Appl. (9), **76**(6) (1997), 477–498.

H. Guillard and C. Viozat. On the behaviour of upwind schemes in the low Mach number limit. Comput. Fluids, **28** (1999), 63–86.

J. P. Guiraud and R. Kh. Zeytounian. Geophys. Astrophys. Fluid Dyn., **15**(3/4) (1980), 283–295.

J. C. Hardin and M. Y. Hussaini (eds.). Computational Aeroacoustics. Springer, New York, 1993.

K. Hunter, A. Majda and R. R. Rosales. Resonantly interacting weakly nonlinear hyperbolic waves II. Several space variables. Stud. Appl. Math., **75** (1986), 187–226.

I. Imai. *Approximation Methods in Compressible Fluid Dynamics.* Institute for Fluid Dynamics and Applied Mathematics, University of Maryland, Technical Note BN-104, March 1957.

O. Janzen. Beitrag zu einer Theorie der stationären Strömungen Kompressibler Flüssiigkeiten. Phys. Z. **14** (1913), 639–643.

J. L. Joly, G. Métivier and J. Rauch. Resonant one-dimensional nonlinear geometric optics. J. Funct. Anal., **114** (1993),

S. Klainerman and A. Majda. Singular limits of quasilinear hyperbolic system with large parameters and the incompressible limit of compressible fluids. Commun. Pure Appl. Math., **34** (1981), 481–524.

S. Klainerman and A. Majda. Compressible and incompressible fluids. Commun. Pure Appl. Math., **35** (1982), 629–651.

R. Klein and N. Peters. Cumulative effects of weak pressure waves during the induction period of thermal explosion in a closed cylinder. J. Fluid Mech., **187** (1988), 197–230.

R. Klein, N. Botta, L. Hofmann, A. Meister, C. D. Munz, S. Roller and T. Sonar. Asymptotic adaptive methods for multi-scale problems in fluid mechanics. J. Engng. Math., **39** (2001), 261–343.

H.-O. Kreiss and J. Lorenz. *Initial-Boundary Value Problems and the Navier-Stokes Equations.* Academic Press, New York, 1989.

V. P. Kuznetsov. Sov. Phys. Acoust. **16** (1970), 467–470.

V. R. Lauvstad. On non-uniform Mach number expansion of the Navier– Stokes equations and its relation to aerodynamically generated sound. J. Sound Vib., **7** (1968), 90–105.

F. G. Leppington and H. Levine. The sound field of a pulsating sphere in unsteady rectilinear motion. Proc. Roy. Soc. A, **412** (1987), 199–221.

M. B. Lesser and D. G. Crighton. Physical acoustics and the method of matched asymptotic expansions. In: *Physical Acoustics*, edited by W. P. Mason and R. N. Thurson, Vol. 11, Academic Press, New York, 1975.

M. J. Lighthill. On sound generated aerodynamically. I. General theory. Proc. Roy. Soc. Lond. A, **211** (1952), 564–514. See also M. J. Lighthill. Proc. Roy. Soc. Lond. A, **267** (1962), 147–182.

P.-L. Lions. Limites incompressibles et acoustiques pour des fluides visqueux, compressibles et isentropiques. C. R. Acad. Sci. Paris, Série I, **316** (1993), 1335–1340; **317** (1993), 115–120; **317** (1993), 1197–1202.

P.-L. Lions. *Mathematical Topics in Fluid Mechanics,* Vol. 2, *Compressible Models.* Oxford Lecture Series in Mathematics and its Applications, No. 10, Clarendon Press, Oxford, 1998.

A. Majda. *Compressible Fluid Flow and Systems of Conservation Laws in Several Space Variables.* Springer, New York, 1984.

A. Majda and R. R. Rosales. Resonantly interacting weakly nonlinear hyperbolic waves I. A single space variable. Stud. Appl. Math., **71** (1983), 149.

A. Majda and J. Sethian. The derivation and numerical solution of the equations for zero Mach number combustion. Combust. Sci. Technol., **42** (1985), 185–205.

N. J. Masmoudi. Asymptotic problems and compressible–incompressible limits. In: *Advances in Mathematical Fluid Mechanics,* edited by J. Malek, J. Necas and M. Rokyta. Springer, Berlin, 2000, pp. 119–156.

M. Matalon and P. Metzener. The propagation of premixed flames in closed tubes. J. Fluid Mech., **336** (1997), 331–350.

W. H. Matthaeus and M. R. Brown. Nearly incompressible magneto-hydrodynamics at low Mach number. Phys. Fluids, **31** (1988), 3634–3644.

A. Meister. Asymptotic single and multiple scale expansions in the low Mach number limit. SIAM J. Appl. Math., **60** (1999), 256–271.

A. Meister. Asymptotic based preconditioning technique for low Mach number flows. Z. Angew. Math. Mech., **83**(1) (2003a), 3–25.

A. Meister. Viscous fluid flow at all speeds: analysis and numerical simulation. Z. Angew. Math. Phys., **54** (2003b), 1010–1049.

G. Métivier and S. Schochet. The incompressible limit of the non-isentropic Euler equations. Arch. Ration. Mech. Anal., **158** (2001), 61–90.

A. S. Monin. Izv. Akad. Nauk SSSR, Ser. Geophys., **4** (1961), 602.

A. S. Monin. *Weather Forecasting as a Problem in Physics.* MIT Press, Cambridge, MA, 1972.

B. Müller. *Computation of Compressible Low Mach Number Flow.* Habilitation thesis, Institut für Fluiddynamik, ETH Zürich, September 1996.

B. Müller. Low-Mach-number asymptotics of the Navier–Stokes equations. J. Engrg. Math., **34** (1998), 97–109.

B. Müller. *Low Mach Number Asymptotics of the Navier–Stokes Equations and Numerical Implications.* Von Karman Institute for Fluid Dynamics, Lecture Notes 1999-03, March 1999.

F. Obermeier. *The Application of Singular Perturbation Methods to Aerodynamic Sound Generation.* Lecture Notes in Mathematics, No. 594, Springer, Berlin, 1976, pp. 400–421.

J. Oliger and A. Sundström. Theoretical and practical aspects of some initial boundary value problems in fluid dynamics. SIAM J. Appl. Math. **35**(3) (November 1978), 419–446.

R. Pego. Some explicit resonating waves in weakly nonlinear gas dynamics. Stud. Appl. Math., **79** (1988), 263–270.

N. A. Phillips. Models for weather prediction. Annu. Rev. Fluid Mech., **2** (1970), 251–292.

Lord Rayleigh, On the flow of compressible fluid past an obstacle. Philos. Mag., **32** (1916), 1–6.

R. G. Rehm and H. R. Baum. J. Res. Nat. Bur. Stand., **83** (1978), 297–308.

R. H. Rhadawan and D. R. Kassoy. The response of a confined gas to a thermal disturbance: rapid boundary heating. J. Engng. Math., **18** (1984), 133–156.

W. Schneider. *Mathematische Methoden in der Strömungsmechanik.* Vieweg, Braunschweig, 1978.

S. Schochet. Resonant nonlinear geometric optics for weak solutions of conservation laws. J. Differential Equations, **113** (1994a), 473–504.

S. Schochet. Fast singular limits of hyperbolic PDEs. J. Differential Equations, **114** (1994b), 476–512.

R. Sery Baye. *Contribution à l' étude de certains écoulements à faibles nombres de Mach.* Thèse de Doctorat de Mécanique, Université de Paris VI, 1994.

A. Shajii and J. P. Freidberg. J. Fluid Mech., **313** (1996), 131–145.

F. Tiberi Timperi. *Simulazione numerica della compressione di un pistone e confroto con l'approssimazione adiabatica.* Testi di laurea, Università dell'Aquila, Laquila, Italy, 1997.

E. K. Tracey. The initial and ultimate motions of an impulsively started cylinder in a compressible fluid. Proc. Roy. Soc. Lond. A, **418** (1988), 209–228.

M. Van Dyke. Long series in mechanics: Jansen–Rayleigh expansion for a circle. Meccanica, **33** (1998), 517–522.

M. G. Velarde and R. Kh. Zeytounian (eds.), *Interfacial Phenomena and the Marangoni Effect.* CISM Courses and Lectures, No. 428, Springer, Vienna, 2002.

C. Viozat. *Calcul d'écoulement stationnaires et instationnaires à petit nombre de Mach, et en maillage étirés.* Thèse de Doctorat en Sciences (de l'ingénieur), Université de Nice-Sophia Antipolis, 1998.

H. Viviand. Etude des écoulements instationnaires à faibles nombre de Mach avec application au bruit aérodynamique. J. de Mécanique, **9** (1970), 573–599.

C. Wilcox. *Scattering Theory for the d'Alembert Equation in Exterior Domain.* Springer, Berlin, 1975.

G. P. Zank and W. H. Matthaeus. The equations of nearly incompressible fluids. I. Hydrodynamics, turbulence and waves. Phys. Fluids A, **3** (1991), 69–82.

G. P. Zank and W. H. Matthaeus. Nearly incompressible fluids. II. Magnetohydrodynamics, turbulence and waves. Phys. Fluids A, **5** (1993), 257–273.

R. Kh. Zeytounian. Arch. Mech. (Warsaw), **26**(3) (1974), 499–509.See also R. Kh. Zeytounian, *Notes sur les Ecoulements Rotationnels de Fluide Parfait.* Lecture Notes in Physics, No. 27, Springer, Heidelberg, 1974.

R. Kh. Zeytounian. On models for lee waves in a baroclinic compressible troposphere. Izv. Akad. Nauk SSSR, Atm. Ocean. Phys., **15**(5) (1979), 498–507.

R. Kh. Zeytounian. *Asymptotic Modeling of Atmospheric Flows.* Springer, Heidelberg, 1990.

R. Kh. Zeytounian. *An asymptotic derivation of the initial condition for the incompressible and viscous external unsteady fluid flow problem.* Int. J. Engng. Sci., **38** (2000), 1983–1992.

R. Kh. Zeytounian. *Asymptotic Modelling of Fluid Flow Phenomena.* Fluid Mechanics and Its Applications, Vol. 64, Series Ed., R. Moreau, Kluwer Academic, Dordrecht, 2002.

R. Kh. Zeytounian. Joseph Boussinesq and his approximation: a contemporary view. C. R. Mecanique., **331** (2003a), 575–586.

R. Kh. Zeytounian. On the foundation of the Boussinesq approximation applicable to atmospheric motions. Izv. Atm. Ocean. Phys., **39** Suppl. 1 (2003b), S1–S14.

R. Kh. Zeytounian and J.-P. Guiraud. Evolution d'ondes acoustiques dans une enceinte et concept d'incompressibilité. IRMA-Lille, **2**, Fasc. 4(4) (1980a), IV-1–IV-16.

R. Kh. Zeytounian and J.-P. Guiraud. C. R. Acad. Sci. Paris, **290B** (1980b), 75–78.

R. Kh. Zeytounian and J.-P. Guiraud. Asymptotic features of low Mach number flows in aerodynamics and in the atmosphere. In: *Advances in Computational Methods for Boundary and Interior Layers, BAIL III*, edited by J. J. H. Miller. Boole Press, Dublin, 1984, pp. 95–100.

3

A First Approach to the Asymptotics of Low-Mach-Number Flows

In this chapter, we describe an initial (rather 'naive') approach to the asymptotics of low-Mach-number flows, based mainly on consideration of the equations, and show for various cases why the initial and boundary conditions (linked to a physically realistic fluid flow problem) are very important in the framework of a consistent theory for the asymptotic modelling of low-Mach-number flows.

3.1 Constant-Density Euler Inviscid Flow

The mathematical model of *constant-density* Euler inviscid flow (in fact, a model where $\rho \equiv 1$, since we work with dimensionless variables), where we have simply the constraint of a divergence-free velocity vector

$$\nabla \cdot \boldsymbol{v} \equiv \operatorname{div} \boldsymbol{v} = 0 \tag{3.1}$$

in place of the compressible continuity equation (1.9), has been and still is very popular for computations of flow. For a long time, this was due to the possibility of using the theory of functions of a complex variable for two-dimensional steady flows. As a matter of fact, the theory of steady two-dimensional flow was one of the most important chapters in any textbook on fluid dynamics, and it was also a very important field for the application of mathematical theory to fluid dynamics. However, nowadays the point of view must be somewhat different. Namely, the possibility of using very rapid computers puts the emphasis no longer on closed-form solutions but rather on mathematical models, derived asymptotically, that allow efficient computer codes to be developed. The constraint (3.1) has a number of consequences (see, for instance, the survey paper by Gresho 1992); in particular, it appears to be very difficult to devise numerical codes which enforce this divergence-free condition with high accuracy unless the code is specially designed for that purpose. However, such

R. Kh. Zeytounian: *Topics in Hyposonic Flow Theory*, Lect. Notes Phys. **672**, 45–75 (2006)
www.springerlink.com

codes which enforce div $\boldsymbol{v} = 0$ automatically actually exist and are very powerful, and this fact argues in favour of the point of view that we intend to emphasize in this book.

It is necessary to observe also that we know, from the mathematical theory of vector fields, how to compute a divergence-free vector field when we know the field of its rotational, namely its vorticity field

$$\omega = \nabla \wedge \boldsymbol{v} \equiv \operatorname{curl} \boldsymbol{v} \ , \tag{3.2}$$

where we are dealing with a velocity field $\boldsymbol{v}$ (for a quite comprehensible treatment of this subject see, for instance, Chap. 2 of Batchelor 2000). Here we mention merely that the velocity field $\boldsymbol{v}$ may be expressed through the formula

$$\boldsymbol{v} = \mathsf{L}\omega + \nabla\varphi \ , \tag{3.3}$$

where

$$\mathsf{L}\omega(\boldsymbol{x}) = -\frac{1}{4\pi}\iiint_D \frac{(\boldsymbol{x}-\boldsymbol{y})\wedge\omega(\boldsymbol{y})}{|(\boldsymbol{x}-\boldsymbol{y})|^3}\, dv_y \ . \tag{3.4}$$

In the *Biot–Savart* formula (3.4), the domain of integration D being the domain occupied by the part of the velocity field which is rotational. As a result of (3.3) together with (3.4), once the vorticity field is known one may apply the methods of potential theory and, in particular, numerical codes derived from it, in order to achieve the computation of the velocity field. Our purpose here is not to consider this matter in any detail but only to emphasize again the considerable interest associated with flows which are divergence-free; readers interested in the use of the above Biot–Savart formula in the computation of flows may refer again to the survey paper by Gresho (1992).

Now that we are convinced of the value of considering divergence-free velocity flows (as a consequence of $\rho \equiv 1$), the next question to ask is: under what circumstances may they be used for the computation of actual flows? The answer is obvious if one is dealing with fluids which are, according to their thermodynamics, of constant density $\rho \equiv 1$. This answer is much less obvious if one is dealing with fluids such as gases which are, intrinsically, compressible. Nevertheless, the matter is undoubtedly important, as can be seen if one considers the numerous applications of constant-density aerodynamics (hydrodynamics). The reader interested in the foundations of the models he/she is using should understand to what extent it is semantically inconsistent to speak about constant-density aerodynamics. As a matter of fact, aerodynamics deals basically with the motion of gases, which are obviously compressible (with a variable density!). On the other hand, nobody would argue against the view that constant-density aerodynamics is a quite useful body of knowledge, at least under appropriate circumstances.

Our purpose in this book is to investigate those circumstances. Any modern texbook on the thermodynamics of fluid flows emphasizes the difference in meaning between the pressure in a liquid on the one hand and the pressure in a gas on the other hand. In a liquid, the pressure is not a thermodynamic quantity, i.e its value is not directly related to the state of the liquid but is instead determined such that the condition $\operatorname{div} \boldsymbol{v} = 0$ is satisfied. In a *gas*, the pressure is a thermodynamic quantity which defines the state of the gas. This difference has far-reaching consequences, which are at the heart of our whole subject. According to the above remarks, a liquid should be considered as a substance whose state is unaffected by the value of the pressure. This is exactly what we mean by saying that a liquid has a constant density. In the case of gases, the situation is completely different, and the density is strongly dependent on the pressure. However, a gas may behave as if it were approximately of a constant density–that is to say, that constant-density aerodynamics may give good answers to a flow (hydrodynamic) problem as a consequence of an approximation. It is the nature of this approximation that we intend to investigate in this book. Our main objective below is to present a serious but relatively simple argument in order to convince the reader that the suitability of divergence-free-velocity flow as a mathematical model has something to do with the smallness of the Mach number.

3.2 Incompressible (Isochoric) Eulerian Fluid Flow

In the case of constant-density aerodynamics, where $\rho \equiv 1$, we obtain the constraint of a divergence-free velocity vector (3.1) as a consequence of the compressible continuity equation (1.9). On the other hand, whatever the pressure, so long as the condition $\nabla \cdot \boldsymbol{v} = 0$ is satisfied, the density, according to (1.9), is invariable for any flowing fluid particle of small volume, i.e.

$$\frac{\mathrm{D}\rho}{\mathrm{D}t} \equiv \frac{\partial \rho}{\partial t} + \boldsymbol{v} \cdot \nabla \rho = 0 . \tag{3.5}$$

At the same time, the whole density field (which is a Lagrangian invariant) may have considerable gradients. The density ρ of the fluid particle is thereby supposed to be independent of any pressure variations. This key assumption, although seeming too strong (as noticed by Fedorchenko 1997, p.136), represents in fact a rigorous definition of an incompressible fluid medium, and, as a result, enables us to derive a closed system of governing equations (for $\boldsymbol{v}$, ρ and p) without using any equation of state. Thus, it is not necessary to regard the system (written with dimensional variables)

$$\begin{gathered} \frac{\partial \rho}{\partial t} + \boldsymbol{v} \cdot \nabla \rho = 0 , \\ \nabla \cdot \boldsymbol{v} = 0 , \\ \rho \left[\frac{\partial \boldsymbol{v}}{\partial t} + (\boldsymbol{v} \cdot \nabla) \boldsymbol{v} \right] + \nabla p = 0 \end{gathered} \tag{3.6}$$

as only a reduction of the general model (1.30) of Euler compressible fluid flow when the characteristic Mach number M tends to zero.

3.2.1 From Adiabaticity to Isochoricity

The Euler non-viscous, incompressible model system (3.6) is called 'isochoric' by Zeytounian (2002, Sect. 4.6). In fact, in the framework of the full Euler compressible, non-viscous adiabatic model system, the *adiabaticity equation* is written as

$$C_v \frac{\mathrm{D}T}{\mathrm{D}t} + p \frac{\mathrm{D}(1/\rho)}{\mathrm{D}t} = 0 \,, \tag{3.7}$$

where the first term in (3.7) represents the part of the heat input expended for an increase in the temperature T, and the second term corresponds to the work done by expansion. Now, if we perform the isochoric limiting process

$$C_v \downarrow 0 \Rightarrow \gamma \uparrow \infty \tag{3.8}$$

(but with $C_p = R = O(1)$), then we derive, in place of (3.7),

$$\frac{\mathrm{D}\rho}{\mathrm{D}t} = 0 \Rightarrow \text{the work done by expansion is zero} \,, \tag{3.9}$$

and, from the continuity equation (1.9), we have also $\nabla \cdot \boldsymbol{v} = 0$. With (3.9), we recover the isochoric (incompressible) Eulerian non-viscous model (3.6).

3.2.2 The Fedorchenko Approach

Some authors, without going into detail, identify (as is observed in Fedorchenko 1997) the term 'incompressible fluid flow' with constant density all over the flow region. The form of the closed system (3.6) for $\boldsymbol{v}, p$ and ρ is regarded as absolutely independent of the boundary conditions (the above 'incompressible model' can be applied to the study of any bounded flow without any changes in (3.6)), in spite of the fact that we are considering so exotic a medium as an incompressible fluid characterized by an infinitely large speed of sound. However, a series of simple illustrations (see Fedorchenko 1997) shows that the usual local model of incompressible fluid flow does not work when applied to the solution of some initial–boundary-value problems. In Fedorchenko (1997), a new non-local mathematical model, where $\operatorname{div} \boldsymbol{v} \neq 0$ in the general case, is proposed for the simulation of unsteady 'hyposonic' flows in a bounded domain with continuously distributed sources of mass, momentum and entropy, also taking into account the effects of viscosity and heat conductivity when necessary (see, Sect. 3.3 below). The exclusion of sound waves is one of the most important features of this Fedorchenko model, which represents a fundamental (but ad hoc!) extension of the conventional model of incompressible fluid flow. The model is built up by modifying both the general system of equations for the motion of a compressible fluid (viscous or inviscid as required) and the appropriate set of boundary conditions. Finally, some particular cases of this model are discussed in Fedorchenko (1997), and a series of exact time-dependent solutions, one- and two-dimensional, is presented to illustrate the model. The present author's opinion is that the Fedorchenko model seems very interesting for numericians who are confronted during numerical simulations with difficulties in accurately resolving complex acoustic phenomena and in assigning the proper boundary conditions. Curiously, in two recent (voluminous) report papers, by Klein et al. (2001) and Meister (2003), devoted to numerical simulation of various fluid flows at low Mach numbers, the paper Fedorchenko (1997) is absent from the list of references.

3.3 Small-Mach-Number Non-Viscous (Eulerian) Models

3.3.1 Euler Compressible Equations and the Limit $M \downarrow 0$

We start from the full dimensionless Euler compressible, non-viscous, adiabatic equations for a thermally perfect gas (see (1.30)):

$$\frac{\mathrm{D}\rho}{\mathrm{D}t} + \rho\nabla \cdot \boldsymbol{v} = 0\,, \tag{3.10a}$$

$$\rho\frac{\mathrm{D}\boldsymbol{v}}{\mathrm{D}t} + \frac{1}{\gamma M^2}\nabla p = 0\,, \tag{3.10b}$$

$$\rho\frac{\mathrm{D}T}{\mathrm{D}t} + (\gamma - 1)p\nabla \cdot \boldsymbol{v} = 0\,, \tag{3.10c}$$

$$p = \rho T\,. \tag{3.10d}$$

Our purpose is to investigate the behaviour of the solutions of these compressible Euler equations when the characteristic Mach number M is much smaller than unity (but γ is fixed), in order to search for some kind of approximation which might be appropriate from the point of view both of accuracy and of suitability for performing computations. First, in place of the equation (3.10c) for the temperature T, we can use the entropy equation for S in the adiabatic case (when the dissipative coefficients k, μ and λ are zero); namely, we use the following Lagrangian invariant:

$$\frac{\mathrm{D}S}{\mathrm{D}t} = 0\,. \tag{3.11a}$$

In this case, in place of the equation of state $p = \rho T$, it is necessary to use the following dimensionless equation of state between the three thermodynamic functions p, S and ρ:

$$p = \rho^{\gamma}\exp(S)\,; \tag{3.11b}$$

in this case we have a set of four dimensionless equations, (3.10a,b) and (3.11a,b), for the four functions $\boldsymbol{v}, p, \rho$ and S. Our objective is not to find an approximate solution of (3.10a–d) or of (3.10a,b) and (3.11a,b) when the Mach number tends to zero, but to derive an approximate (model) system of incompressible equations, in place of (3.10a–d) or (3.10a,b) and (3.11a,b), that are applicable to low-Mach-number flows.

Small-M Limiting Process

Before embarking on this subject, we emphasize that arguments relying on the above Euler compressible equations written in dimensionless form are very useful and often quite significant, but may turn out to be misleading. One should keep in mind that quite a strong hypothesis is at the root of the reasoning, namely that all the terms which occur in the equations, except the dimensionless, parameter M (the Mach number) and the ratio γ, are of order unity and, in particular, that this order-of-magnitude assumption remains valid when differentials operate on the quantities at hand; we shall have to deal with this situation later. In an analogous but slightly different setting, we may remark that our argument will rely on the assumption that the quantities at hand do not change, as far as their order of magnitude is concerned, when one goes from place to place or from time to time within the domain of the flow in space–time. For instance, in the framework of the 'small-M' limiting process, we assume here that

$$M \downarrow 0, \quad \text{with } t \text{ and } \boldsymbol{x} \text{ fixed} , \tag{3.12}$$

the ratio γ also being fixed, and we emphasize that, when the Mach number tends to zero, the time t and the position vector $\boldsymbol{x}$ are both fixed – this a strong constraint and has some important consequences. First, obviously, independently of any sophisticated argument relying on asymptotic expansions, a matter that we shall consider soon, we may argue quite simply that the pressure gradient ∇p in (3.10b) must be very small, and indeed be $O(M^2)$. Otherwise, the rather high term $(1/\gamma M^2)\nabla p$, with γ fixed, would be unbalanced in the momentum equation, leading, quite obviously, to a contradiction. The only way out of this contradiction is to assume that, in the limiting process (3.12), the pressure may be rewritten as

$$p\,(t, x) = p_0(t) + M^2\pi(t, \boldsymbol{x}) , \tag{3.13}$$

where $p_0(t)$ may be an arbitrary function of time but must be independent of the spatial variables. In (3.13), the function $\pi(t, \boldsymbol{x})$ is a pseudo-pressure, which is introduced for convenience. From the equation of state (3.10d), we obtain

$$\rho T = p_0(t) + O(M^2) \tag{3.14}$$

and, as a first approximation, we may neglect the term $O(M^2)$ in accordance with (3.12).

3.3.2 Small-M Non-Viscous Euler Equations

Assuming that the approximation (3.13), together with (3.14), is an adequate approximation, the reader may find easily from (3.10a) and the equation (3.10c) for T that the following equation holds:

$$\gamma p_0(t)(\nabla \cdot \boldsymbol{v}) = -\frac{\mathrm{d}p_0(t)}{\mathrm{d}t} . \tag{3.15}$$

The momentum equation (3.10b) for $\boldsymbol{v}$ and the entropy equation (3.11a) for S, taking account of the equation of state (3.11b), lead to the following set of two approximate equations:

$$[p_0(t)]^{\frac{1}{\gamma}} \exp\left(-\frac{S_0}{C_p}\right) \frac{D\boldsymbol{v}}{\mathrm{D}t} + \nabla\left(\frac{\pi}{\gamma}\right) = 0 , \tag{3.16a}$$

$$\frac{\mathrm{D}S_0}{\mathrm{D}t} = 0 , \tag{3.16b}$$

where

$$S_0(t, x) = C_v \log\left(\frac{p_0(t)}{\rho^\gamma}\right) \tag{3.16c}$$

is the specific entropy (but written with $p_0(t)$ and ρ) and C_p is the specific heat cpacity at constant pressure (see (1.27)). Obviously, in place of the two equations (3.10a,b), we can also write

$$\rho\frac{\mathrm{D}v}{\mathrm{D}t} + \nabla\left(\frac{\pi}{\gamma}\right) = 0 , \tag{3.16d}$$

$$\frac{\mathrm{D}}{\mathrm{D}t}\left(\frac{[p_0(t)]^{\frac{1}{\gamma}}}{\rho}\right) = 0 . \tag{3.16e}$$

It can be readily seen that we can recover from (3.15) and (3.16) the model of constant-density inviscid flow.

Constant-Density, Non-Viscous Euler Equations for an External Problem

It can be readily seen that we can recover from (3.15) and (3.16) the non-viscous Euler model equations for a constant density with a (divergence-free velocity vector) in the case of an external flow around a bounded solid body in an unbounded atmosphere, provided we add two statements. The first statement is that the function $p_0(t)$ is a constant (p_{00}), and the second is that the specific entropy $S_0(t, \boldsymbol{x})$ is also a constant (S_{00}); in such a case, the density is also constant:

$$\rho = (p_{00})^{1/\gamma} \exp\left(-\frac{S_{00}}{C_p}\right) \equiv \rho_{00} = \text{const} . \tag{3.17}$$

The first statement holds under quite general conditions, namely that the approximation (3.13), together with (3.14), is valid up to infinity (which is the case for external aerodynamics) and that the pressure is constant-valued there ($p_0(t) \equiv p_{00} = \text{const}$). The second statement is merely a matter of boundary/initial conditions and may be considered to hold for problems of external flows over bodies in an unbounded atmosphere. The first statement is a matter of the uniform validity of an approximation (for instance at infinity, when $|\boldsymbol{x}| \to \infty$), and we shall have a lot to say about this matter in Chap. 4. Concerning the *constancy* of the specific entropy, the reader experienced in compressible-flow theory knows the argument: the conservation equation (3.16b) states that entropy is constant along particle trajectories, and our second statement holds whenever the particles come from a region of space where the entropy may be considered the same for all of the particles. On the other hand, any reader familiar with the basic elements of gas dynamics knows that the jump in the dimensionless entropy across a (very weak) shock wave is of the order of M^3, and this reinforces the above conclusion concerning the constancy of the entropy. Finally, when both of the above statements hold, we obtain the usual form of the constant-density Euler equations, namely

$$\nabla \cdot \boldsymbol{v} = 0 \,, \tag{3.18a}$$

$$\rho_{00}\frac{D\boldsymbol{v}}{Dt} + \nabla\left(\frac{\pi}{\gamma}\right) = 0 \,, \tag{3.18b}$$

where ρ_{00} must be considered as a constant (independent of both space and time and, in fact, equal to unity since we are working with dimensionless quantities).

We find that the equations (3.18) are the ones appropriate to the flow of a constant-density fluid but that the 'pressure' occurring in the non-viscous Euler momentum equation (3.17b), using ρ_{00}, is not the true pressure of the gas but the quantity

$$\frac{\pi}{\gamma} = \frac{p-1}{\gamma M^2} \,, \tag{3.18c}$$

since (in the present case, in (3.13)) $p_0(t) \equiv p_{00} = 1$; this has been introduced through manipulation of the starting Euler compressible equations (3.10) by our use of (3.12). The reader may ask what is the physical meaning of this fictitious and somewhat mysterious quantity, which looks like a pressure (more precisely, a pressure perturbation) but is not the actual pressure in the gas. As long as the flow extends to a infinity, where all quantities tend to a constant value, we may choose (as has been observed above for the constants ρ_{00} and p_{00}) the characteristic quantities in such a way that the constant values at infinity are $p^\infty = 1, \rho^\infty = 1$ and $|\boldsymbol{v}^\infty| = 1$. A fairly simple computation then shows that

$$\frac{2\pi}{\gamma} = C_p \,, \tag{3.18d}$$

where C_p is the *pressure coefficient*. We have thus found that, as a consequence of our approximation, the pressure coefficient has nothing to do with the thermodynamics of the fluid; instead, it is adjusted to an appropriate value such that the flow velocity remains divergence-free.

Small-Mach-Number Equations for a Slow, Non-Viscous Motion in a Time-Dependent Bounded Container $\Omega(t)$

In the case of a slow, non-viscous motion in a bounded time-dependent container $\Omega(t)$, it is necessary to consider the small-M model equations (3.15) and (3.16). In a such case, as a consequence of the conservation of the ratio $[p_0(t)]^{1/\gamma}/\rho$, and (again) because the jump in the dimensionless specific entropy across a shock wave is of the order of the cube of its strength, which means here that it must be of the order of M^3, we obtain the result that

$$p_0(t) = \rho_0(t)^\gamma \, , \tag{3.19}$$

and the following two equations, for the velocity vector $\boldsymbol{v}$ and the pseudo-pressure function $\Pi = \pi/\gamma\rho_0(t)$, can be derived:

$$\nabla \cdot \boldsymbol{v} + \frac{1}{\rho_0(t)} \frac{\mathrm{d}\rho_0(t)}{\mathrm{d}t} = 0 \, , \tag{3.20a}$$

$$\frac{\mathrm{D}\boldsymbol{v}}{\mathrm{D}t} + \nabla \Pi = 0 \, . \tag{3.20b}$$

More precisely, if we consider small-Mach-number slow motion in a bounded container $\Omega(t)$ with a time-dependent deformation of the wall $\partial\Omega(t)$, then it is easy to show that $p_0(t)$and$\rho_0(t)$ belong to a family of functions corresponding to adiabatic thermostatic evolution of a perfect gas in a container, with the following relation for the temperature $T_0(t)$:

$$T_0(t) = [p_0(t)]^{(\gamma-1)/\gamma} \, , \tag{3.20c}$$

and are determined from the overall conservation of the mass M^0 in the bounded container. Namely, for the density we have:

$$\rho_0(t) = \frac{1}{\mathsf{V}(t)} \, , \tag{3.20d}$$

where

$$\mathsf{V}(t) = \frac{\rho_c|\Omega(t)|}{\mathsf{M}^0} \quad \text{and} \quad \mathsf{V}(0) = 1 \, ,$$

and $|\Omega(t)|$ denotes the volume of the bounded container $\Omega(t)$, which is a known function of the time t. The above equations (3.20a,b), together with the entropy equation (3.16b) and also (3.20c,d), were used in an investigation of the propagation of premixed flames in closed tubes by Matalon and Metzener (1997). In such a case, if the flame front is described by the equation $F(t, \boldsymbol{x}) = 0$, where the unit normal, $\boldsymbol{n}$, points towards the burt gas ($\boldsymbol{n} = -\nabla F/|\nabla F|$), the clasical Rankine–Hugoniot jump relations across the flame can be used and the above relations are valid when $F(t, \boldsymbol{x}) > 0$ in the region of unburnt gas.

3.3.3 Various Small-Mach-Number Systems in a Time-Dependent Bounded Container $\Omega(t)$

In a more general case, we assume that

$$[\boldsymbol{v}(t,\boldsymbol{x};M),\rho(t,\boldsymbol{x};M),T(t,\boldsymbol{x};M)] \to [\boldsymbol{v}_0(t,\boldsymbol{x}),\rho_0(t,\boldsymbol{x}),T_0(t,\boldsymbol{x})] \tag{3.21}$$

when $M \to 0$ with t and $\boldsymbol{x}$ fixed. Then, from the Euler compressible equations (3.10), we derive the following small-Mach-number quasi-compressible, dimensionless model equations for the velocity vector $\boldsymbol{v}_0(t,\boldsymbol{x})$, the perturbation of the pressure $\pi(t,\boldsymbol{x})$ and the two thermodynamic functions $\rho_0(t,\boldsymbol{x})$ and $T_0(t,\boldsymbol{x})$:

$$\begin{gathered}
\nabla\,.\,\boldsymbol{v}_0 = -\frac{1}{\rho_0}\frac{\mathrm{D}\rho_0}{\mathrm{D}t}\,, \\
\rho_0\frac{\mathrm{D}\boldsymbol{v}_0}{\mathrm{D}t} + \nabla\left(\frac{\pi}{\gamma}\right) = 0\,, \\
\rho_0\frac{\mathrm{D}T_0}{\mathrm{D}t} + (\gamma-1)p_0(t)\nabla\,.\,\boldsymbol{v}_0 = 0\,, \\
p_0(t) = \rho_0 T_0\,,
\end{gathered} \tag{3.22}$$

where
$$\frac{\mathrm{D}}{\mathrm{D}t} \equiv \frac{\partial}{\partial t} + (\boldsymbol{v}_0\,.\,\boldsymbol{D})$$

The Rohm–Baum and Majda Systems

In place of (3.22), we can write the following system of small-Mach-number dimensionless equations:

$$\begin{gathered}
\frac{\partial\rho_0}{\partial t} + \nabla\,.\,(\rho_0\boldsymbol{v}_0) = 0\,, \\
\frac{\partial(\rho_0\boldsymbol{v}_0)}{\partial t} + (\rho_0\boldsymbol{v}_0\,.\,\nabla)\boldsymbol{v}_0 + \boldsymbol{v}_0\nabla\,.\,(\rho_0\boldsymbol{v}_0) + \nabla\left(\frac{\pi}{\gamma}\right) = 0\,, \\
\gamma\rho_0\left(\frac{\partial T_0}{\partial t} + \boldsymbol{v}_0\,.\,\nabla T_0\right) - (\gamma-1)\frac{\mathrm{d}P(t)}{\mathrm{d}t} = 0\,, \\
P(t) = \rho_0 T_0\,.
\end{gathered} \tag{3.23}$$

If, in place of the third equation (for T_0), we use the following non-adiabatic (Sect. 3.4) equation for the temperature T_0,

$$\frac{\gamma}{\gamma-1}\rho_0\left(\frac{\partial T_0}{\partial t} + \boldsymbol{v}_0\,.\,\nabla T_0\right) - \frac{\mathrm{d}P(t)}{\mathrm{d}t} = Q_0\,, \tag{3.24}$$

where Q_0 is the energy-source (non-adiabatic) term, then the first two equations (3.23) become, when we take into account the buoyancy force (proportional to Fr^{-2}) in the equation of motion for $\boldsymbol{v}_0$, toghether with (3.24), the low-Mach-number equations of Rehm and Baum (1978) (see Müller 1998, p. 107). Rehm and Baum obtained these equations by an asymptotic analysis of the compressible Euler equations for slow heat addition to describe flows in fires, the zeroth-order temperature of a particle being changed by heat conduction and the heat release rate (both contained in Q_0) and by the global rate of change of pressure with time $\mathrm{d}P(t)/\mathrm{d}t$. The pressure $P(t) \equiv p_0(t)$ in the system of equations (3.23) can be determined by integration of the third equation of (3.23) over $\Omega(t)$ (the computational bounded container), knowing $w_0 \equiv \boldsymbol{v}_0 \cdot \boldsymbol{n}$ at the boundary $\partial\Omega = \Sigma$ and taking into account the first equation of (3.23) with $P(t) = \rho_0 T_0$. Namely, since

$$\gamma\rho_0 \frac{\mathrm{D}T_0}{\mathrm{D}t} = \gamma \frac{\mathrm{d}P(t)}{\mathrm{d}t} + \gamma P(t)\nabla \cdot \boldsymbol{v}_0 \,, \tag{3.25}$$

we obtain the following equation for the function $P(t)$:

$$\left[\iiint_{\Omega} \mathrm{d}\Omega\right] \frac{\mathrm{d}P(t)}{\mathrm{d}t} + \left[\gamma \iint_{\Sigma} w_0 \,\mathrm{d}\Sigma\right] P(t) = 0 \,. \tag{3.26}$$

We may solve this ordinary differential equation to determine $P(t)$, starting from a given initial condition $P(0)$. We note that a system of equations similar to (3.23) can be used in the case of low-Mach-number *combustion* (see, for instance, Majda 1984, Majda and Sethian 1985, and Dwyer 1990) in a idealized special case, where in place of the third equation of (3.23) for T_0, we write the following two equations in accordance with Sect. 1.4.1:

$$\frac{\mathrm{D}Z_0}{\mathrm{D}t} = -K \exp\left(-\frac{A_0}{T_0}\right) Z_0 \,, \tag{3.27a}$$

$$\gamma\rho_0 \left(\frac{\partial T_0}{\partial t} + \boldsymbol{v}_0 \cdot \nabla T_0\right) - (\gamma - 1)\frac{\mathrm{d}P(t)}{\mathrm{d}t} = (\gamma - 1)K q_0 \rho_0 \exp\left(-\frac{A_0}{T_0}\right) Z_0 \,. \tag{3.27b}$$

The Fedorchenko System in Which Acoustics is Excluded

In Fedorchenko (1997), as a non-viscous approximate system of equations, with acoustics excluded, the following system of equations is considered:

$$\frac{\partial\rho_0}{\partial t} + \nabla \cdot (\rho_0 \boldsymbol{v}_0) = 0 \,,$$

$$\frac{\partial(\rho_0 \boldsymbol{v}_0)}{\partial t} + (\rho_0 \boldsymbol{v}_0 \cdot \nabla)\boldsymbol{v}_0 + \boldsymbol{v}_0 \nabla \cdot (\rho_0 \boldsymbol{v}_0) + \nabla\left(\frac{\pi}{\gamma}\right) = 0 \,,$$

$$\frac{\partial S_0}{\partial t} + \boldsymbol{v}_0 \cdot \nabla S_0 = 0 \,, \tag{3.28a}$$

with an implicit equation of state

$$F(S_0, P(t), \rho_0) = 0\ . \tag{3.28b}$$

In place of (3.28a), we can obtain an equivalent system in which entropy is excluded:

$$\frac{\partial \rho_0}{\partial t} + \nabla \cdot (\rho_0 \boldsymbol{v}_0) = 0\ ,$$
$$\frac{\partial(\rho_0 \boldsymbol{v}_0)}{\partial t} + (\rho_0 \boldsymbol{v}_0 \cdot \nabla)\boldsymbol{v}_0 + \boldsymbol{v}_0 \nabla \cdot (\rho_0 \boldsymbol{v}_0) + \nabla\left(\frac{\pi}{\gamma}\right) = 0\ ,$$
$$\nabla \cdot \boldsymbol{v}_0 = -\frac{1}{\mathsf{V}(t)} \int\!\!\int_{\Sigma} w_0\, \mathrm{d}\Sigma \tag{3.29}$$

where we take into account (3.26), with (3.20c,d).

The last equation in the Fedorchenko system (3.29) reflects the most remarkable feature which distinguishes the model (3.29) from any traditional non-viscous approaches – the system of equations (3.29) governing the motion of the fluid in any small internal volume is instantaneously non-locally related to the boundary conditions. This is not the usual instantaneous influence of boundary conditions on the local motion of the fluid that is inherent in the classical incompressible flow model owing to its partially elliptic nature; it is an explicit non-local dependence of the basic equations on the boundary conditions. One can see that only in the particular case when $\mathrm{d}P(t)/\mathrm{d}t = 0$ does the above model (3.29) reduce to the classical model of incompressible fluid flow (see (3.6)), where $\nabla \cdot \boldsymbol{v}_0 = 0$, and $S_0 \equiv S_0(\rho_0)$. Mainly as a consequence of the reduced equation of state (3.28b), where in place of π we have $P(t)$, and of the term $\nabla(\pi/\gamma)$ in the second of the equations (3.28a), the model (3.28a,b) filters all sound waves.

In Fedorchenko (1997), the reader can find a local characteristic analysis of a hyposonic flow model with acoustics excluded, following an approach developed in Courant and Hilbert (1962). For this analysis the quasi-linear version of the small-Mach-number approximate equations for independent flow variables $(\boldsymbol{v}_0, \rho_0, \pi/\gamma = \Pi)$ is considered:

$$\frac{\partial v_i}{\partial t} + v_k \frac{\partial v_i}{\partial x_k} + \frac{1}{\rho}\frac{\partial \Pi}{\partial x_i} = 0\ ,$$
$$\frac{\partial \rho}{\partial t} + v_k \frac{\partial \rho}{\partial x_k} = -\rho m\ ,$$
$$\frac{\partial v_k}{\partial x_k} = m\ , \tag{3.30a}$$

where $i, k = 1, 2, 3$, and $m = -(1/\gamma P(t))\,\mathrm{d}P(t)/dt$ is a zero-order forcing term which does not change the main local properties of (3.28a). To classify this system, we need to investigate the matrix differential operator corresponding to (3.30a). In doing, so we find a characteristic algebraic equation (characteristic cone) for new variables (τ, θ_i) instead of (t, x_i),

$$(\tau + v_k \theta_k)^3 \left({\theta_k}^2 + \tau^2\right) = 0\,. \tag{3.30b}$$

This characteristic equation shows that the above small-Mach-number system of three equations displays combined local properties: it is partially hyperbolic (there are three sets of characteristcs responsible for convection by flow) and partially elliptic (there is an instantaneous global connection between the fields of pressure and velocity). These properties resemble features of the traditional system of equations applied to the simulation of incompressible (inviscid) fluid flows. In fact, the small-Mach-number model permits us to study the evolution of two types of disturbances, namely vorticity and entropy (and in problems of hydrodynamic stability too), but with sound waves excluded. The absence of sound waves seems obvious just from the modified equation of state (3.28b), where π is omitted, i.e. we have no explicit connection between local variations of π and ρ_0, and this results in an infinitely large speed of sound (or $M \ll 1$).

However, the significant aspect (observed by Fedorchenko 1997, p. 144) of this small-Mach-number (hyposonic) model with acoustics excluded is that it is possible to estimate values of $a^2(t, \boldsymbol{x}) = \gamma P(t)/\rho_0(t, \boldsymbol{x})$ and hence to calculate the characteristic value of the low-Mach number M. Obviously, if acoustics is not the main research goal, it is not helpful to simulate simultaneously both the evolution of the unsteady mean flow and the attendant acoustic processes, since these phenomena, when considered separately, usually have very different characteristic times. For a flow in a bounded spatial domain this approach would mean that the medium would be compressible as a whole, with a considerable change of the average pressure with time being allowed, but the existence of sound waves would be completely precluded. In using this model, it is assumed that the sound speed is infinitely large even if we retain something like a thermodynamic equation of state. As a consequence a time-dependent model system would have no hyperbolic characteristics responsible for the propagation of sound; only convection by the flow is possible. The example given by Fedorchenko (1997, pp. 145–147) demonstrate some specific effects which may be observed in a bounded volume under non-steady boundary conditions which change the average pressure $P(t)$, in particular. We observe that changes in $P(t)$ with time could be a reason for the appearance of vorticity sources, which in turn could have a radical influence on the stability of all the flow.

3.3.4 The Problem of Adjustment to the Initial Conditions in a Time-Dependent Bounded Container

Here we consider the two equations (3.20a,b) and take into account (3.20d) in the case of a time-ependent bounded container $\Omega(t)$. In a such case, when at the initial time $t \leq 0^-$ we start from rest, i.e. $\boldsymbol{v} = 0$, we can assume as a consequence of the equation (3.20b) for $\boldsymbol{v}$ that $\boldsymbol{v}(= \nabla\phi)$ is irrotational, and for ϕ we obtain the following elliptic equation from (3.20a):

$$\Delta\phi = \frac{\mathrm{d}\log \mathsf{V}(t)}{\mathrm{d}t}, \tag{3.31a}$$

with the condition

$$\frac{\mathrm{d}\phi}{\mathrm{d}n}|_{\Sigma(t)} = W_{\Sigma}(t, \boldsymbol{P}), \tag{3.31b}$$

derived from the slip condition

$$(\boldsymbol{v} \cdot \boldsymbol{n})|_{\Sigma(t)} = W_{\Sigma}(t, \boldsymbol{P}), \quad \text{on the wall } \partial\Omega = \Sigma(t), \tag{3.31c}$$

where $W_{\Sigma}(t, \boldsymbol{P})$ is the normal component of the boundary velocity, and $\boldsymbol{P} \in \Sigma$. When ϕ is known, as a solution of the problem (3.31a,b), we can deduce Π thanks to the Bernoulli integral:

$$\Pi = -\left(\frac{\partial\phi}{\partial t} + \frac{1}{2}|\nabla\phi|^2\right). \tag{3.31d}$$

Obviously, for the validity of the above Neumann problem (3.31a,b)), derived when $M \downarrow 0$, with the time t fixed, it is necessary that compatibility with the initial conditions

$$t \leq 0^-, \quad \boldsymbol{v} = 0, \quad p = \rho = T = 1 \tag{3.32}$$

is ensured. For this, it is necessary that

$$\frac{\mathrm{d}\phi}{\mathrm{d}n}|_{\Sigma(0)} = 0. \tag{3.33}$$

The Case When the Deformable Time-Dependent Wall of the Bounded Container is Started Impulsively from Rest

However, (3.33) is not valid when the deformable wall is started *impulsively from rest* (or very rapidly during a time interval $O(M)$), and in such a case we have an incompatibility between our present approximate small-Mach-number model (3.31a,b), derived when $M \downarrow 0$ with t fixed, and the given initial conditions (3.32). Indeed, when the wall $\Sigma(t)$ is started impulsively from rest (or even started rapidly, during a dimensionless time $t = O(M)$) we shall show below that we have an incompatibility of the above approximate, compressible, small-Mach-number model (3.31a,b) and (3.33) with the Eulerian initial conditions (3.32), the compressible, small-Mach-number inviscid Neumann problem (3.31a,b) being singular close to $t = 0$. The low-Mach-number compressible inviscid flow valid close to the initial time $t = 0$, with the initial conditions (3.32), is characterized by a short acoustic time

$$\tau = \frac{t}{M}, \tag{3.34}$$

since the initial acoustic transition stage accurs mainly during a time interval $O(M)$; for the derivation of the associated equations of the acoustics, it is necessary to consider, in place of the limiting process (3.12), a new low-Mach-number limiting process, namely

$$M \downarrow 0, \quad \text{with } \tau \text{ and } \boldsymbol{x} \text{ fixed}\,, \tag{3.35}$$

together with (in place of (3.21)) the asymptotic expansion

$$\boldsymbol{v} = \boldsymbol{v}_{\mathrm{a}}(\tau, \boldsymbol{x}), \quad p = 1 + M\gamma\rho_{\mathrm{a}}(\tau, \boldsymbol{x}) \quad \text{and} \quad \rho = 1 + M\rho_{\mathrm{a}}(\tau, \boldsymbol{x})\,, \tag{3.36}$$

leaving $\boldsymbol{v}$ (denoted by $\boldsymbol{v}_{\mathrm{a}}$) unscaled in the acoustic theory. In such a case, $\boldsymbol{v}_{\mathrm{a}}(\tau, \boldsymbol{x})$ and $\rho_{\mathrm{a}}(\tau, \boldsymbol{x})$ are governed by the classical linear equations of acoustics,

$$\frac{\partial \rho_{\mathrm{a}}}{\partial \tau} + \nabla \cdot \boldsymbol{v}_{\mathrm{a}} = 0, \quad \frac{\partial \boldsymbol{v}_{\mathrm{a}}}{\partial \tau} + \nabla \rho_{\mathrm{a}} = 0\,. \tag{3.37}$$

Using

$$\boldsymbol{v}_{\mathrm{a}} = \nabla \varphi \quad \text{and} \quad \rho_{\mathrm{a}} = -\frac{\partial \varphi}{\partial \tau} \tag{3.38}$$

for the potential function $\varphi(t, \boldsymbol{x})$, we derive the classical hyperbolic wave equation

$$\frac{\partial^2 \varphi}{\partial \tau^2} - \Delta \varphi = 0\,, \tag{3.39}$$

and an acoustic problem of adjustment to the initial conditions must be considered. Unfortunately, the matching according to the MAE technique,

$$\lim_{\tau \to \infty} [M \downarrow 0 \text{ with } \tau \text{ fixed}] = \lim_{t \to 0} [M \downarrow 0 \text{ with } t \text{ fixed}] \tag{3.40}$$

is not possible, as is shown below, because the time-dependent domain $\Omega(t)$ is bounded. From the given initial conditions (3.32), together with the above expansion (3.36) for $\boldsymbol{v}$ and ρ we can write the initial conditions for the acoustic system (3.37)

$$\tau = \leq 0^-, \quad \boldsymbol{v}_{\mathrm{a}} = 0 \text{ and } \rho_{\mathrm{a}} = 0\,. \tag{3.41}$$

Concerning the slip condition (3.31c) for this inviscid acoustic system (3.37), we observe that, for $t = M\tau$ and $\tau = O(1)$, we have $\Sigma(t) = \Sigma(0) + O(M)$ and, as a consequence, we can write for $\boldsymbol{v}_{\mathrm{a}}$

$$(\boldsymbol{v}_{\mathrm{a}} \cdot \boldsymbol{n})|_{\Sigma(0)} = \mathsf{H}(\tau) w_{\Sigma}(\boldsymbol{P})\,. \tag{3.42a}$$

This simulates an impulsive motion of the wall on a timescale $\tau = O(1)$; $\mathsf{H}(\tau)$ is the unit Heaviside function such that

$$\mathsf{H}(\tau \leq 0^-) \equiv 0, \quad \text{but } \mathsf{H}(\tau \uparrow \infty) \equiv \mathsf{H}(t \to 0^+) \equiv 1\,. \tag{3.42b}$$

Obviously, when the short time τ tends to $+\infty$ (which is equivalent, if matching is performed, to the statement that the time t tends to 0^+), we obtain a steady slip condition

$$(\boldsymbol{v}_{\mathrm{a}} \cdot \boldsymbol{n})|_{\Sigma(0)} = w_{\Sigma}(\boldsymbol{P})\,. \tag{3.43}$$

A straightforward analysis via a Laplace transform shows that the solution of the above acoustic problem (3.37), (3.41), (3.42) and (3.43), in a bounded time-dependent container, is

$$\boldsymbol{v}_{\mathrm{a}} = \nabla\varphi^{\infty} + \Sigma_{n\geq 1}[\alpha_n C_n - \beta_n S_n]\,\boldsymbol{U}_n\,, \tag{3.43a}$$

$$\rho_{\mathrm{a}} = \Sigma_{n\geq 1}[\alpha_n S_n + \beta_n C_n] R_n\,, \tag{3.43b}$$

where $S_n = \sin(\omega_n\tau), C_n = \cos(\omega_n\tau)$ and $\boldsymbol{U}_n, R_n$ are the normal modes of vibration of the container $\Sigma(0)$ with eigenfrequencies ω_n. The velocity potential φ^{∞}, in the solution (3.43a) for $\boldsymbol{v}_{\mathrm{a}}$ satisfies the following steady-state problem:

$$\Delta\varphi^{\infty} = \iint_{\Sigma(0)} w_{\Sigma}(\boldsymbol{P})\,\mathrm{d}s, \quad \text{with } \frac{\mathrm{d}\varphi_{\infty}}{\mathrm{d}n} = w_{\Sigma}(\boldsymbol{P})\,. \tag{3.44}$$

We observe that, in reality, we have the following Neumann problem for φ^{∞}:

$$\Delta\varphi^{\infty} + f^0 = 0, \quad \frac{\mathrm{d}\varphi^{\infty}}{\mathrm{d}n} = w_{\Sigma}(\boldsymbol{P})\,, \tag{3.45a}$$

where f^0 is a constant, with the compatibility relation

$$f^0 + \iint_{\Sigma(0)} w_{\Sigma}(\boldsymbol{P})\,\mathrm{d}s = 0\,. \tag{3.45b}$$

The reader can find a detailed derivation in Zeytounian and Guiraud (1980). We note also that when $\omega_n \neq 0$,

$$\iiint_{\Omega(0)} \left[|\boldsymbol{U}_n|^2 + |R_n|^2\right]\,\mathrm{d}v = 1\,, \tag{3.46}$$

and in the case of an impulsive motion of the wall we can derive the following two relations for the anplitudes α_n and β_n, in the solution (3.43):

$$\alpha_n = \frac{1}{2}\frac{\iint_{\Sigma(0)} R_n w_{\Sigma}(\boldsymbol{P})\,\mathrm{d}s}{\iiint_{\Omega(0)} |\boldsymbol{U}_n|^2 \mathrm{d}v}\,, \quad \beta_n \equiv 0\,. \tag{3.47}$$

As a consequence of (3.47), the above solution (3.43) of the acoustic problem (3.37), (3.41) and (3.42) does not tends to a defined limit when τ tends to infinity. The constraint (3.33) is not satisfied at $t = 0$, because the matching (3.40) is not possible. We observe that the initial value $\phi(0, \boldsymbol{x})$ can be identified with φ^{∞}. Naturally, if $\alpha_n \equiv 0$ also in (3.47), then in this case the matching (3.40) of the two solutions is possible, but this is the case only when the setting in motion of the wall is *progressive*, on a timescale $t = O(1)$.

Proposition *Unfortunately, when the wall of the bounded container (in the case of internal aerodynamics) is set in motion impulsively from rest, then at the end of the initial transition stage, at a time* $t = O(M) \Leftrightarrow \tau = O(1)$, *the acoustic oscillations remain undamped and continue in the stage when $t = O(1)$, and a multiple-scale technique is necessary for the investigation of the long-time evolution of these rapid oscillations. The MAE technique and the two-timescale, single-space-scale low-Mach-number asymptotic analyses of Müller (1998) and Ali (2003) do not work.*

In Chap. 5, we consider again the above internal-aerodynamics problem via a multiple-timescale approach when, together with the slow time t linked to the boundary velocity, we consider also in the solution an infinity of fast times designed to cope the infinity of periods of free vibrations of the time-dependent bounded domain $\Omega(t)$.

3.4 Small-Mach-Number Viscous and Heat-Conducting (NSF) Models

Below, in accordance with the NSF system (1.29) but using the Stokes relation (1.4) and assuming that the two viscosity coefficients μ and λ are both constant and that the heat conduction coefficient k is independent of ρ and T, we consider the following NSF equations:

$$\frac{\mathrm{D}\rho}{\mathrm{D}t} + \rho(\nabla \cdot \boldsymbol{v}) = 0 , \tag{3.48a}$$

$$\rho\frac{\mathrm{D}\boldsymbol{v}}{\mathrm{D}t} + \frac{1}{\gamma M^2}\nabla p = \frac{1}{Re}\left[\Delta \boldsymbol{v} + \frac{1}{3}\nabla(\nabla \cdot \boldsymbol{v})\right] , \tag{3.48b}$$

$$\rho\frac{\mathrm{D}T}{\mathrm{D}t} + (\gamma - 1)p\nabla \cdot \boldsymbol{v} = \frac{\gamma}{Pr\,Re}\Delta T + \gamma(\gamma - 1)\frac{M^2}{Re}\Phi(\boldsymbol{v}) , \tag{3.48c}$$

$$p = \rho T . \tag{3.48d}$$

where $\mathrm{D}/\mathrm{D}t = \partial/\partial t + (\boldsymbol{v} \cdot \nabla)$. When $Re \equiv \infty$, with $Pr = O(1)$, we obtain again the Euler equations (3.10). We assume that the NSF equations (3.48) are valid within a domain (bounded or unbounded) Ω, and on the boundary $\partial\Omega$ it is necessary to write a no-slip condition for $\boldsymbol{v}$ and also a condition for the temperature T, but for the moment we shall not write any initial(at $t = 0$) or boundary (on $\partial\Omega$) conditions. Our results below are valid (with some obvious modifications) also when the viscosity and heat conduction coefficients are dependent on ρ and T and when the two viscosity coefficients do not satisfy Stokes' hypothesis. We consider again the limiting process (3.12), with the Reynolds and Prandtl numbers Re and Pr and the ratio of specific heats γ also fixed and $O(1)$. If we write the asymptotic expansion

$$U = [\boldsymbol{v}, p, \rho, T](\boldsymbol{x}, t; M) = U_0 + MU_1 + M^2U_2 + \dots , \tag{3.49}$$

where

$$M \to 0, t \text{ and } \boldsymbol{x} \text{ fixed},$$

then we have the following expansion for the pressure $p(\boldsymbol{x}, t)$:

$$p = p_0(t) + Mp_1(t) + M^2 p_2(\boldsymbol{x}, t) + \dots, \tag{3.50a}$$

where $p_0(t) = \rho_0 T_0$; we assume that

$$[\boldsymbol{v}, \rho, T](\boldsymbol{x}, t; M) \to [\boldsymbol{v}_0, \rho_0, T_0](\boldsymbol{x}, t) \quad \text{when } M \to 0 \text{ with } (t, \boldsymbol{x}) \text{ fixed}. \tag{3.50b}$$

In this case, we can derive a 'small-Mach-number dissipative model' for $\boldsymbol{v}_0, \rho_0, T_0$ and p_2, where $p_0(t)$ is also present. Namely, in place of the NSF full system (3.48), we obtain

$$\begin{aligned}
&\frac{\mathrm{D}\rho_0}{\mathrm{D}t} + \rho_0 \nabla \cdot \boldsymbol{v}_0 = 0, \\
&\rho_0 \frac{\mathrm{D}\boldsymbol{v}_0}{\mathrm{D}t} + \frac{1}{\gamma}\nabla p_2 = \frac{1}{Re}\left[\Delta \boldsymbol{v}_0 + \frac{1}{3}\nabla(\nabla \cdot \boldsymbol{v}_0)\right], \\
&\rho_0 \frac{\mathrm{D}T_0}{\mathrm{D}t} + (\gamma - 1)p_0(t)(\nabla \cdot \boldsymbol{v}_0) = \frac{\gamma}{Pr\, Re}\Delta T_0,
\end{aligned} \tag{3.51}$$

where $\mathrm{D}/\mathrm{D}t = \partial/\partial t + (\boldsymbol{v}_0 \cdot \nabla)$. An obvious consequence of the first and last equations of the small-Mach-number dissipative model system (3.51) is the following relation for the unknown leading-order pressure $p_0(t)$:

$$\frac{\mathrm{d}p_0(t)}{\mathrm{d}t} = -\gamma \nabla \cdot \left[p_0(t)\boldsymbol{v}_0 - \frac{1}{Pr\, Re}\nabla T_0\right]. \tag{3.52}$$

In Fröhlich's thesis (Fröhlich 1990), the reader can find various methods for the computation of the thermodynamic pressure $p_0(t)$. From (3.52), we arrive at the equation

$$\nabla \cdot \boldsymbol{v}_0 = -\frac{1}{\gamma p_0(t)}\frac{\mathrm{d}p_0(t)}{\mathrm{d}t} + \frac{1}{Pr\, Re}\nabla\left(\frac{T_0}{p_0(t)}\right), \tag{3.53}$$

which represents a logical extension of the last equation of (3.29). If we now use (3.53) instead of the third equation of (3.51) for T_0, then we obtain an approximate small-Mach-number dissipative system, which is strictly equivalent to the system (3.51) but now excludes the temperature. Namely, for the temperature T_0 we have the relation

$$T_0 = \frac{p_0(t)}{\rho_0}, \tag{3.54a}$$

and for $\boldsymbol{v}_0, p_2$ and ρ_0 we have the following small-Mach-number dissipative system:

$$\frac{\mathrm{D}\rho_0}{\mathrm{D}t} + \rho_0 \nabla . \boldsymbol{v}_0 = 0,$$
$$\rho_0 \frac{\mathrm{D}\boldsymbol{v}_0}{\mathrm{D}t} + \frac{1}{\gamma}\nabla p_2 = \frac{1}{Re}\left[\Delta \boldsymbol{v}_0 + \frac{1}{3}\nabla(\nabla . \boldsymbol{v}_0)\right],$$
$$\nabla . \boldsymbol{v}_0 = -\frac{1}{\gamma p_0(t)}\frac{\mathrm{d}p_0(t)}{\mathrm{d}t} + \frac{1}{Pr\,Re}\nabla\left(\frac{T_0}{p_0(t)}\right) . \tag{3.54b}$$

Here, $p_0(t)$ is also present.

It seems, according to Embid (1987), that the system (3.51) is well-posed for a judicious choice of initial and boundary conditions. However, the equation of state $p_0(t) = \rho_0 T_0$ implies a singular behaviour close to the initial time $t = 0$, because the system (3.51) or (3.54) does not take acoustic phenomena into account. The system (3.54) is a radically new non-local model, which includes the Navier constant-density viscous model; the latter is derived when $T_0 \equiv 1$ and $\rho_0 \equiv 1$, such that $p_0(t) \equiv 1$. For the divergence-free Navier velocity vector $\boldsymbol{v}_{\mathrm{N}}$ and the pseudo-pressure p_{N}, we obtain the following classical viscous dynamic model, which is usually valid in external hypo-aerodynamics problems:

$$\nabla . \boldsymbol{v}_{\mathrm{N}} = 0, \quad \frac{\mathrm{D_N}\,\boldsymbol{v}_{\mathrm{N}}}{\mathrm{D}t} + \frac{1}{\gamma}\nabla p_{\mathrm{N}} = \frac{1}{Re}\Delta \boldsymbol{v}_{\mathrm{N}} , \tag{3.55}$$

where $\mathrm{D_N}/\mathrm{D}t = \partial/\partial t + \boldsymbol{v}_{\mathrm{N}} . \nabla$, and $p_{\mathrm{N}} = (p-1)/M^2$. In fact, the small-Mach-number model (3.51), with $\rho_0 = p_0(t)/T_0$, is a 'pseudo-dilatable' model, but it seems that is not obvious that one can derive from the equations (3.51) a shallow Rayleigh–Bénard thermal-convection (Boussinesq) simplified system, as is claimed in Viozat (1998) (see Sect. 1.2.2), when the reference difference of temperatures is small. In Sect. 7.2, we give a detailed asymptotic derivation of these equations for the Rayleigh–Bénard problem (see also Sect. 2.5.1).

3.4.1 The Case of a Time-Dependent Bounded Domain

By integrating (3.52) over a time-dependent bounded domain $\Omega(t)$, we obtain

$$\frac{\mathrm{d}p_0(t)}{\mathrm{d}t} = -\frac{\gamma}{\mathsf{V}(t)}\iint_{\Sigma(t)}\left[p_0(t)\boldsymbol{v}_0 - \frac{1}{Pr\,Re}\nabla T_0\right] . \boldsymbol{n}\,\mathrm{d}\Sigma , \tag{3.56}$$

where $\Sigma(t) = \partial\Omega(t)$ and where $\mathsf{V}(t)$ is the reduced, dimensionless volume of the time-dependent bounded domain (in accordance with (3.20d)). The above relation (3.56) must be solved with an independently assigned set of both initial and boundary conditions. Bearing in mind the inviscid version of the model, we can ignore (3.56) only if we are considering an external flow in an unbounded atmosphere. Thus we have derived a closed system of unsteady equations (3.54) and (3.56) – of integro-differential type in the general case – for the set of flow variables $(\boldsymbol{v}_0, p_2, \rho_0, T_0, p_0(t))$ in the bounded (closed) time-dependent domain $\Omega(t)$. One can confirm that this model is characterized by the complete exclusion of any acoustic effects, as in the case of its inviscid version. A local analysis of the system (3.54) and (3.56) also excludes characteristics responsible for the propagation of sound disturbances, because of the new form of the equation of state (3.28b). Indeed, we now have an incompletely parabolic system, where the second spatial derivatives cannot lead to the existence of sound waves. If $\Sigma(t) \equiv \Sigma_0$ is a non-moving, impermeable wall,the average pressure $p_0(t)$ depends only on the total heat flux through the whole time-independent boundary Σ_0. This flux could be found either directly from a boundary condition on the wall or in an implicit manner from the solution of the initial–boundary–value problem at the time under consideration. The introduction of the 'global'variable $p_0(t)$ results in the appearance of a new degree of freedom when the total mass balance in a closed volume with heat conduction through the walls is considered.

This model has notable advantages (first, we now have a better form of the continuity equation, ensuring mass conservation) over the traditional models of incompressible fluid flow which are commonly used in various problems of heat convection. The important distinctive feature of the above small-Mach-number dissipative model (as is observed by Fedorchenko 1997, p. 149, pertinently)is that $\operatorname{div} \boldsymbol{v}_0 \neq 0$ even if $\mathrm{d}p_0(t)/\mathrm{d}t = 0$. This reflects the conformity of our model to the natural fact that any change in the temperature field due to heat conduction must be accompanied by flow. For instance, such a change can induce a streaming, although this may be rather slow, caused by a non-equilibrium temperature field imposed on an initially static fluid under a steady pressure. Finally, it should be recalled that within the above model, the smoothing of the pressure field by sound waves is supposed to occur much faster than the above changes in the fields of both temperature and density. In Fedorchenko (1997, pp. 149–153), the reader can find a set of exact solutions that demonstrates how the above approach could be used for the simulation of diverse hyposonic viscous flows. All these solutions represent a necessary supplement to the general description of the small-Mach-number dissipative model. Additionally, they can serve as tests if we decide to apply that model to the computational study of multi-dimensional unsteady problems. Moreover, every exact solution of a nonlinear problem in fluid mechanics, expecially by the time-dependent solution of the NSF equations, has a high intrinsic value irrespective of its practical interpretation. In Meister (2003), the case of an adiabatic boundary condition ($\partial T_0/\partial n|_{\Sigma(t)} = 0$) with vanishing compression and expansion over the boundary ($\iint_{\Sigma(t)} (\boldsymbol{v}_0 \cdot \boldsymbol{n})\, \mathrm{d}\Sigma = 0$) is considered, and in this case $p_0(t) = \text{const}$ ($\equiv 1$, since we are working with dimensionless quantities).

Unfortunately, in combustion problems it is necessary to take account of the fast acoustic waves which emerge in the initial stage of the unsteady flow, and their infuence on the averaged quasi-incompressible flow. In the dissipative case, this is a very difficult problem, and we are actually only able to derive a fully consistent asymptotic model only for the non-viscous case (see Sect. 5.3). However, for the dissipative case, some new interesting results are presented in this book: in Sect. 5.4.1, we derive an average Navier–Fourier model when

$Re = O(1)$ and $Pr = O(1)$, and in Sect. 5.4.2, we investigate the damping process of the acoustic oscillations when $Re \gg 1$, via the introduction of a Stokes layer of thickness $\Delta = \sqrt{M/Re}$.

The Viscous Case when $T_0 = T_0(t)$

The equation for $T_0(\boldsymbol{x}, t)$, in the small-Mach-number model (3.51) can be rewritten (for the specific entropy S_0) in the form

$$\left(\frac{\partial}{\partial t} + \boldsymbol{v}_0 \,.\, \nabla\right) \log\left[\frac{p_0(t)}{(\rho_0)^\gamma}\right] = \frac{\gamma}{Pr\, Re}\Delta\left(\frac{T_0}{p_0(t)}\right) . \tag{3.57}$$

If we assume that in (3.50b)

$$T_0 = T_0(t) , \tag{3.57a}$$

then we obtain the following conservation (Lagrangian-invariant) equation for the specific entropy S_0 from (3.57):

$$\left(\frac{\partial}{\partial t} + \boldsymbol{v}_0 \,.\, \nabla\right) S_0 = 0, \quad \text{with } p_0(t) = (\rho_0(t))^\gamma \exp(S_0) . \tag{3.58}$$

This is in accordance with (3.54a). It seems that we can, *as a conjecture*, assume that when $M \downarrow 0$, the shocks are transferred to a distance $O(1/M)$ in space and in time (but such a statement necessitates a rigorous mathematical proof), and if this is really the case, then (we note that the additive constant in the definition of the specific entropy S is not taken into account)

$$S_0 \equiv 0 \quad \text{and } p_0(t) = (\rho_0(t))^\gamma . \tag{3.59}$$

This is also true in the dissipative case when

$$\partial T_0/\partial n|_{\Sigma(t)} = 0 , \tag{3.57b}$$

since this thermal condition on the boundary $\Sigma(t)$ (the wall $\partial\Omega(t)$) is compatible with the solution (3.57a). The above conclusion is also true if we write, for the full NSF starting system of equations (3.48), the following thermal boundary condition on the wall $\partial\Omega(t) \equiv \Sigma(t)$:

$$T = T_0(t) + \Lambda^* M^\alpha \Theta(t, \boldsymbol{P}), \quad \alpha > 0, \quad \Lambda^* = O(1), \quad \boldsymbol{P} \in \Sigma(t) , \tag{3.57c}$$

where Λ^* is a measure of the given thermal field $\Theta(t, \boldsymbol{P})$ on $\Sigma(t)$, and $a > 0$ is a positive scalar.We observe that the thermal condition (3.57c) is compatible with the following thermodynamic initial conditions:

$$\text{at } t = 0: \quad p = 1, \quad T = 1, \quad \rho = 1 \quad \text{and } S = 0 . \tag{3.60}$$

This reinforces (3.59). We now assume that the boundary $\Sigma(t)$ is deformed over time with the given 'boundary velocity'

$$\boldsymbol{U}_\Sigma(t,\boldsymbol{P}) = \boldsymbol{V}_\Sigma(t,\boldsymbol{P}) + W_\Sigma(t,\boldsymbol{P})\boldsymbol{n}\ , \tag{3.61}$$

and we write first a *no-slip* condition for the velocity vector $\boldsymbol{v}$ (for the viscous equation of motion (3.48b)),

$$\boldsymbol{v}|_{\Sigma(t)} = \boldsymbol{U}_\Sigma(t,\boldsymbol{P})\ . \tag{3.62}$$

As initial condition for $\boldsymbol{v}$, the solution of (3.48b), we write

$$\boldsymbol{v}|_{t=0} = \boldsymbol{v}^0(\boldsymbol{x})\ . \tag{3.63}$$

In this case, with the no-slip condition (3.62) and the thermal condition (3.57c), where $\Lambda^* = O(1)$, we can derive, from the relation (3.56)(which is valid for small-Mach-number flows, i.e. for $M \downarrow 0$ with t and $\boldsymbol{x}$ fixed), an integro-differential relation between $p_0(t)$ and $W_\Sigma(t,\boldsymbol{P})$, namely

$$\frac{\mathrm{d}\,\log p_0(t)}{\mathrm{d}t} = -\frac{\gamma}{\mathsf{V}(t)}\iint_{\Sigma(t)} W_\Sigma(t,\boldsymbol{P})\,\mathrm{d}\Sigma\ . \tag{3.64}$$

When we observe that the conservation of the global mass of the time-dependent bounded body $\Omega(t)$ gives, in dimensionless form,

$$\rho_0(t) = \frac{1}{\mathsf{V}(t)} = [p_0(t)]^{1/\gamma}\ , \tag{3.65a}$$

we obtain

$$\frac{\mathrm{d}\mathsf{V}(t)}{\mathrm{d}t} = -\iint_{\Sigma(t)} W_\Sigma(t,\boldsymbol{P})\,\mathrm{d}\Sigma \tag{3.65b}$$

in place of (3.64).

Proposition 1. *With the thermal boundary condition (3.57c) and the thermodynamic initial conditions (3.60), we can derive asymptotically, from the full starting NSF equations (3.48) and as a consequence of the low-Mach-number limiting process $M \downarrow 0$, with t and $\boldsymbol{x}$ fixed, the following viscous model equations for the limiting functions $\boldsymbol{v}_0(t,\boldsymbol{x})$ and $p_2(\boldsymbol{x},t)$:*

$$\nabla\,.\,\boldsymbol{v}_0 = \frac{\mathrm{d}\,\log \mathsf{V}(t)}{\mathrm{d}t}\ , \tag{3.66a}$$

$$\frac{\mathrm{D}\boldsymbol{v}_0}{\mathrm{D}t} + \nabla\Pi = \nu^0(t)\,\Delta\boldsymbol{u}_0\ , \tag{3.66b}$$

where

$$\Pi = \frac{\mathsf{V}(t)}{\gamma}p_2(t,\boldsymbol{x}),\quad \nu^0(t) = \frac{1}{Re\,\mathsf{V}(t)}\quad \text{and}\quad \frac{\mathrm{D}}{\mathrm{D}t} = \frac{\partial}{\partial t} + \boldsymbol{v}_0\,.\,\nabla\ . \tag{3.66c}$$

Here, the no-slip condition on $\Sigma(t)$,

$$\boldsymbol{v}_0|_{\Sigma(t)} = \boldsymbol{U}_\Sigma(t,\boldsymbol{P}),\quad \boldsymbol{P}\in\Sigma(t)\ , \tag{3.66d}$$

is applied. These equations are valid inside the time-dependent domain $\Omega(t)$ bounded by the wall $\Sigma(t)$.

When $\nu^0(t) = 0$ ($Re \equiv \infty$) we recover the inviscid model (3.20a,b) in place of (3.66a,b). However, because the above system (3.66a–d) governs an unsteady hyposonic flow with acoustics excluded, we do not have the possibility to apply the initial condition

$$\boldsymbol{v}_0|_{t=0} = \boldsymbol{v}^0(\boldsymbol{x}) \,, \tag{3.67}$$

which is a direct consequence of the initial condition (3.63) for the velocity vector $\boldsymbol{v}(t, \boldsymbol{x})$, in the NSF system (3.48). As a consequence, close to the initial time $t = 0$, the above model problem (3.66) is certainly singular, and it is necessary to introduce again a short time $\tau = t/M$, together with the limiting inner-in-time process (3.35) and the asymptotic acoustic expansion (3.36). At the leading order, from the NSF equations (3.48), with Re and Pr fixed in the limiting inner-in-time process (see (3.34)–(3.36)), we derive the following classical (inviscid) linear equations of acoustics:

$$\begin{aligned} \frac{\partial \rho_{\rm a}}{\partial \tau} + \nabla \cdot \boldsymbol{v}_{\rm a} = 0, \\ \frac{\partial \boldsymbol{v}_{\rm a}}{\partial \tau} + \frac{1}{\gamma} \nabla p_{\rm a} = 0, \\ \frac{\partial T_{\rm a}}{\partial \tau} + (\gamma - 1) \nabla \cdot \boldsymbol{v}_{\rm a} = 0 \,, \end{aligned} \tag{3.68}$$

where $p_{\rm a} = \rho_{\rm a} T_{\rm a}$, and $T_{\rm a} = (T - 1)/M$, when M tends to zero with $\tau = \frac{t}{M}$ fixed. In the case of a time-dependent bounded domain, it is obvious that the acoustic waves generated at the intial time $t \geq 0^+$ by the deformation (as a function of time) of the bounded time-dependent domain under the effect of the 'boundary velocity' do not, in general, vanish when $\tau \uparrow \infty$. This is obviously the case when the motion of the wall $\Sigma(t)$ of the time-dependent bounded domain is accelered from rest to a finite velocity in a time $O(M)$. Finally, we can formulate the following second proposition.

Proposition 2. *At $t \geq 0^+$, as a consequence of the deformation (as a function of time) of the bounded/closed domain, under the effect of the 'boundary velocity', we expect that the (acoustic) solution of (3.66a,b) will oscillate on a timescale $O(M)$. On the other hand, the solution of the limiting low-Mach-number NSF flow inside the time-dependent bounded domain evolves on a timescale $O(1)$ owing to the time evolution of the wall. Therefore, we must build into the structure of the solution of the NSF equations a multiplicity of timescales:a slow time t and an infinity of fast times designed to cope with the infinity of periods of free vibrations of the bounded domain.*

3.4.2 The Case of an External Dissipative, Unsteady Flow

In the external dissipative case, we can see (according to Sects. 2.1.1 and 2.1.2) that it is necessary to investigate two main singular local, inviscid regions and match both these local regions with the outer Navier region.

Far Field

In the case of an external flow, it is necessary to consider first the far field, where $|\boldsymbol{x}| \to \infty$, this region is governed by the equations of acoustics. More precisely, these acoustic equations emerge via the outer (far-field) limiting process (2.1), $\lim^{0\text{ff}}$, mentioned in Chap. 2, when we consider the following far-field expansion:

$$\boldsymbol{v} = M^{\beta}\boldsymbol{v}^{0\text{ff}} + \dots, \quad (p, \rho, T) = 1 + M^{\beta+1}(p^{0\text{ff}}, \rho^{0\text{ff}}, T^{0\text{ff}}) + \dots , \tag{3.69}$$

where the far-field terms (with the superscript '0ff') are dependent on the time–space variables $(t, \boldsymbol{\xi})$, and $\boldsymbol{\xi} = M\boldsymbol{x}$. An analysis of the far field shows that $\beta \geq 2$ (see, for instance, Sery Baye, 1994). With (3.69) we can derive, for the solutions, $\boldsymbol{v}_{\mathrm{N}}$ and p_{N} of the usual divergence-free (constant-density) viscous Navier model equations (3.55), the following behaviour conditions at infinity:

$$|\boldsymbol{v}_{\mathrm{N}}| \to 0 \quad \text{and } p_{\mathrm{N}} \to 0, \quad \text{when } |\boldsymbol{x}| \to \infty . \tag{3.70}$$

In Chap. 4, we consider again the above far field in the framework of an external unbounded flow around a solid bounded body. Here we note only that an asymptotic investigation of the far field, in relation to its application to aerodynamic sound generation (à la Lighthill 1952), has been developed by Obermeier (1976). In Sery Baye's thesis (Sery Baye 1994), the far-field flow and its matching with the outer (Eulerian inviscid) flow are considered in the framework of the Steichen equation for a compressible, potential, unsteady fluid flow with $\boldsymbol{v} = \nabla\Phi$. The Steichen equation is

$$\begin{aligned}\Delta\Phi - SM^2\frac{\partial^2\Phi}{\partial t^2} = M^2\Bigg\{S\frac{\partial}{\partial t}\left(\frac{1}{2}|\nabla\Phi|^2\right)\\ + \left[(\gamma-1)\,\Delta\Phi + \nabla\Phi\,.\,\nabla)\right]\left(S\frac{\partial\Phi}{\partial t} + \frac{1}{2}|\nabla\Phi|^2\right)\Bigg\},\end{aligned} \tag{3.71}$$

where $\Delta = \nabla^2$. This equation governs an Eulerian irrotational potential flow, which gives the Laplace equation $\Delta\Phi = 0$ when the main limiting process, $M \downarrow 0$ with t and $\boldsymbol{x}$ fixed, is considered. In such a case, it is necessary to introduce a 'distal' (distant) potential function $\Phi^*(t, \xi)$ for the far field, with the following distal asymptotic expansion:

$$\Phi = M^q\Phi^*(t, \boldsymbol{\xi}) = \sum_{p=0}^{6-q} M^p\Phi_p^*(t, \boldsymbol{\xi}) + \sum_{s=4-q}^{6-q} (M^s \operatorname{Log} M)\Phi_{s,1}^*(t, \boldsymbol{\xi}) , \tag{3.72}$$

where $\boldsymbol{\xi} = M\boldsymbol{x}$. For monopole-like, dipole-like and quadrupole-like behaviour of the leading-order 'proximal' potential function $\Phi_0(t, \boldsymbol{x}) \approx O(1/r^q)$, $q > 0$, such that $\Delta\Phi_0 = 0$ when $r = |\boldsymbol{x}| \to \infty$, we have $q = 1, q = 2$ and $q = 3$, respectively. The proximal asymptotic expansion associated with the distal expansion (3.72) is

$$\Phi(t, \boldsymbol{x}) = \sum_{p=0}^{6} M^p \Phi_p(t, \boldsymbol{x}) + \sum_{s=4}^{6} (M^s \operatorname{Log} M) \Phi_{s,1}(t, \boldsymbol{x}) \ , \tag{3.73}$$

according to Sery Baye (1994).

The Region Close to $t = 0$ and the Acoustic d'Alembert Equation

Obviously, the Navier viscous model (3.55), with (3.70), for an external dissipative flow is also singular close to the initial time, because the Navier velocity vector $\boldsymbol{v}_{\mathrm{N}}$ is divergence-free. Close to $t = 0$, the inner-in-time acoustic limiting process (2.4), $\lim^{\mathrm{Ac}}$. must be considered, with the following local-in-time inner (acoustic) expansion:

$$\boldsymbol{v} = \boldsymbol{v}_{\mathrm{a}0} + M \boldsymbol{v}_{\mathrm{a}1} + \dots \ , \tag{3.74a}$$

$$(p, \rho, T) = 1 + M(p_{\mathrm{a}1}, \rho_{\mathrm{a}1}, T_{\mathrm{a}1}) + M^2(p_{\mathrm{a}2}, \rho_{\mathrm{a}2}, T_{\mathrm{a}2}) + \dots \ , \tag{3.74b}$$

where the acoustic terms (with a subscript 'a') are dependent on the short acoustic time $\tau = t/M$ and the position vector $\boldsymbol{x}$. The equations derived from the full NSF unsteady equations for $\boldsymbol{v}_{\mathrm{a}0}$ and $(p_{\mathrm{a}1}, \rho_{\mathrm{a}1}, T_{\mathrm{a}1})$ are again (when $Re = O(1)$ and $Pr = O(1)$) the classical inviscid equations of acoustics (3.68), but written for $\boldsymbol{v}_{\mathrm{a}0}$ and $(p_{\mathrm{a}1}, \rho_{\mathrm{a}1}, T_{\mathrm{a}1})$. Via the usual formulae for an exterior unbounded domain,

$$\boldsymbol{v}_{\mathrm{a}0} = \boldsymbol{v}_{\mathrm{N}}(t = 0, \boldsymbol{x}) + \nabla \phi_{\mathrm{a}0}, \quad (p_{\mathrm{a}1}, \rho_{\mathrm{a}1}, T_{\mathrm{a}1}) = (-1, -\gamma, (\gamma - 1)) \frac{\partial \phi_{\mathrm{a}0}}{\partial \tau} \ , \tag{3.75a}$$

we can solve these acoustic equations, provided $\phi_{\mathrm{a}0}$ is a solution of the dimensionless d'Alembert equation of acoustics,

$$\frac{\partial^2 \phi_{\mathrm{a}0}}{\partial \tau^2} - \Delta \phi_{\mathrm{a}0} = 0 \ . \tag{3.75b}$$

We observe that in the formula for $\boldsymbol{v}_{\mathrm{a}0}$ (3.75a), we have, according to (3.55),

$$\nabla \cdot \boldsymbol{v}_{\mathrm{N}}(t = 0, \ \boldsymbol{x}) = 0 \ . \tag{3.76a}$$

Also, necessarily (since in (3.75a) $\boldsymbol{v}_{\mathrm{a}0}$ is derived from a potential function, as a consequence of the second equation of acoustics (3.68)), we have

$$\boldsymbol{v}_{\mathrm{N}}(t = 0, \ \boldsymbol{x}) = \nabla \psi_{\mathrm{N}}^0 \ , \tag{3.76b}$$

where the Navier initial velocity vector, $\boldsymbol{v}_{\mathrm{N}}(t = 0, \ \boldsymbol{x}) \equiv \boldsymbol{v}_{\mathrm{N}}^0$, is an unknown.

Now the main problem is the study of the behaviour of the d'Alembert equation (3.75b) when

$$\tau \to \infty \ . \tag{3.77}$$

Although matching is not always possible in the case of a bounded domain (for instance, when we consider an impulsive motion of the wall from rest), matching is, in contrast, possible in the case of *external aerodynamics* in the framework of the unsteady adjustment problem, and the MAE technique works very well, because the behaviour of the solution of the hyperbolic equation (3.75b) for the acoustic waves is such that it vanishes for a large short time $\tau = t/M$. In fact, in the case of external aerodynamics, a rigorous investigation of the behaviour at large distance (or large time) of solutions of the wave equation (3.75b) is a quite difficult question, as far as the mathematical part of it is concerned; the reader can find in Wilcox (1975) a rigorous scattering theory for the d'Alembert equation (3.75b) in an exterior domain. Namely, in an exterior domain, we can derive, as an asymptotic result (for a large time $\tau \to \infty$, with $|\boldsymbol{x}| - \tau = O(1)$), the following behaviour for ϕ, the solution of the d'Alembert acoustic equation $\partial^2\phi/\partial\tau^2 - \Delta\phi = 0$.

$$\phi \approx \frac{1}{\tau}\,\mathsf{F}\left(|\boldsymbol{x}| - \tau, \frac{\boldsymbol{x}}{|\boldsymbol{x}|}\right) + o\left(\frac{1}{\tau}\right), \quad \tau \to \infty, \tag{3.78}$$

where the function F depends on the various conditions which contribute to define the mathematical problem.

It may be of some interest to the reader to know a quite simple derivation of (3.78), via a multiple-scaling technique, if he/she is not interested in mathematically rigorous arguments. For this purpose, we write the solution of the acoustic equation in the following form:

$$\phi(\tau, \boldsymbol{x}) = \varphi * (f(\tau, \boldsymbol{x}); \tau, \boldsymbol{x}), \tag{3.79}$$

where the dependence of φ^* on τ and $\boldsymbol{x}$ through $f(\tau, \boldsymbol{x})$ is much more rapid than the direct dependence. The physical reason for using such an a priori representation is that acoustic waves are, basically, propagated to great distances without being much distorted. If they have some wavelengths or frequencies built in these must persist, as far as the order of magnitude is concerned, up to infinity. Such a behaviour is precisely accounted for by the dependence through $f(\tau, \boldsymbol{x})$. On the other hand, the waves evolve in space and time over spatial distances and time intervals which involve very many wavelengths or frequencies. This feature is accounted for by the direct dependence through τ and $\boldsymbol{x}$. From (3.79), to the level of the above acoustic approximation and keeping only the leading approximation near infinity, we obtain the result that for $r = -\partial\varphi\partial\tau$ and $\boldsymbol{v} = \nabla\varphi$ we have

$$r = r^* = -\frac{\partial\varphi^*}{\partial f}\frac{\partial f}{\partial\tau} \quad \text{and } \boldsymbol{v} = \boldsymbol{v}^* = \frac{\partial\varphi^*}{\partial f}\nabla f, \tag{3.80}$$

and the acoustic equation is equivalent to the following two equations:

$$\frac{\partial f}{\partial\tau}\frac{\partial r^*}{\partial f} + \nabla f \cdot \left(\frac{\partial\boldsymbol{v}^*}{\partial f}\right) = -\left[\frac{\partial r^*}{\partial\tau} + \nabla\cdot\boldsymbol{v}^*\right], \tag{3.81a}$$

$$\frac{\partial f}{\partial\tau}\frac{\partial\boldsymbol{v}^*}{\partial f} + \frac{\partial r^*}{\partial f}\nabla f = -\left[\frac{\partial\boldsymbol{v}^*}{\partial\tau} + \nabla r^*\right]. \tag{3.81b}$$

Concurrently, φ^* may be an arbitrary function of f, while $\partial f/\partial\tau$ and ∇f have to be related by the dispersion relation between the wave vector and the frequency, where we have to choose the sign appropriate to waves going out towards infinity. We obtain a first-order partial differential equation for $f(\tau, \boldsymbol{x})$, which is solved by the standard method of characteristics. The solutions of the *characteristic strips* are solutions to the following set of equations:

$$\frac{\mathrm{d}\tau}{-\omega} = \frac{\mathrm{d}x_j}{k_j} = \frac{\mathrm{d}\omega}{0} = \frac{\mathrm{d}k_j}{0}\,, \tag{3.82}$$

where ω stands for $-(\partial f/\partial\tau)$ and k_j for $\partial f/\partial x_j$. From (3.82) we obtain the result that along the acoustic rays $\boldsymbol{x} = \boldsymbol{x}^0 + (\tau - \tau^0)\boldsymbol{e}, \omega = -(\partial f/\partial\tau)$ and $\boldsymbol{k} = \nabla f$ are both constant, with $\omega = |\boldsymbol{k}|$ and $\boldsymbol{e} = \boldsymbol{k}/|\boldsymbol{k}|$. In place of (3.81), we can write

$$\begin{aligned} -\frac{\partial f}{\partial\tau}\left[\frac{\partial r^*}{\partial\tau} + \nabla\,.\,\boldsymbol{v}^*\right] + \nabla f\,.\left[\frac{\partial\boldsymbol{v}^*}{\partial\tau} + \nabla r^*\right] \\ = \left\{\left(\frac{\partial f}{\partial\tau}\right)^2 - |\nabla f|^2\right\}\left(\frac{\partial r^*}{\partial f}\right) \equiv 0\,, \end{aligned} \tag{3.83}$$

if we want to improve the approximation (3.80). This approximation (3.80) must satisfy the compatibility condition

$$\frac{\partial f}{\partial\tau}\left[\frac{\partial r^*}{\partial\tau} + \nabla\,.\,\boldsymbol{v}^*\right] - \nabla f\left[\frac{\partial\boldsymbol{v}^*}{\partial\tau} + \nabla r^*\right] = 0\,. \tag{3.84}$$

Continuing the computation, we find that

$$2\left(\frac{\partial}{\partial\tau} + \boldsymbol{e}\,.\,\nabla\right)\frac{\partial\varphi^*}{\partial f} + \frac{\partial\varphi^*}{\partial f}\nabla\,.\,\boldsymbol{e} = 0\,, \tag{3.85}$$

because $\partial\omega/\partial\tau + \nabla\,.\,\boldsymbol{k} = \nabla\,.\,\boldsymbol{e}$ and $\omega\,\partial/\partial\tau + \boldsymbol{k}\,.\,\nabla = \partial/\partial\tau + \boldsymbol{e}\,.\,\nabla$, where the unit vector $\boldsymbol{e}$ is that which occurs in the above definition of the acoustic rays. In the far field, we may obviously use $\boldsymbol{x} \approx \tau\boldsymbol{e}$, from which we derive

$$\nabla\,.\,\boldsymbol{e} \approx \frac{2}{\tau}\,. \tag{3.86}$$

As a consequence, (3.85) shows that

$$\tau\frac{\partial\varphi^*}{\partial f} \text{ is constant along each ray}\,. \tag{3.87}$$

But this result is precisely what is stated after (3.78), and we hope that the above argument will have been of some help to the reader in understanding one of the deepest results of acoustic theory. A direct consequence of (3.78) is that, for the solution of the acoustic equation (3.75b), we have as the behaviour for large time

$$\tau \to \infty\,:\,\phi_{\mathrm{a}0}(\tau, \boldsymbol{x}) \to 0\,, \tag{3.88}$$

which is guaranted by the mathematically rigorous theory of d'Alembert's equation (3.75b), according to Wilcox (1975). More precisely, according to this theory, we have the result that

$$\tau \to \infty : \left|\frac{\partial \phi_{\mathrm{a}0}}{\partial \tau}\right| \to 0, \quad |\nabla \phi_{\mathrm{a}0}| \to 0 . \tag{3.89}$$

The interested reader may find a proper account in Wilcox (1975), but some sophisticated mathematical arguments are involved. In Chap. 4, we consider again (3.75a)–(3.76b) and derive a well-posed problem for the function ψ^0_{N}, in (3.76b), which gives the possibility to obtain an asymptotically consistent initial value $\boldsymbol{v}_{\mathrm{N}}(t = 0, \boldsymbol{x}) \equiv \boldsymbol{v}^0_{\mathrm{N}}$ for the Navier viscous dynamic system of equations (3.55).

3.5 The Weakness of Our First Approach

It is not difficult to see that our first approach, which we have discussed in Sects. 3.1–3.4, fails in a number of circumstances. The first obvious example arises when there is some local region of the flow where the Mach number cannot be considered as small, even though in the main part of the flow this kind of approximation is fully valid. Suppose, for instance, that we force the flow to leave the main domain by passing through a small hole; in this case the velocity may change by an order of magnitude, and it may happen that compressibility has to be taken care of in the subdomain under consideration. In such a case we may easily see that the situation is manageable, because the failure of the low-Mach-number approximation is localized to a region of small extent and we may expect it to have only a small influence on the main part of the flow–in fact, the hole may be accounted for by simply writing that it is a sink in the incompressible approximation. A much more difficult situation is encountered when we are dealing with low-Mach-number flow over a bluff body. The reason is that compressibility may have to be brought in over a rather extended region; if we are dealing with a flow over a circular cylinder with a fairly low value of the Mach number at infinity $M = 0.1$, for example, this leads to rather high subsonic condition in a region which may not be considered as localized. Considering again a hole, the flow may enter rather than leave, and then we may be faced with a high subsonic jet which penetrates quite deeply into the low-Mach-number region of the flow.

In the preceding paragraph, we were briefly discussing circumstances under which the first approach fails because the velocity changes by an order of magnitude in such a way that the local Mach number does not remain small. Remembering that the low-Mach-number limit may lead to the acoustic equations rather than the incompressible equations, we may suspect that the first approach will fail whenever we encounter changes so rapid that they occur on a short timescale which is of the order of M^{-1}. This happens in a number of circumstances, and a well-known example is that of *impulsive motion*, which is considered in incompressible-flow theory and is dealt with at various places in Lamb's book (Lamb 1945). Even when the motion is not produced impulsively, the changes that take place at the beginning, provided that they occur at a sufficiently high rate, send out sound waves, which, in our dimensionless world, travel at a speed which is of the order of $1/M$. Whenever we need to take into account such a small time lapse, our first approach will fail. Of course, when we are considering external aerodynamics, these sound waves are very rapidly–in a time $O(M)$–propagated out of the domain of interest, and generally we may simply ignore them; indeed, if this were not true, as is shown in Chap. 4, this problem would be well known and a proper technique for dealing with would be available. Nevertheless, the phenomenom is worth some consideration, and it may even sometimes be troublesome, for example if the sound waves under consideration are reflected from a body that is not too remote. If the motion is repeated from one cycle to another, as occurs in engines, any rapid changes occurring at some phase of the cycle generate sound waves that are not easily damped out, except by turbulence, and our first approach is again expected to fail.

In the above, we were considering changes which occur so rapidly so that the time derivative changes by an order of magnitude. Another facet of the same failure occurs when the space derivatives change by an order of magnitude, becoming $O(M)$ instead of remaining $O(1)$. If, simultaneously, the time derivative preserves its order of magnitude, exactly the same imbalance as the one alluded to before will occur between the space and time derivatives. The most obvious situation concerns the *far field*, at a distance $O(M^{-1})$ from the main region of flow. If the flow happens to be more or less periodic, the far field will also be periodic to the same extent, with the same pseudo-frequency. This means that the time derivative will retain the same order of magnitude when one goes from the main region to the far field, while the space derivatives will be changed in the inverse ratio of the distances. This situation is very well known and has been very much studied because it accounts for what is called 'sound generated aerodynamically'. This situation was touched on very briefly in Sect. 3.4.2 and will be considered again in Chap. 4 in the framework of the derivation of the Navier–Fourier constant-density, viscous and heat-conducting model.

A quite analogous kind of failure of our first approach will occur any time that in a domain of diameter $O(1)$ the space derivatives happen to be $O(M)$. Assume, for example, that in a channel we encounter a long, straight part; in this case we expect that the space derivatives will be lowered to a small level of $O(M)$ or less. On the other hand, as we are dealing with a constant-density flow, the time derivatives are transmitted instantaneously, so that they will generally be lowered to the same extent as the space derivatives, and we shall again encounter the same kind of imbalance, which will lead to the acoustic limit.

An interesting problem corresponds to what happens when two flows generated by bodies moving well within the hyposonic regime come close together; an obvious application is linked to the meeting of two high-speed trains (as the TGV in France) and see the paper by Schetz (2001). In the case of the diffraction of sound by a rigid body, two parameters appear in the formulation of the problem: a small Mach number and the ratio of the two length scales linked to the rigid body and the wavelength of the sound wave. The impact of a rigid, rounded body with water is also a low-Mach-number situation when we consider the initial stage, linked to a very tiny period of time; see, for instance, H. Lamb (1932). Tracey (1988) investigated the motion of an infinitely long, rigid circular cylinder in air considered as an inviscid perfect gas, but did not investigates the formation of a shock, because her main concern was with the slowing down of the cylinder by unsteady aerodynamic forces – for a brief discussion of this problem, see the review paper by Guiraud (1995, pp. 280–281)). Finally, the present author thinks that a very good test problem is Rayleigh's problem for the flow of a weakly compressible (low-Mach-number) viscous, heat-conducting fluid in the framework of the one-dimensional unsteady NSF equations (see Zeytounian 2004, Sect. 3.7, pp. 78–88 and Sect. 7.4 of the present book). This problem has been considered, in particular, by Howarth (1951) and Van Dyke (1952). We end here this account of some situations which require the low-Mach-number flows to be considered in a less naive manner than might seem appropriate at the first sight.

References

G. Ali, Low Mach number flows in time-dependent domain. SIAM J. Appl. Math., **63**(6) (2003), 2020–2041.

G. K. Batchelor, *An Introduction to Fluid Dynamics.* Cambridge University Press, Cambridge, 2000.

R. Courant and D. Hilbert, *Methods of Mathematical Physics,* Vol. 2, Interscience, New York, 1962.

H. A. Dwyer, *Calculation of low Mach number flows.* AIAA J. **28** (1990), 98–105.

P. Embid. Commun. Partial Differential Equations, **12** (1987), 1227–1284.

A. T. Fedorchenko, J. Fluid Mech., **334** (1997), 135–155.

J. Fröhlich, *Résolution numérique des équations de Navier-Stokes à faible nombre de Mach par méthode spectrale.* Thèse de Doctorat, Université de Nice-Sophia-Antipolis, 1990.

P. M. Gresho, Adv. Appl. Mech. **28** (1992), 45–140.

J. P. Guiraud, Going on with asymptotics. In: *Asymptotic Modelling in Fluid Mechanics,* Lecture Notes in Physics, No. 442, pp. 257–307. P. A. Bois, E. Deriat, R. Gatignol and A. Rigolot eds. Springer, Berlin, 1995.

L. Howarth, Some aspects of Rayleigh's problem for a compressible fluid. Q. J. Mech. Appl. Math., **4** (1951), 157.

R. Klein, N. Botta, L. Hofmann, A. Meister, C. D. Munz, S. Roller and T. Sonar. Asymptotic adaptive methods for multi-scale problems in fluid mechanics. J. Engng. Math., **39** (2001), 261–343.

A. Korobkin, J. Fluid Mech., **244** (1992), 437–453.

H. Lamb, Hydrodynamics. Cambridge Univ. Press, 6th ed. (1932); Reprint, Dover New York 1945.

M. J. Lighthill, On sound generated aerodynamically. I. General theory. Proc. Roy. Soc. Lond. A **211** (1952), 564–514.

A. Majda, *Compressible Fluid Flow and Systems of Conservation Laws in Several Space Variables.* Springer, New York, 1984.

A. Majda and J. Sethian, The derivation and numerical solution of the equations for zero Mach number combustion. Combust. Sci. and Technol., **42** (1985), 185–205.

M. Matalon and P. Metzener, The propogation of premixed flames in closed tubes. J. Fluid Mech., **336** (1997), 331–350.

A. Meister, Viscous fluid flow at all speeds: analysis and numerical simulation. Z. Angew. Math. Phys., **54** (2003), 1010–1049.

B. Müller, *Computation of Compressible Low Mach Number Flow.* Habilitation thesis, Institut für Fluiddynamik, ETH Zürich, September 1996.

B. Müller, *Low-Mach-number asymptotics of the Navier–Stokes equations.* J. Engng. Math., **34** (1998), 97–109.

F. Obermeier, *The Application of Singular Perturbation Methods to Aerodynamic Sound Generation.* Lecture Notes in Mathematics, No. 594, Springer, Berlin, 1976, pp. 400–421.

R. G. Rehm and H. R. Baum. J. Res. Nat. Bur. Stand., **83** (1978), 297–308.

J. A. Schetz, Aerodynamics of High-Speed Trains. Ann. Rev. Fluid Mech. **33** (2001), 371–414.

R. Sery Baye, *Contribution à l'étude de certains écoulements à faibles nombres de Mach.* Thèse de Doctorat de Mécanique, Université de Paris VI, 1994.

E. K. Tracey, The initial and ultimate motions of an impulsively started cylinder in a compressible fluid. Proc. Roy. Soc. Lond. A, **418** (1988), 209–228.

M. Van Dyke. ZAMP **3**(5) (1952), 343–353.

C. Viozat, *Calcul d'écoulement stationnaires et instationnaires à petit nombre de Mach, et en maillage étirés.* Thèse de Doctorat en Sciences (de l'ingénieur) Université de Nice-Sophia Antipolis, 1998.

C. Wilcox, *Scattering Theory for the d'Alembert Equation in Exterior Domain.* Springer, Berlin, 1975.

R. Kh. Zeytounian, *Theory and Applications of Nonviscous Fluid Flows.* Springer, Heidelberg, 2002.

R. Kh. Zeytounian, *Theory and Applications of Viscous Fluid Flows.* Springer, Heidelberg, 2004.

R. Kh. Zeytounian and J.-P. Guiraud, Evolution d'ondes acoustiques dans une enceinte et concept d'incompressibilité. IRMA-Lille, **2**, Fasc. 4(4) (1980), IV-1–IV-16. See also R. Kh. Zeytounian and J.-P. Guiraud. C. R. Acad. Sci. Paris, **290B** (1980), 75–78.

4

Some Aspects of Low-Mach-Number External Flows

In this chapter, we consider first, in Sect. 4.1, a detailed derivation of the Navier–Fourier initial–boundary-value problem for the Navier velocity vector $\boldsymbol{v}_{\mathrm{N}}$, the pseudo-pressure p_{N} and the perturbation of the temperature T_F, which are the solutions of the system of the Navier equations (3.55) and an approximate equation for T_F, derived from the full NSF equation (3.48c), for the temperature T. Then, in Sect. 4.2, a derivation of Burgers' dissipative model equation is given from the full NSF equations for large Strouhal and Reynolds numbers, in the context of low-Mach-number asymptotics, when a thin dissipative region of thickness $O(M)$ close to an acoustic wavefront is considered. Section 4.3 is also related to a large Reynolds number, in the context of low-Mach-number asymptotics, when a similarity rule is taken into account between the two small parameters $1/Re$ and M; as an application, the classical Blasius (1908) problem in investigated in accordance with Godts and Zeytounian (1990). The last section, Sect. 4.4, is related to a small-Reynolds-number, weakly compressible, low-Mach-number fluid flow, in the framework of the Stokes–Oseen exterior problem, where we take into account a judicious similarity rule such that the Knudsen number $Kn = M/Re \ll 1$, as this is the case for a continuous, macroscopic, Newtonian fluid flow; the corresponding steady Stokes and Oseen compressible-model equations are asymptotically derived.

4.1 The Navier–Fourier Initial–Boundary-Value Model Problem

In this section, we consider a detailed asymptotic derivation of the Navier–Fourier initial–boundary-value problem for the Navier velocity $\boldsymbol{v}_{\mathrm{N}}$, the pseudo-pressure p_{N} and the perturbation of the temperature T_F, which are the solutions of the system of the Navier equations (3.55) and an equation for T_F derived from the NSF equation (3.48c) for the temperature T. As a physical

R. Kh. Zeytounian: *Topics in Hyposonic Flow Theory*, Lect. Notes Phys. **672**, 77–113 (2006)
www.springerlink.com

problem, we consider unsteady, slightly compressible, viscous, heat conducting laminar flow around a rigid, bounded, body in an unbounded atmosphere when the Coriolis and gravitational forces are neglected, as is the case in classical aerodynamics. The starting 'exact' problem is governed by the NSF system of equations (3.48) with some initial conditions (a state of rest), a no-slip condition for the velocity vector, a condition for the temperature and some conditions on the behaviour at infinity. These conditions are specified below, and we consider the case when the solid body starts from a state of rest impulsively or during a dimensionless interval of time $O(M)$.

4.1.1 The Incompressible (Navier) Main Limit

An undesirable consequence of the incompressible (Navier) main limit (2.2),

$$\mathrm{Lim}^{\mathrm{Nav}} = [M \to 0 \text{ with } t \text{ and } \boldsymbol{x} \text{ fixed}] , \tag{4.1}$$

when we assume that in the NSF starting equations (3.48) the Reynolds and Prandtl numbers Re and Pr are $O(1)$, is the filtering of the time derivative of the density in the Navier model system (3.55), which leads, in leading order, to a divergenceless constraint on the Navier velocity vector $\boldsymbol{v}_\mathrm{N}$,

$$\nabla \cdot \boldsymbol{v}_\mathrm{N} = 0 . \tag{4.2a}$$

Because of this, we encounter the fundamental problem of deciding what initial conditions one may prescribe for $\boldsymbol{v}_\mathrm{N}$ in the incompressible/constant-density approximate Navier model equation

$$\frac{\mathrm{D}_\mathrm{N}\boldsymbol{v}_\mathrm{N}}{\mathrm{D}t} + \frac{1}{\gamma}\nabla p_\mathrm{N} = \frac{1}{Re}\Delta \boldsymbol{v}_\mathrm{N} , \tag{4.2b}$$

where $\mathrm{D}_\mathrm{N}/\mathrm{D}t = \partial/\partial t + \boldsymbol{v}_\mathrm{N} \cdot \nabla$. We observe again that in (4.2b), the pseudopressure p_N is given by

$$p_\mathrm{N} = \frac{p-1}{M^2} , \tag{4.3}$$

where $p(t, x)$ is the dimensionless 'true' pressure in the NSF system of equations (3.48). However, it is important to note that the above Navier system of two equations (4.2) is asymptotically consistent only when $M \downarrow 0$ with t and $\boldsymbol{x}$ both fixed and $O(1)$, and also that, in the case of an external problem, we have the following far-field expansion (see (3.69)) for the NSF pressure when $|\boldsymbol{x}| \to \infty$ with $\boldsymbol{\xi} = M\boldsymbol{x}$ fixed (see (4.5)):

$$p(t, \boldsymbol{x}) = 1 + M^{\beta+1} p^{0\mathrm{ff}}(t, \boldsymbol{\xi}) + \dots , \quad \text{with } \beta \geq 2 . \tag{4.4}$$

4.1.2 The Far-Field (Lighthill) Local Limit

It seems that the first paper devoted to aerodynamic sound emission as a singular perturbation problem was a preprint by Crow (1966) (see also Lauvstad 1968), which investigated the singular far-field region in the framework of low-Mach-number asymptotics. As a consequence, in place of the main limit (4.1), it is necessary to consider the following 'à la Lighthill (1952)' local, far-field limit (see (2.1)):

$$\mathrm{Lim}^{0\mathrm{ff}} = [M \to 0 \text{ with } t \text{ and } \xi \text{ fixed}] . \tag{4.5}$$

In such a case, we know that (see (3.69)), with the expansion (4.4) for the pressure p, we have the following far-field expansions for $\boldsymbol{v}$, T and ρ:

$$\boldsymbol{v} = M^{\beta}\boldsymbol{v}^{0\mathrm{ff}} + \dots, \quad \rho = M^{\beta+1}\rho^{0\mathrm{ff}} + \dots, \quad T = M^{\beta+1}T^{0\mathrm{ff}} + \dots . \tag{4.6}$$

Using (4.4) and (4.6), we can derive from the NSF system (3.48) a system of two acoustic equations for $\boldsymbol{v}^{0\mathrm{ff}}(t,\xi)$ and $\rho^{0\mathrm{ff}}(t,\xi)$:

$$S\frac{\partial \rho^{0\mathrm{ff}}}{\partial t} + \nabla^{0\mathrm{ff}} \,.\, \boldsymbol{v}^{0\mathrm{ff}} = 0, \quad S\frac{\partial \boldsymbol{v}^{0\mathrm{ff}}}{\partial t} + \nabla^{0\mathrm{ff}}\rho^{0\mathrm{ff}} = 0 , \tag{4.7}$$

where $p^{0\mathrm{ff}} = \gamma\rho^{0\mathrm{ff}}$ and $\nabla^{0\mathrm{ff}} = (1/M)\nabla$, if we assume that Re and Pr are fixed and $O(1)$.

The Results of Obermeier (1977)

In a paper by Obermeier (1977), the reader can find a very pertinent discussion concerning the application of low-Mach-number asymptotics to aerodynamic sound generation, where the starting equation (4.8) is a rearranged version of the Lighthill inhomogeneous wave equation (derived from the equations of fluid motion) for the flow density ρ, with a formal source distribution on the right-hand side, namely

$$\begin{aligned}\nabla\rho - \frac{1}{c^2}\frac{\partial^2\rho}{\partial t^2} = \frac{\partial^2}{\partial x_i\,\partial x_j}\Bigg\{&\rho v_i v_j + (p - c^2\rho)\delta_{ij} \\ &- \mu\left[\frac{\partial v_i}{\partial x_j} + \frac{\partial v_j}{\partial x_i} - \frac{2}{3}\frac{\partial v_k}{\partial x_k}\delta_{ij}\right]\Bigg\} ,\end{aligned} \tag{4.8}$$

where $v_i (i = 1,2,3)$ are the components of the flow velocity $\boldsymbol{v}$ in the direction x_i, δ_{ij} is the usual Kronecker symbol and $c^2 = \gamma p/\rho$. However, it is necessary to put in place of c^2 the square of the speed of sound in the medium at rest, c_0^2, and to assume in addition that the right-hand side of (4.8) decreases sufficiently fast outside the actual flow region D. The effects on sound generation by solid bodies inside the flow region D were discussed by Curle (1955) in a generalization of the Lighthill (1952) theory, at about the same time as jet-powered aircraft entered the commercial air traffic scene. In Obermeier (1977), the inner flow region and the outer flow region (where the sound is observed) are analysed and matching via an intermediate region is discussed; however, we shall not go into the details of the above-mentioned papers, and, instead, the reader is referred to the literature. Here, we note only that if we assume that the Strouhal number St is equal to 1, one must deal, in the general case, with the smallness of both the Mach number M and the inverse Reynolds number $1/Re$, and a way of dealing with such a situation is to enforce a relation such that $1/Re = O(M^\alpha)$ with a properly chosen exponent α.

The Results of Viviand (1970)

In the *inviscid* case, from the work of Lauvstad (1968), Crow (1970) and Viviand (1970), one may infer that the low-Mach-number expansion should proceed as follows, if we do not take into account the influence of the Reynolds number in the isentropic case:

$$p = 1 + M^2 p_2 + M^3 p_3(t) + M^4 p_4 + \cdots , \qquad \boldsymbol{v} = \boldsymbol{v}_0 + M^2 \boldsymbol{v}_2 + M^3 \boldsymbol{v}_3 + \cdots . \tag{4.9}$$

These formulae exhibit some peculiar features of the low-Mach-number expansion which show that it is not as straightforward as might be expected. The first point is that $p_3(t)$ is a function of time only. This was shown by Crow (1970) and Viviand (1970), who discovered that the expansion (4.9) is a 'proximal' one (valid only when $|\boldsymbol{x}| = O(1)$), which is not valid at large distances from the wall bounding the unsteady, exterior fluid motion. In fact, according to Viviand (1970), we have the following Poisson equation for the term $p_s + 2$:

$$\Delta p_{s+2} = \frac{\partial^2 p_s}{\partial t^2} - \gamma \nabla \cdot [\nabla \cdot (\boldsymbol{v}_0 \boldsymbol{v}_s + \boldsymbol{v}_s \boldsymbol{v}_0 + \rho_s \boldsymbol{v}_0 \boldsymbol{v}_0)] \, . \tag{4.10}$$

It is clear that it is not possible to find a solution of (4.10) which vanishes at a large distance from the bounded solid body. For instance, we note that the behaviour of p_2 at infinity is $O(1/|\boldsymbol{x}|)$ and, as consequence, p_4 is $O(|\boldsymbol{x}|)$ for $|\boldsymbol{x}| \to +\infty$. Following Lauvstad (1968), Crow and Viviand were able, independently and almost simultaneously in 1970, to match the proximal expansion (4.9), with t and $|\boldsymbol{x}|$ fixed, to the 'distal' expansion (4.4) and (4.6), where $\xi = M\boldsymbol{x}$, which is valid when $\xi = O(1)$. With an error of $O(M^{\beta+1})$, we find in this case that $p^{0\text{ff}}$, $\boldsymbol{v}^{0\text{ff}}$ and $\rho^{0\text{ff}} = (1/\gamma) p^{0\text{ff}}$ are solutions of the equations of acoustics; namely, for $p^{0\text{ff}}$, we derive the classical linear wave equation

$$\frac{\partial^2 p^{0\text{ff}}}{\partial t^2} - \Delta^{0\text{ff}} p^{0\text{ff}} = 0, \quad \Delta^{0\text{ff}} \equiv (\nabla^{0\text{ff}})^2 \, . \tag{4.11}$$

The solution of (4.11) must satisfy the (Sommerfeld) radiation condition at infinity and the matching conditions when $|\xi|$ tends to zero, and this solution must be singular at $|\xi| = 0$ (or else this solution is identically zero). Viviand (1970) utilizes a composite expansion for the pressure, namely

$$C(p) = 1 + M^2 P_2(t, \boldsymbol{x}, \xi) + M^3 P_3(t, \boldsymbol{x}, \xi) + M^4 P_4(t, \boldsymbol{x}, \xi) + \dots , \qquad (4.12)$$

and for each P_s in (4.12) we have

$$P_s = p_s(t, \boldsymbol{x}) - \Pi_s(t, \boldsymbol{x}) + p_s^{0\mathrm{ff}}(t, \xi) - \Phi_s^{0\mathrm{ff}}(t, \xi) , \qquad (4.13)$$

where $\Pi_s(t, \boldsymbol{x})$ is the part of the proximal expansion of p_s which does not tend to zero when $|\boldsymbol{x}| \to +\infty$, and $\Phi_s^{0\mathrm{ff}}(t, \xi)$ is an unknown function of ξ which must be determined from the condition of the uniform validity of (4.12). In Viviand (1970, p. 590), the reader can find this composite expansion up to the term proportional to M^5. On the other hand, in Viviand (1970, Sect. 6) the results are also compared with Lighthill's theory in the case of small Mach number, and are found to agree with it up to terms of order M^5. We observe that in (4.13), $p_3^{0\mathrm{ff}}(t, \xi) \equiv p^{0\mathrm{ff}}(t)$. Expansion (4.12) matches either with a *source*, if $\beta = 2$ and $p^{0\mathrm{ff}}(t)$ in (4.4) is proportional to the second derivative of the mass flow; with a *dipole*, if $\beta = 3$ and $p^{0\mathrm{ff}}(t) = 0$; or with a *quadrupole*, if $\beta = 4$ and again $p^{0\mathrm{ff}}(t) = 0$. The quadrupole situation is the one appropriate for sound generation; for a pertinent account, see Ffowcs Williams (1984). If we go back to (4.9), we find that the pair $[(1/\gamma)p_2, \boldsymbol{v}_0]$ is governed by the equations of incompressible/constant-density inviscid aerodynamics, while the pairs $[(1/\gamma)p_4, \boldsymbol{v}_2]$ and so on exhibit some slight compressibility. Sligthly compressible (inviscid) external aerodynamics becomes the aerodynamics of compressible flow at large distances from the sources which create the motion and, furthermore, this flow is in fact acoustically dominated; this conclusion is also true for a viscous, compressible, heat-conducting flow, if in place of the Euler equations we consider the full NSF equations, and it is true even at moderate Reynolds numbers. It is in this apparently simple remark that the theory of sound propagation is profoundly rooted.

The Guiraud (1983)–Sery Baye (1994) Investigation of the Far Field for the Steichen Equation

Following Guiraud's unpublished notes (Guiraud, 1983), Sery Baye (1994) constructed an asymptotic algorithm (in the inviscid potential case) for the scalar velocity potential Φ, which satisfies the unsteady Steichen equation (3.71), written in dimensionless form. For this hyperbolic equation (3.71), which is second-order in time, it is necessary to impose two conditions for $t = 0$, namely (with $St \equiv 1$)

$$\text{at } t = 0: \quad \Phi = \Phi^0 \quad \text{and} \quad \frac{\partial \Phi}{\partial t} = \Phi_1 . \qquad (4.14)$$

Guiraud's asymptotic algorithm is expressed in terms of integer powers of M, including products of such powers with $\operatorname{Log} M$, and the calculation is truncated at M^6. Terms in M^p, where p is an odd integer, and terms in $M^q \operatorname{Log} M$, where $q \geq 4$, appear in the asymptotic expansions for the far field because of the behaviour in the far field of the classical Janzen–Rayleigh (M^2) expansion in the unsteady case when $|\boldsymbol{x}|$ goes to infinity. In fact, one of the difficulties is that the expansion cannot be generated by iterative application of a single limiting process $M \downarrow 0$, keeping the time t and position $\boldsymbol{x}$ fixed.

As a consequence, two asymptotic expansions are needed: one (proximal) for $|\boldsymbol{x}| = O(1)$, and another (distal) for $|\boldsymbol{x}| = O(1/M)$. In relation to the position $\boldsymbol{x}$, Sery Baye (1994) took great care with the matching conditions. For the purpose of this matching (see Chap. 3), the proximal expansion is written in the form (3.73) for the solution of the unsteady Steichen equation (3.71), and the associated distal asymptotic expansion for the far field is written as (3.72). In the distal expansion (3.72), the scalar $q > 0$ is determined by the behaviour at large distances, i.e. when $r = |\boldsymbol{x}| \to +\infty$, of the first term of (3.73), $\Phi_0(t, \boldsymbol{x})$, which is a solution of the Laplace equation $\Delta\Phi_0 = 0$. In fact, at large distances we have the behaviour $\Phi_0(t, \boldsymbol{x}) \approx O(1/r^q)$; more precisely, if we write only the three first terms (which are necessary for the discussion below), we have the following formula:

$$\Phi_0 = \frac{m^0}{r} - \boldsymbol{D}^0 \cdot \nabla\left(\frac{1}{r}\right) - \mathbf{Q}^0 : \nabla\nabla\left(\frac{1}{r}\right) + O\left(\frac{1}{r^4}\right), \quad \text{when } r \to +\infty\,. \tag{4.15}$$

In Viviand (1970), the reader can find a more precise formula, with an error of $O(1/r^7)$. In (4.15), m^0, $\boldsymbol{D}^0$ and $\boldsymbol{Q}^0$ are arbitrary functions (a scalar, a vector and a second-order tensor, respectively), depending on the time t. As is noticed in Sery Baye (1994, p. 37), we have the following:

$$\begin{aligned}
&\text{if } m^0 \neq 0, \quad \text{then } \mathrm{q} = 1 \\
&\quad \text{(monopole-like behaviour of } \Phi_0 \text{ when } r \to +\infty)\,; \\
&\text{if } m^0 = 0 \text{ but } \boldsymbol{D}^0 \neq 0, \quad \text{then } q = 2 \\
&\quad \text{(dipole-like behaviour of } \Phi_0 \text{ when } r \to +\infty); \\
&\text{if } m^0 = \boldsymbol{D}^0 = 0 \text{ but } \mathbf{Q}^0 \neq 0, \text{then} \quad q = 3 \\
&\quad \text{(quadrupole-like behaviour of } \Phi_0 \text{ when } r \to +\infty)\,.
\end{aligned} \tag{4.16}$$

When the behaviour of Φ_0 is quadrupole-like for $r \to +\infty$, the terms proportional to $M^q \operatorname{Log} M$ are zero in (3.72) and (3.73). The results obtained by Sery Baye (1994) extend the results of Viviand (1970) and also those of Leppington and Levine (1987). Finally, thanks to far field expansion for $\boldsymbol{v}$ (associated with (4.6) for $\beta \geq 2$) and by matching with the main Navier expansion (associated with (4.1)), we have the possibility to write the behaviour conditions at infinity (3.70) for the Navier leading-order system of two equations (3.55) (for any finite time).

4.1.3 The Initial-Time (Acoustic) Local Limit

In the region close to the initial time $t = 0$, we can consider the local acoustic limit (3.35) and the following initial conditions (state of rest) for the starting dimensionless NSF equations (3.48):

$$\text{at } \tau < 0^- : \quad p = \rho = T = 0 \quad \text{and } \boldsymbol{v} = 0 , \tag{4.17}$$

where $\tau = t/M$. This gives us the possibility to assume the existence of a local-in-time acoustic expansion (3.74) together with the formulae (3.75a) for an exterior unbounded domain (outside a bounded rigid body). As a consequence, we obtain the d'Alembert equation (3.75b),

$$\frac{\partial^2 \phi_{\mathrm{a}0}}{\partial \tau^2} - \Delta \phi_{\mathrm{a}0} = 0 , \tag{4.18}$$

for $\phi_{\mathrm{a}0}$ together with the Laplace equation (as a consequence of the two relations (3.76a,b))

$$\Delta \psi^0_{\mathrm{N}} = 0 \tag{4.19}$$

for $\boldsymbol{v}_{\mathrm{N}}(t = 0, \boldsymbol{x}) = \nabla \psi^0_{\mathrm{N}}$, where the Navier initial-velocity vector $\boldsymbol{v}_{\mathrm{N}}$ $(t = 0, \boldsymbol{x}) \equiv \boldsymbol{v}^0_{\mathrm{N}}$ is an unknown vector. For (4.18) and (4.19), we have two inviscid problems: one for $\phi_{\mathrm{a}0}(\tau, \boldsymbol{x})$ and another for $\psi^0_{\mathrm{N}}(\boldsymbol{x})$.

The Flow Around a Rigid, Bounded Body

We consider a test problem below, but our asymptotic approach is also suitable for more realistic external problems. For the dimensionless velocity vector $\boldsymbol{v}$ (rendered dimensionless through $\boldsymbol{U}_{\mathrm{c}}$, linked to the characteristic value of the velocity of displacement of the body), we write the usual no-slip condition for the NSF equations (3.48).

$$\boldsymbol{v} = \boldsymbol{U}_{\Sigma}(t, \boldsymbol{P}) \quad \text{all along } \Sigma, \text{for } t \geq 0^+ . \tag{4.20}$$

We assume that the fluid is set in motion from rest (i.e. (4.17) applies for $t < 0^-$) by the displacement of a bounded, non-deformable body Ω in an unbounded atmosphere. We associate with each point $\boldsymbol{P}(s_1, s_2)$ of the wall $\partial\Omega = \Sigma$ a dimensionless velocity $\boldsymbol{U}_{\Sigma}(t, \boldsymbol{P})$, and we denote by $z = \boldsymbol{P} \,.\, \boldsymbol{n}$ the coordinate normal to the wall Σ directed into the fluid. Close to the initial time, as a consequence of the limiting acoustic process (3.35), together with (3.74) and (3.75a), we obtain the inviscid equations of acoustics (3.68) for the leading-order terms, and then we derive the above d'Alembert equation (4.18) for $\phi_{\mathrm{a}0}$ (see (3.75b) in Sect. 3.4.2). As a consequence, close to the initial time, in place of the no-slip boundary condition (4.20), we have the possibility to write only a *slip* condition on Σ,

$$\boldsymbol{v} = \boldsymbol{U}_{\Sigma} \,.\, \boldsymbol{n} \equiv W_{\Sigma}(t, \mathbf{P}) \quad \text{all along } \Sigma, \text{ for } t \geq 0^+ . \tag{4.21}$$

More precisely, we assume below that in (4.21),

$$\boldsymbol{U}_{\Sigma}(t,\boldsymbol{P}) = H(t)\boldsymbol{u}_{\Sigma}(\boldsymbol{P}) \quad \text{and } w_{\Sigma}(\boldsymbol{P}) = \boldsymbol{n}\,.\,\boldsymbol{u}_{\Sigma}(\boldsymbol{P})\,, \tag{4.22}$$

where $H(t)$ is the Heaviside function (or unit function; $H(t \leq 0^-) = 0$ and $H(t > 0^+) \equiv 1$), which simulates the situation where the bounded body starts its motion impulsively from rest when $t \leq 0$. Using (4.21) and (4.22), we write the boundary condition for $\phi_{\mathrm{a}0}$, the solution of the wave equation (4.18),

$$\frac{\partial \phi_{\mathrm{a}0}}{\partial n} = 0 \quad \text{all along } \Sigma, \text{for } t \geq 0^+. \tag{4.23}$$

The initial conditions are

$$\phi_{\mathrm{a}0} = -\psi_{\mathrm{N}}^0 \quad \text{and } \frac{\partial \phi_{\mathrm{a}0}}{\partial \tau} = 0, \quad \text{at } \tau = 0\,, \tag{4.24}$$

when we take into account (3.75a). In this case, together with (4.23), for ψ_{N}^0, the solution of (4.19), we must write the boundary condition (when $t > 0^+$) as

$$\frac{\partial \psi_{\mathrm{N}}^0}{\partial n} = w_{\Sigma}(\boldsymbol{P}) \text{ all along } \Sigma\,, \tag{4.25}$$

when we take into account (4.21) and (4.22). Finally, (4.18), and (4.19), with the conditions (4.23) and (4.24) for $\phi_{\mathrm{a}0}$ and (4.25) for ψ_{N}^0, lead to two well-posed problems for $\phi_{\mathrm{a}0}$ and ψ_{N}^0, respectively, provided we add the following behavior conditions at infinity:

$$\phi_{\mathrm{a}0} \to 0 \quad \text{and } \psi_{\mathrm{N}}^0 \to 0, \quad \text{as } |\boldsymbol{x}| \to \infty\,. \tag{4.26}$$

We observe that this added information (4.26), namely that no perturbations are coming from infinity towards the body, is a physical rather than a mathematical condition.

The Results of Wilcox (1975), and Matching

We are not actually interested in finding $\phi_{\mathrm{a}0}$; all that we want to know is that (see, for instance, (3.88) and (389))

$$\phi_{\mathrm{a}0} \to 0 \text{ when } \tau \to \infty\,, \tag{4.27}$$

which is guaranteed by the mathematically rigorous theory of d'Alembert's equation presented by Wilcox (1975), where a simple approach based on a self-adjoint extension of the Lapacian in Hilbert space is presented. The principal results are the construction of eigenfunction expansions for the Laplacian and the calculation of the asymptotic form of solutions of d'Alembert's equation for large values of the time parameter: every wave with finite energy is asymptotically equal, for $\tau \to \infty$, to a diverging spherical wave. Moreover, it is shown how the profile of this wave can be calculated from the initial state.

These results are then used to calculate the asymptotic distribution of the energy for $\tau \to \infty$. More precisely we have, according to Wilcox (1975, p. 6),

$$\frac{\partial \phi_{a0}(\tau, \boldsymbol{x})}{\partial x_k} \approx u_k^{\infty}(\tau, \boldsymbol{x}), \quad \tau \to \infty, \quad k = 0, 1, 2, 3, \quad x_0 = \tau , \tag{4.28a}$$

where $\boldsymbol{x} = (x_1 = s_1, x_2 = s_2, x_3 = z)$, and

$$u_k^{\infty}(\tau, \boldsymbol{x}) \equiv \frac{1}{|\boldsymbol{x}|} F_k(|\boldsymbol{x}| - \tau, \boldsymbol{x}/|\boldsymbol{x}|) , \tag{4.28b}$$

in the sense that, for $k = 0, 1, 2, 3$,

$$\lim_{\tau \to \infty} \int_{\Omega} \left[\frac{\partial \phi_{a0}(\tau, \boldsymbol{x})}{\partial x_k} - u_k^{\infty}(\tau, \boldsymbol{x}) \right]^2 \mathrm{d}x = 0 . \tag{4.28c}$$

In such a case, we obtain automatically

$$\boldsymbol{v}_{a0}(\tau, \boldsymbol{x}) \to \boldsymbol{v}_{\mathrm{N}}(t = 0, \boldsymbol{x}) \quad \text{when } \tau \to \infty \tag{4.29}$$

from (3.75a), so that matching (according to the MAE technique) between the Navier solution (at $t = 0$) and the local-in-time acoustic solution (when $\tau \to \infty$) is realized, and we have found what was missing, namely that the function ψ_{N}^0 (when we assume that $\boldsymbol{v}_{\mathrm{N}}(t = 0, \boldsymbol{x}) = \nabla \psi_{\mathrm{N}}^0$) is now completely determined by the following Neumann problem:

$$\Delta \psi_{\mathrm{N}}^0 = 0, \tag{4.30a}$$

$$\frac{\partial \psi_{\mathrm{N}}^0}{\partial n} = w_{\Sigma}(\boldsymbol{P}) \text{ all along } \Sigma, \quad \text{and } \psi_{\mathrm{N}}^0 \to 0 \text{ as } |\boldsymbol{x}| \to \infty . \tag{4.30b}$$

4.1.4 The Navier–Initial–Boundary-Value Model Problem

To achieve a complete formulation of the Navier–initial–boundary-value (Navier–IBV) model problem (since the bounded body starts its motion impulsively from at rest according to (4.22) when $t > 0$), we write the no-slip condition for the Navier model equations (3.55) as

$$\boldsymbol{v}_{\mathrm{N}} = \boldsymbol{u}_{\Sigma}(\boldsymbol{P}), \quad \text{all along} \quad \Sigma \text{ for all } t > 0^+ , \tag{4.31a}$$

and, together with (4.30) and (3.70), namely

$$|\boldsymbol{v}_{\mathrm{N}}| \to 0 \text{ and } p_{\mathrm{N}} \to 0 \quad \text{when } |\boldsymbol{x}| \to \infty, \tag{4.31b}$$

we obtain a well-posed Navier–IBV model problem for the Navier equations (3.55).

Unfortunately, the Navier–IBV model problem derived here does not replace completely our NSF–IBV problem, because, first, all information linked to the temperature T (solution of (3.48c)) and its associated thermal condition (3.57c), is missing. In Sect. 4.1.5, devoted to the Fourier equation for the perturbation of the temperature, this question is discussed, and we show the role played by the thermal condition (3.57c) in this asymptotic derivation; in fact, it is necessary to consider three cases, $\alpha < 2, \alpha = 2$ and $\alpha > 2$, when the similarity parameter $\Lambda* = O(1)$. A second, curious feature of the Navier–IBV model problem, with the initial condition $\boldsymbol{v}_{\mathrm{N}}(t = 0, \boldsymbol{x}) = \nabla\psi^0_{\mathrm{N}}$, where ψ^0_{N}, is a solution of the above Neumann problem (4.30), is linked to the fact that the slip condition on Σ in that Neumann problem is written in relation to normal steady component $w_\Sigma(\boldsymbol{P}) = \boldsymbol{n} \,.\, \boldsymbol{u}_\Sigma(\boldsymbol{P})$ of the velocity $\boldsymbol{U}_\Sigma = H(t)\boldsymbol{u}_\Sigma(\boldsymbol{P})$. The reason is that, close to the initial time, where τ is the significant time, the acoustic local limiting process (3.35), together with (3.34), which leads, at the leading order, to linear acoustic equations ((3.68) with a slip condition on Σ) and, from (3.74) and (3.75a), to d'Alembert's equation (4.18), appears as singular when we consider simultaneously the region close to the initial time ($t = 0$) and a thin dissipative layer near the wall Σ. As a result, on Σ we have the possibility to write a no-slip condition for the velocity vector and a condition for the temperature which follows from (3.57c) for small Mach number. As a consequence, it seems necessary to consider, in place of the *slip* condition (4.21) on the boundary with (4.22) on the wall Σ, a *matching* condition. This matching is performed between the classical linear, inviscid, acoustic equations (for instance (3.68)) and a new set of unsteady *dissipative* equations that are significant simultaneously close to the initial time and near the wall Σ, such that no-slip boundary conditions for the velocity vector and a condition for the temperature can be prescribed on Σ. This problem has not yet been completely solved. The consideration of this problem, which has theoretical interest in itself and is a necessary step for the obtaining of a significant model close to the initial time for low-Mach-number external flows, is still work in progress.

4.1.5 The Role of Thermal Effects

The main limiting process $\lim^{\mathrm{Nav}}$, (2.2), in the case of a external flow, can be associated with the following full asymptotic expansion of the NSF unsteady, dissipative equations (3.48):

$$\boldsymbol{v} = \boldsymbol{v}_{\mathrm{N}} + M^a \boldsymbol{v}_a + \cdots\,, \quad p = 1 + M^2 p_{\mathrm{N}} + M^b p_b + \dots\,, \tag{4.32a}$$

$$T = 1 + M^c T_c + \cdots\,, \quad \rho = 1 + M^d \rho_d + \cdots\,. \tag{4.32b}$$

In (4.32), the scalars a, b, c and d are positive arbitrary real numbers, and for $\boldsymbol{v}_{\mathrm{N}}$ and p_{N} we obtain, again, at the leading order from the NSF equations

(3.48), the Navier system of two equations (3.55). Then, from the NSF equation of continuity (3.48a), we derive the following continuity equation for the function ρ_d in (4.32b):

$$\frac{\partial \rho_d}{\partial t} + \boldsymbol{v}_{\mathrm{N}} \,.\, \nabla \rho_d + M^{a-d} \nabla \,.\, \boldsymbol{v}_a = 0 \,. \tag{4.33}$$

Obviously the case $a = d$ is the most significant (the least degenerate), but for the moment $d > 0$ is indeterminate. Next, from the equation (3.48c) for the temperature T in the NSF dissipative system, we derive an equation for T_c (since $b > 2$):

$$\frac{\partial T_c}{\partial t} + \boldsymbol{v}_{\mathrm{N}} \,.\, \nabla T_c - \frac{\gamma - 1}{\gamma} M^{2-c} \left[\frac{\partial p_{\mathrm{N}}}{\partial t} + \boldsymbol{v}_{\mathrm{N}} \,.\, \nabla p_{\mathrm{N}} \right] = \frac{1}{Pr\, Re} \Delta T_c + \frac{\gamma - 1}{Re} M^{2-c} \Phi(\boldsymbol{v}_{\mathrm{N}}) \,, \tag{4.34}$$

where $\Phi(\boldsymbol{v}_{\mathrm{N}})$ is the dissipation function written relative to the Navier velocity vector $\boldsymbol{v}_{\mathrm{N}}$. From the thermal condition (3.57c), when $T_0(t) \equiv 1$, we derive the boundary condition for T_c, the solution of the temperature equation (4.34), as

$$T_c = \Lambda * M^{q-c} \Theta(t, \mathbf{P}) \text{ all along } \Sigma \,. \tag{4.35}$$

In fact, from the dimensionless equation of state $p = \rho T$ together with (4.32), we see that three cases are possible, namely (i) $d = c = 2$, (ii) $d = c < 2$ and (iii) $d = 2$. Respectively, for q in (4.35), we can consider the cases (Zeytounian 1977) $0 < q < 2, q = 2$ and $4 > q > 2$. Below, we restrict ourselves to the cases $q = 1$ and $q = 2$, and for both of these cases we have the possibility to formulate a Fourier–initial–boundary-value (F–IBV) problem for T_c, with $c = q$, which takes thermal effects into account.

The Case $q = 2$

In the most significant case, when in the expansion (4.32) we have

$$0 < a = c = d = 2 \text{ and } b = 4 \,, \tag{4.36}$$

together with $q = c = 2$ in (4.35), we obtain the following equation for T_2 from the above temperature equations (4.34):

$$\frac{\partial T_2}{\partial t} + \boldsymbol{v}_{\mathrm{N}} \,.\, \nabla T_2 - \frac{\gamma - 1}{\gamma} \left[\frac{\partial p_{\mathrm{N}}}{\partial t} + \boldsymbol{v}_{\mathrm{N}} \,.\, \nabla p_{\mathrm{N}} \right] = \frac{1}{Pr\, Re} \Delta T_2 + \frac{\gamma - 1}{Re} \Phi(\boldsymbol{v}_{\mathrm{N}}) \,. \tag{4.37}$$

With (4.37) it is necessary to associate, first, an initial condition at $t = 0$, which is determined by the relation (obtained by matching – see Sect. 4.1.7)

$$T_2(0, \boldsymbol{x}) = -(\gamma - 1)\nabla^2[\lim_{\tau\to\infty}\phi_{A1}] , \tag{4.38}$$

where ϕ_{A1} is the solution of an inhomogeneous d'Alembert equation, derived below in Sect. 4.1.7. At infinity we write, as a consequence of (4.6) for T_2,

$$T_2 \to 0 \quad \text{when } |\boldsymbol{x}| \to \infty , \tag{4.39}$$

because $\beta \geq 2$ in the far-field expansion (4.6). Finally, on the wall Σ of the rigid body, we obtain the following thermal condition from (4.35):

$$T_2 = \Lambda * \Theta(t, \boldsymbol{P}) \text{ all along } \Sigma, t > 0 . \tag{4.40}$$

The Case $q = 1$

When $q = c = 1$ in (4.35), the following 'reduced' temperature equation can be derived from (4.34):

$$\frac{\partial T_1}{\partial t} + \boldsymbol{u}_{\mathrm{N}} \cdot \nabla T_1 = \frac{1}{Pr\, Re}\Delta T_1 . \tag{4.41}$$

On the wall Σ of the bounded rigid body we write:

$$T_1 = \lambda^*\Theta(t, \mathbf{P}) , \tag{4.42a}$$

and at infinity:

$$\text{when } |\boldsymbol{x}| \to \infty : T_1 \to 0 . \tag{4.42b}$$

At $t = 0$, we have, by matching, in accordance with (4.32b), (3.75a) and (3.89),

$$T_1(0, \boldsymbol{x}) = -(\gamma - 1)\lim_{\tau\to\infty}\left[\frac{\partial \phi_{\mathrm{A0}}}{\partial \tau}\right] = 0 . \tag{4.42c}$$

Finally, we observe that, in the expansion $\rho = 1 + M\rho_1 + \cdots$, we have

$$\rho_1 = T_1 . \tag{4.42d}$$

4.1.6 The Navier–Fourier Quasi-Compressible Model Problem

On the other hand, in the more significant case $q = 2$, when $\boldsymbol{v}_{\mathrm{N}}, p_{\mathrm{N}}$ and T_2 are known, we have the possibility to derive also a system of two equations for the terms $\boldsymbol{v}_2$ and p_4 in the expansion (4.32a) which take into account some effects of weak compressibility, linked mainly to the temperature perturbation T_2 and the associated thermal condition for the wall, and also to the effect of the fact that there is no divergence-free condition for $\boldsymbol{v}_2$. Indeed, we have the possibility to supplement the classical Navier–IBV model problem for $\boldsymbol{v}_{\mathrm{N}}$ and p_{N} with a second-order 'quasi-compressible' model equations relating to the functions $\boldsymbol{v}_2$ and p_4. Namely, we obtain from the NSF equations (3.48), together with the expansion

$$\boldsymbol{v} = \boldsymbol{v}_\mathrm{N} + M^2\boldsymbol{v}_2 + \cdots , \quad p = 1 + M^2 p_\mathrm{N} + M^4 p_4 + \cdots , \quad T = 1 + M^2 T_2 + \ldots , \tag{4.43}$$

a system of two equations:

$$\nabla \cdot \boldsymbol{v}_2 = -\left(\frac{\partial \rho_2}{\partial t} + \boldsymbol{v}_\mathrm{N} \cdot \nabla \rho_2\right) , \tag{4.44a}$$

$$\frac{\partial \boldsymbol{v}_2}{\partial t} + (\boldsymbol{v}_\mathrm{N} \cdot \nabla)\boldsymbol{v}_2 + (\boldsymbol{v}_2 \cdot \nabla)\boldsymbol{v}_\mathrm{N} + \frac{1}{\gamma}\nabla p_4 - \frac{1}{Re}\Delta \boldsymbol{v}_2$$
$$= -\frac{1}{3Re}\nabla\left(\frac{\partial \rho_2}{\partial t} + \boldsymbol{v}_\mathrm{N} \cdot \nabla \rho_2\right) - \rho_2\left[\frac{\partial \boldsymbol{v}_\mathrm{N}}{\partial t} + (\boldsymbol{v}_\mathrm{N} \cdot \nabla)\boldsymbol{v}_\mathrm{N}\right] , \tag{4.44b}$$

together with

$$\rho_2 = p_\mathrm{N} - T_2 . \tag{4.44c}$$

Obviously, only our above systematic asymptotic approach ('fluid dynamics inspired by asymptotics') gives the possibility to take into account, in the above second-order, quasi-compressible system for $\boldsymbol{v}_2$ and p_4, the well-balanced terms in the right-hand sides of (4.44a) and (4.44b) which are responsible for the effects of slight compressibility and the thermal effects. We observe also that the above system of two equations (4.44a,b), together with (4.44c), is completely determined only when the model problem for the three functions $\boldsymbol{v}_\mathrm{N}$, p_N and T_2, governed by the Navier equations (3.55) and the equation for the temperature perturbation (4.37), has been solved. Concerning the initial and boundary conditions for $\boldsymbol{v}_2$, the solution, together with p_4, of the system of two equations (4.44a,b) with (4.44c), we write the following homogeneous initial and boundary conditions:

$$\boldsymbol{v}_2 = 0 \quad \text{all along } \Sigma, \text{ at } t = 0 \text{ and when } |\boldsymbol{x}| \to \infty . \tag{4.45}$$

Nevertheless, when we have a *source* as $|\boldsymbol{x}| \to \infty$ and, as a consequence, $\beta = 2$ in the equation for $\boldsymbol{v}$ in far-field expansion (4.6), the behaviour condition at infinity must be replaced by a matching between the terms $M^2\boldsymbol{v}_2$ and $M^\beta \boldsymbol{v}^{0\mathrm{ff}}$. On the other hand, the rigorous justification of the zero initial condition at $t = 0$ for $\boldsymbol{v}_2$ in (4.45) is again linked to an unsteady acoustic adjustment problem close to the initial time for the third-order acoustic equations and for matching between $M^2\boldsymbol{v}_2$ and $M^2\boldsymbol{v}_{\mathrm{a}2}$, when, in place of the asymptotic expansions (3.74), we write the following acoustic expansions (local in time, where the terms are dependent on τ and $\boldsymbol{x}$):

$$\boldsymbol{v} = \boldsymbol{v}_{\mathrm{a}0} + M\boldsymbol{v}_{a1} + M^2\boldsymbol{v}_{\mathrm{a}2} + \ldots , \tag{4.46a}$$

$$(p, \rho, T) = (1, 1, 1) + M(p_{\mathrm{a}1}, \rho_{\mathrm{a}1}, T_{\mathrm{a}1}) + M^2(p_{\mathrm{a}2}, \rho_{\mathrm{a}2}, T_{\mathrm{a}2})$$
$$+ M^3(p_{\mathrm{a}3}, \rho_{\mathrm{a}3}, T_{\mathrm{a}3}) + \ldots . \tag{4.46b}$$

We observe that, despite the homogeneous (zero) initial and boundary conditions (4.45), the solution of the two second-order equations (4.44a,b) is not zero, thanks to the source terms on the right-hand side arising from the above Navier–Fourier model problem for $\boldsymbol{v}_{\mathrm{N}}, p_{\mathrm{N}}$ and T_2. It is also necessary to state that when we have $\beta = 4$ in the far-field expansions (4.4) and (4.6), which represents a 'quadrupole situation' appropriate to sound generation à la Lighthill (see Ffowcs Williams 1984), then the far field does not interact with the above Navier–Fourier model problem complemented by the problem of the second-order quasi-compressible effects governed by the equations (4.44a,b) with the conditions (4.45).

4.1.7 The Second-Order Acoustic Equations and the Initial Condition for T_2

Going back to the initial-time (acoustic) local limit (3.4) and the acoustic expansion (3.74), we can derive from the NSF equations (3.48) the following second-order inhomogeneous acoustic equations, for the functions $\boldsymbol{v}_{\mathrm{a}1}, p_{\mathrm{a}2}, T_{\mathrm{a}2}$ and $\rho_{\mathrm{a}2}$;

$$
\begin{aligned}
\frac{\partial \rho_{\mathrm{a}2}}{\partial \tau} + \nabla \cdot \boldsymbol{v}_{\mathrm{a}1} &= \frac{\partial Q_{\mathrm{a}0}}{\partial \tau} , \\
\frac{\partial \boldsymbol{v}_{\mathrm{a}1}}{\partial \tau} + \nabla \left(\frac{p_{\mathrm{a}2}}{\gamma} \right) &= \nabla P_{\mathrm{a}0} , \\
\frac{\partial T_{\mathrm{a}2}}{\partial \tau} + (\gamma - 1) \nabla \cdot \boldsymbol{v}_{\mathrm{a}1} &= \frac{\partial R_{\mathrm{a}0}}{\partial \tau} , \\
p_{\mathrm{a}2} - (\rho_{\mathrm{a}2} + T_{\mathrm{a}2}) &= (\gamma - 1) \left(\frac{\partial \phi_{\mathrm{a}0}}{\partial \tau} \right)^2 .
\end{aligned}
\tag{4.47}
$$

From (3.75a), $\boldsymbol{v}_{\mathrm{N}}(t = 0, \boldsymbol{x}) = \nabla \psi_{\mathrm{N}}^0$, and when we assume that Re and Pr are both $O(1)$, we have

$$
Q_{\mathrm{a}0} = \nabla \psi_{\mathrm{N}}^0 \cdot \nabla \phi_{\mathrm{a}0} + \frac{1}{2} (\nabla \phi_{\mathrm{a}0})^2 + \frac{1}{2} \left(\frac{\partial \phi_{\mathrm{a}0}}{\partial \tau} \right)^2 , \tag{4.48a}
$$

$$
P_{\mathrm{a}0} = \frac{1}{2} \left(\frac{\partial \phi_{\mathrm{a}0}}{\partial \tau} \right)^2 - \frac{1}{2} \mid \nabla \psi_{\mathrm{N}}^0 + \nabla \phi_{\mathrm{a}0} |^2 + \frac{2}{3\,Re} \nabla^2 \phi_{\mathrm{a}0} ,
$$

$$
R_{\mathrm{a}0} = (\gamma - 1) \left[\nabla \psi_{\mathrm{N}}^0 \cdot \nabla \phi_{\mathrm{a}0} + \frac{1}{2} (\nabla \phi_{\mathrm{a}0})^2 + \frac{\gamma}{2} \left(\frac{\partial \phi_{\mathrm{a}0}}{\partial \tau} \right)^2 \right. \tag{4.48b}
$$

$$
\left. - \left(\frac{\gamma}{Pr\,Re} \right) \nabla^2 \phi_{\mathrm{a}0} \right] . \tag{4.48c}
$$

It is easily checked that the following formulae solve (4.47):

$$\boldsymbol{v}_{\mathrm{a1}} = \nabla\left(\frac{\partial \phi_{\mathrm{a1}}}{\partial \tau}\right), \quad \rho_{\mathrm{a2}} = -\nabla^2\phi_{\mathrm{a1}} + Q_{\mathrm{a0}} \,, \tag{4.49a}$$

$$T_{\mathrm{a2}} = -(\gamma-1)\nabla^2\phi_{\mathrm{a1}} + R_{\mathrm{a0}}, \quad p_{\mathrm{a2}} = -\gamma\frac{\partial^2\phi_{\mathrm{a1}}}{\partial\tau^2} + \gamma P_{\mathrm{a0}} \,, \tag{4.49b}$$

provided ϕ_{a1} is a solution to the inhomogeneous d'Alembert' equation

$$\frac{\partial^2\phi_{\mathrm{a1}}}{\partial\tau^2} - \nabla^2\phi_{\mathrm{a1}} = P_{\mathrm{a0}} - \frac{1}{\gamma}[Q_{\mathrm{a0}} + R_{\mathrm{a0}}] - \frac{\gamma-1}{\gamma}\left(\frac{\partial\phi_{\mathrm{a0}}}{\partial\tau}\right)^2 , \tag{4.50}$$

to which we must add the conditions

$$\text{at } \tau = 0: \quad \phi_{\mathrm{a1}} = 0, \quad \frac{\partial\phi_{\mathrm{a1}}}{\partial\tau} = 0, \quad \text{everywhere outside the body;} \tag{4.51a}$$

$$\frac{\partial\phi_{\mathrm{a1}}}{\partial n} = 0, \text{ all along the wall } \Sigma \,. \tag{4.51b}$$

From the values of P_{a0}, Q_{a0} and R_{a0} according to (4.48), we conclude that the right-hand side of the inhomogeneous d' Alembert equation (4.50) tends to zero when $|\boldsymbol{x}| \to \infty$ with τ fixed. Relying on the same physical argument as used for (4.26) for ϕ_{a0}, we may add

$$\phi_{\mathrm{a1}} \to 0 \text{ when} |\boldsymbol{x}| \to \infty \,, \tag{4.51c}$$

and obtain with (4.51), a well-posed initial–boundary-value problem for the inhomogeneous d'Alembert equation (4.50) relating to the function ϕ_{a1}. A rough argument suggests that the solution of the inhomogeneous equation (4.50) for $\phi_{\mathrm{a1}}(\tau, \boldsymbol{x})$ tends to zero as $\tau \to \infty$,(and as a consequence of (4.38), $T_2(0, \boldsymbol{x}) = 0$!). However, such a statement necessitates a rigorous proof, and the above conclusion is, in fact, only a *conjecture* for the moment. On the other hand, from known results concerning d'Alembert's equation of acoustics (4.18) for ϕ_{a0}, we may be more precise about the behaviour (4.51c). Namely (see Wilcox 1975), at any fixed point, ϕ_{a0} (together with $\partial\phi_{\mathrm{a0}}/\partial\tau$ and $\nabla\phi_{\mathrm{a0}}$) vanishes identically within some vicinity (depending on the point $\boldsymbol{x}$ considered) of $\tau = \infty$, and the same holds true for P_{a0}, Q_{a0} and R_{a0}.

Considering the above problem, we may construct the solution of the equation (4.50) together with (4.51) for ϕ_{a1} in two steps. The first step is a retarded-potential solution of the inhomogeneous equation (4.50) ignoring the initial and boundary conditions (4.51a,b), corresponding to an extension of the right-hand side of (4.50) to zero for time values $\tau < 0$ and/or within the body. The second step is to solve a homogeneous problem (with zero as the right-hand side in (4.50)) with initial and boundary data according to (4.51a,b). Obviously, the solution of this second-step homogeneous problem is identically zero. Seeing that the right-hand side of the inhomogeneous d'Alembert equation (4.50) tends to zero as τ tends to infinity, we may think that the solution of the first-step inhomogeneous problem also tends to zero when the time τ tends to infinity. In reality, however, this conclusion seems not to be sufficiently rigorous, because the behaviour of this particular solution as τ tends to infinity is strongly related to the nature of the behaviour (the rapidity with which it tends to zero) of the right-hand side of (4.50) when $\tau \to \infty$. But here, we should not be disturbed by this purely mathematical problem.

4.2 From the NSF Equations to Burgers' Model Equation

The equations of gas dynamics (in fact, the non-viscous, compressible, unsteady Euler equations (1.30) derived from the NSF equations (3.48) when $Re \equiv \infty$ and $Pr = O(1)$) constitute a quasi-linear hyperbolic system. These unsteady equations of inviscid fluid dynamics are 'properly embedded' in the full viscous, thermally conducting equations, and one remarkable model equation of fluid dynamics, Burgers' equation, illustrates this embedding. It describes the motion of weak nonlinear (acoustic) waves in gases when one needs to take account of dissipative effects to a first approximation. In the limit of vanishing dissipation, Burger's equation provides the proper interpretation of the inviscid solution. The history of the Burgers' equation is too rich to detail here. It was proposed by Burgers in 1948 (Burgers 1948) as a model equation for one-dimensional turbulence, and Lighthill (1956) showed that, with the proper interpretation, this model equation was appropriate for describing the propagation of weak planar waves. On the other hand, the essential ideas concerning the asymptotic derivation of Burgers' equation from the full NSF equations may be found in publications by Guiraud (see, for instance, Guiraud 1967, pp. 18–20), and in Germain (1971) the reader can find a derivation of Burgers' equation by a method which is essentially equivalent. For a more rigorous mathematical point of view, see the paper by Choquet-Bruhat (1969). The main result in Sect. 4.2.2 below is the asymptotic low-Mach-number derivation of Burgers' dissipative model equation from the full NSF system of equations as a transport equation valid in a thin singular region of thickness $O(M)$ close to an acoustic wavefront inside an acoustic field governed by the classical linear acoustic equations. Before that, in Sect. 4.2.1, we discuss briefly the asymptotic status of Burgers' equation.

4.2.1 A Simple Derivation of Burgers' Equation

Burgers' equation is very simply derived as a compatibility–non-secularity condition via a two-timescale approach, in the framework of low-Mach-number asymptotics, from the one-dimensional unsteady NSF equations for large Reynolds number ($Re \gg 1$) and Strouhal number ($St \gg 1$) such that (Zeytounian 2002a)

$$Re^* = M\,Re = O(1) \quad \text{and } M\,St \equiv 1, \quad \text{when } M \ll 1\,. \tag{4.52}$$

Namely, we obtain, for an unsteady one-dimensional sound wave in which $u = F(\tau, \sigma)$, moving to the right, after a long time τ,

$$\frac{\partial F}{\partial t} + \frac{1}{2}(\gamma + 1)F\left(\frac{\partial F}{\partial \sigma}\right) = 2S^* \frac{\partial^2 F}{\partial \sigma^2}\,, \tag{4.53}$$

with the following dissipative Stokes coefficient (with $Re^* = O(1)$ for zero bulk viscosity and $\tau = M^2 t, \sigma = x - t$):

$$S^* = \frac{1}{Re^*}\left(\frac{4}{3} + \frac{\gamma - 1}{Pr}\right)\,. \tag{4.54}$$

4.2.2 Burgers' Equation as a Dissipative Model Transport Equation in a Thin Region Close to an Acoustic Wavefront

In the NSF equations (3.48), we have assumed that the Strouhal number $St = L_c/t_c U_c \equiv 1$; below, in contrast, we assume that $St \gg 1$ and we write in the NSF equations (3.49), in place of the material derivative $\mathrm{D}/\mathrm{D}t = \partial/\partial t + \boldsymbol{v}\,.\,\nabla$, the material derivative $St\,\mathrm{D}/\mathrm{D}t = St\,\partial/\partial + \boldsymbol{v}\,.\,\boldsymbol{\nabla}$. We then consider the following case, where, in the NSF equations with $St\,\mathrm{D}/\mathrm{D}t$ in place of $\mathrm{D}/\mathrm{D}t$ we assume that

$$M \ll 1, \quad St \gg 1 \quad \text{and } Re \gg 1 \tag{4.55a}$$

Simultaneously, such that

$$St\,M = 1 \quad \text{and } \frac{Re}{St} = 1\,, \tag{4.55b}$$

and in this case our main small parameter is M. We aim to derive, via (4.55), a significant transport equation valid in a thin region close to a wavefront where the classical acoustic inviscid equations are singular. The reader can find in Germain (1971) a general approach related to the concept of 'progressive waves' and a derivation of the most typical canonical form of the transport equation the generalized Burgers equation for purely dissipative effects.

Acoustic Inviscid Region

When $M \ll 1$ together with (4.55b), we consider the following (acoustic) expansions for the solution of the NSF equations:

$$\boldsymbol{v} = \boldsymbol{v}_0 + M\boldsymbol{v}_1 + \ldots, \quad p = 1 + Mp_1 + M^2 p_2 + \ldots\,, \tag{4.56a}$$

$$T = 1 + MT_1 + \ldots, \quad \rho = 1 + M\rho_1 + \ldots\,. \tag{4.56b}$$

When

$$M \to 0 \text{ with } t \text{ and } \boldsymbol{x} \text{ fixed}\,, \tag{4.57}$$

we derive again, for $\boldsymbol{u}_0, p_1, T_1$ and ρ_1, the usual linear equations of acoustics:

$$\frac{\partial \rho_1}{\partial t} + \nabla \cdot \boldsymbol{u}_0 = 0\,,$$
$$\frac{\partial \boldsymbol{u}_0}{\partial t} + \nabla\left(\frac{p_1}{\gamma}\right) = 0\,,$$
$$\frac{\partial T_1}{\partial t} + (\gamma-1)\nabla \cdot \boldsymbol{u}_0 = 0\,,$$
$$p_1 = \rho_1 + T_1\,, \tag{4.58}$$

from these equations of acoustics (4.58), we derive a single equation for $\boldsymbol{v}_0$,

$$\frac{\partial^2 \boldsymbol{v}_0}{\partial t^2} - \nabla^2 \boldsymbol{v}_0 = 0\,. \tag{4.59}$$

For the wave equation (4.59) for $\boldsymbol{v}_0$, we consider the case where the wave train is trapped inside a thin region close to a wavefront. Outside this thin domain (characterized by a thickness of order $O(M)$), in the acoustic region, the above equations (4.58) or (4.59) are valid at the leading order.

Thin Region Close to Wavefront

We assume that the equation of the wavefront can be written as:

$$F(\boldsymbol{x}) - t = 0\,, \tag{4.60}$$

and for the description of a thin region (with a thickness of order $O(M)$) close to the wavefront, we introduce a local variable

$$\theta = \frac{1}{M}[F(\boldsymbol{x}) - t]\,. \tag{4.61}$$

Using (4.61), we consider, in the thin region close to the wavefront, in place of the (acoustic) expansions (4.56), the following local expansions:

$$\boldsymbol{v} = \boldsymbol{v}_0^*(\theta,\boldsymbol{x}) + M\boldsymbol{v}_1^*(\theta,\boldsymbol{x}) + \cdots\,,\; p = 1 + Mp_1^*(\theta,\boldsymbol{x}) + M^2 p_2^*(\theta,\boldsymbol{x}) + \ldots\,, \tag{4.62a}$$
$$T = 1 + MT_1^*(\theta,\boldsymbol{x}) + M^2 T_2^*(\theta,\boldsymbol{x}) + \ldots\,, \tag{4.62b}$$
$$\rho = 1 + M\rho_1^*(\theta,\boldsymbol{x}) M^2 \rho_2^*(\theta,\boldsymbol{x}) + \cdots\,, \tag{4.62c}$$

with the local low-Mach-number limiting process

$$M \to 0 \text{ with } \theta \text{ and } \boldsymbol{x} \text{ fixed} \tag{4.63}$$

in place of (4.57). First, we derive for $\boldsymbol{v}_0^*, p_1^*$, T_1^*, and ρ_1^*, at the order M, the following equations:

$$\frac{\partial}{\partial\theta}[-\rho_1^* + \boldsymbol{k}\,.\,\boldsymbol{v}_0^*] = 0\,,$$
$$\frac{\partial}{\partial\theta}\left[-\boldsymbol{v}_0^* + \frac{1}{\gamma}(\rho_1^* + T_1^*)\right] = 0\,,$$
$$\frac{\partial}{\partial\theta}[-T_1^* + (\gamma-1)\boldsymbol{k}\,.\,\boldsymbol{v}_0^*] = 0\,, \tag{4.63}$$

where

$$\boldsymbol{k} \equiv \boldsymbol{k}(\boldsymbol{x}) = \nabla F(\boldsymbol{x})\,. \tag{4.64}$$

However, we see that the leading-order system (4.63) is linear and homogeneous with respect to the terms $\partial/\partial\theta(\rho_1^*), \partial/\partial\theta(\boldsymbol{v}_0^*)$ and $\partial/\partial\theta(T_1^*)$ and, as a consequence, a non-trivial (different from zero) solution of this system (4.63) exists if and only if the matrix of the coefficients of this system is identically zero. The result is that, necessarily,

$$\boldsymbol{k}^2 = 1, \quad \text{or}\,[\nabla F(\boldsymbol{x})]^2 = 1\,, \tag{4.65}$$

which is a first-order hyperbolic equation (the 'eikonal' equation) for the unknown function $F(\boldsymbol{x})$. More precisely, the characteristic form of this eikonal equation is

$$\frac{\mathrm{d}F(\boldsymbol{x})}{\mathrm{d}\kappa} = 1, \quad \frac{\mathrm{d}\boldsymbol{k}}{\mathrm{d}\kappa} = 0, \quad \text{along}\,\frac{\mathrm{d}\boldsymbol{x}}{\mathrm{d}\kappa} = \boldsymbol{k}\,, \tag{4.66}$$

where κ is a parameter which varies along the characteristics in the direction of increasing F. Usually, $\boldsymbol{k} = 0$ on the initial surface $F(\boldsymbol{x}) = 0$, corresponding to the initial time $t = 0$. In a simple case (no caustics) it is possible to assume that the surface $F(\boldsymbol{x}) = 0$ is smooth (differentiable), and for the determination of the function $F(\boldsymbol{x})$ it is necessary now to consider the Hessian of $F(\boldsymbol{x})$, which is defined by the matrix

$$\mathbf{A} \equiv \mathbf{A}^{\mathrm{T}} = \frac{\partial^2 F}{\partial x_i\,\partial x_j}\,. \tag{4.67a}$$

We observe, as in classical theory, that

$$\mathbf{A}\,.\,\boldsymbol{k} = 0 \quad \text{and}\,\frac{\mathrm{d}\mathbf{A}}{\mathrm{d}\kappa} = -\mathbf{A}^2\,, \tag{4.67b}$$

the null value is an eigenvalue of $\mathbf{A}$, and $\mathbf{A}$ is a symmetric matrix. As a consequence, if λ_1 and λ_2 are two other eigenvalues of the matrix $\mathbf{A}$, we can write

$$\frac{\mathrm{d}\lambda_j}{\mathrm{d}\kappa} = -\lambda_j^2, \quad j = 1 \text{ and } 2\,, \tag{4.67c}$$

and the $\lambda_j \equiv K_j\,(j = 1\,\text{and}\,2)$, are, in fact, the principal curvatures of the wavefront surface $F = \text{const}$. Finally, we derive the following equation:

$$\Delta F = \frac{\sum K_{\mathrm{j}}^0}{1 + K_{\mathrm{j}}^0\kappa}, \quad j = 1 \text{ and } 2\,, \tag{4.68}$$

where K_j^0 corresponds to the initial ($\kappa = 0$) wavefront. The general solution of the system (4.63) of three equations can be written as

$$\rho_1^* = \frac{1}{\gamma - 1}\,\sigma(\theta, \boldsymbol{x}) + R(\boldsymbol{x})\,, \tag{4.69a}$$

$$T_1^* = \sigma(\theta, \boldsymbol{x}) + T(\boldsymbol{x})\,, \tag{4.69b}$$

$$\boldsymbol{v}_0^* = \frac{1}{\gamma - 1}\sigma(\theta, \boldsymbol{x})\boldsymbol{k} + \boldsymbol{U}(\boldsymbol{x})\,. \tag{4.69c}$$

In (4.69), the arbitrary functions $R(\boldsymbol{x}), T(\boldsymbol{x})$ and $\boldsymbol{U}(\boldsymbol{x})$ are determined by the compatibility condition for the equations of order M^2 (see (4.70) below), the conditions for matching with the acoustic solutions valid upstream and downstream of the thin region close to the wavefront, and the initial and boundary conditions prescribed for the physical problem.

At the order M^2, we obtain, after a lengthy but straightforward calculation, the system of equations

$$\frac{\partial}{\partial \theta}[-\rho_2^* + \boldsymbol{k}\,.\,\boldsymbol{v}_1^*] = -\nabla\,.\,\boldsymbol{v}_0^* - \frac{\partial}{\partial \theta}(\rho_1^* \boldsymbol{k}\,.\,\boldsymbol{v}_0^*)\,, \tag{4.70a}$$

$$\begin{aligned}\frac{\partial}{\partial \theta}\left[-\boldsymbol{v}_1^* + \frac{1}{\gamma}(\rho_2^* + T_2^*)\right] &= -\frac{1}{\gamma}\nabla(\rho_1^* + T_1^*) + \rho_1^*\frac{\partial \boldsymbol{v}_0^*}{\partial \theta} \\ &- (\boldsymbol{k}\,.\,\boldsymbol{v}_0^*)\frac{\partial \boldsymbol{v}_0^*}{\partial \theta} - \frac{1}{\gamma}\frac{\partial}{\partial \theta}(\rho_1^* T_1^*)\boldsymbol{k} + \frac{\partial^2}{\partial \theta^2}\left[\frac{1}{3}(\boldsymbol{k}\,.\,\boldsymbol{v}_0^*)\boldsymbol{k} + \boldsymbol{k}^2 \boldsymbol{v}_0^*\right]\,,\end{aligned} \tag{4.70b}$$

$$\begin{aligned}\frac{\partial}{\partial \theta}[-T_2^* + (\gamma - 1)\boldsymbol{k}\,.\,\boldsymbol{v}_1^*] &= -(\gamma - 1)\nabla\,.\,\boldsymbol{v}_0^* + \rho_1^*\frac{\partial T_1^*}{\partial \theta} \\ &- (\boldsymbol{k}\,.\,\boldsymbol{v}_0^*)\frac{\partial T_1^*}{\partial \theta} + \frac{1}{2}\gamma(\gamma - 1)\frac{\partial (\boldsymbol{v}_0^*)^2}{\partial \theta} \\ &- (\gamma - 1)\frac{\partial}{\partial \theta}[(\rho_1^* + T_1^*)(\boldsymbol{k}\,.\,\boldsymbol{v}_0^*)] + \frac{\gamma}{Pr}\boldsymbol{k}^2\frac{\partial^2 T_1^*}{\partial \theta^2}\,.\end{aligned} \tag{4.70c}$$

The second-order system of three equations (4.70) is an inhomogeneous linear system with respect to the terms $\partial/\partial\theta(\rho_2^*), \partial/\partial\theta(\boldsymbol{v}_1^*)$ and $\partial/\partial\theta(T_2^*)$, and we know that a compatibility condition exists necessarily for the solvability of this inhomogeneous system. A very naive, simple way to derive this compatibility condition is to write the combination $\gamma \boldsymbol{k} \times$ (4.70b) + (4.70a) + (4.70c), which eliminates the terms containing $\boldsymbol{v}_1^*, T_2^*$ and ρ_2^*. As a result, we find the following compatibility condition:

$$\begin{aligned}(\boldsymbol{k}\,.\,\nabla)(\rho_1^* + T_1^*) + \gamma\nabla\,.\,\boldsymbol{v}_0^* &+ \frac{\partial}{\partial \theta}\Big[\rho_1^* \boldsymbol{k}\,.\,\boldsymbol{v}_0^* + \frac{\gamma}{2}(\boldsymbol{k}\,.\,\boldsymbol{v}_0^*)^2 + \rho_1^* T_1^* \\ &- \frac{1}{2}\gamma(\gamma - 1)(\boldsymbol{v}_0^*)^2 + (\gamma - 1)(\rho_1^* + T_1^*)(\boldsymbol{k}\,.\,\boldsymbol{v}_0^*)\Big] - \gamma\rho_1^*\frac{\partial}{\partial \theta}(\boldsymbol{k}\,.\,\boldsymbol{v}_0^*) \\ &- \rho_1^*\frac{\partial T_1^*}{\partial \theta} + (\boldsymbol{k}\,.\,\boldsymbol{v}_0^*)\frac{\partial T_1^*}{\partial \theta} = \frac{4\gamma}{3}\frac{\partial^2}{\partial \theta^2}(\boldsymbol{k}\,.\,\boldsymbol{v}_0^*) + \frac{\gamma}{Pr}\frac{\partial^2 T_1^*}{\partial \theta^2}\,.\end{aligned} \tag{4.71}$$

Now, we take into account, in the above compatibility condition, the solution (4.69) for the functions ρ_1^*, T_1^* and $\boldsymbol{v}_0^*$, and also the three relations (4.65), (4.66) and (4.68). As a result, we obtain the following compatibility equation for the amplitude function $\sigma(\theta, \boldsymbol{x})$:

$$\frac{\Delta F}{2}\sigma + \frac{\mathrm{d}\sigma}{\mathrm{d}\kappa} + \left(\boldsymbol{k} \,.\, \boldsymbol{U} + \frac{T}{2}\right)\frac{\mathrm{d}\sigma}{\mathrm{d}\theta} + \frac{1}{2}\frac{\gamma+1}{\gamma-1}\sigma\frac{\mathrm{d}\sigma}{\mathrm{d}\theta}$$
$$= \beta\frac{\partial^2\sigma}{\partial\theta^2} - \frac{1}{2}(\gamma-1)\nabla \,.\, \boldsymbol{U} - \frac{1}{2\gamma}(\gamma-1)\frac{\mathrm{d}(R+T)}{\mathrm{d}\kappa}\,, \tag{4.72}$$

where

$$0 < \beta = \frac{1}{2}\left(\frac{4}{3} + \frac{\gamma-1}{Pr}\right)\,. \tag{4.73}$$

Matching and Burgers' Equation

The most interesting solutions of the above equation (4.72) are those with the following behavior (related to the matching process):

$$\sigma(\theta, \boldsymbol{x}) \to \sigma_{\mathrm{am}}(\boldsymbol{x}) \equiv 0 \quad \text{when}\, \theta \to -\infty\,. \tag{4.74a}$$

In such a case, we obtain

$$\rho_1^* \sim R(\boldsymbol{x}), \quad T_1^* \sim T(\boldsymbol{x}), \quad \boldsymbol{v}_0^* \sim \boldsymbol{U}(\boldsymbol{x}) \quad \text{when}\, \theta \to -\infty\,, \tag{4.75a}$$

in the *upstream* proximity of the wavefront, while

$$\frac{\sigma_{\mathrm{av}}(\boldsymbol{x})}{\gamma-1} + R(\boldsymbol{x}), \quad \sigma_{\mathrm{av}}(\boldsymbol{x}) + T(\boldsymbol{x}), \quad \left[\frac{\sigma_{\mathrm{av}}(\boldsymbol{x})}{\gamma-1}\right]\boldsymbol{k} + \boldsymbol{U}(\boldsymbol{x}) \tag{4.75b}$$

specify the state of the flow in the *downstream* proximity of the wavefront, where

$$\sigma(\theta, \boldsymbol{x}) \to \sigma_{\mathrm{av}}(\boldsymbol{x}) \quad \text{when}\, \theta \to +\infty\,. \tag{4.74b}$$

With the above conditions, we derive, two relations from (4.72):

$$\frac{\mathrm{d}(R+T)}{\mathrm{d}\kappa} + \gamma\nabla \,.\, \boldsymbol{U} = 0, \quad \frac{1}{2}\Delta F\,\sigma_{\mathrm{av}} + \frac{\mathrm{d}\sigma_{\mathrm{av}}}{\mathrm{d}\kappa} = 0\,. \tag{4.76}$$

Now, we can match the local solution (valid close to the wavefront) and the two outer solutions (acoustic, valid in the upstream and downstream regions close to the wavefront). Namely, on the wavefront $t = F(\boldsymbol{x})$, we write for the upstream side

$$\rho_{1\mathrm{am}}(t, \boldsymbol{x}) = R(\boldsymbol{x}), \quad \boldsymbol{v}_{0\mathrm{am}}(t, \boldsymbol{x}) = \boldsymbol{U}(\boldsymbol{x}), \quad T_{1\mathrm{am}}(t, \boldsymbol{x}) = T(\boldsymbol{x})\,, \tag{4.77a}$$

and for the downstream side

$$\rho_{1\mathrm{av}}(t,\boldsymbol{x}) = \frac{\sigma_{\mathrm{av}}(\boldsymbol{x})}{\gamma-1} + R(\boldsymbol{x}), \quad T_{1\mathrm{av}}(t,\boldsymbol{x}) = \sigma_{\mathrm{av}}(\boldsymbol{x}) + T(\boldsymbol{x}) ,$$

$$\boldsymbol{v}_{0\mathrm{va}}(t,\boldsymbol{x}) = \left[\frac{\sigma_{\mathrm{av}}(\boldsymbol{x})}{\gamma-1}\right]\boldsymbol{k} + \boldsymbol{U}(\boldsymbol{x}). \tag{4.77b}$$

Finally, we obtain

$$\frac{\mathrm{d}\sigma}{\mathrm{d}\kappa} + \alpha_1\frac{\mathrm{d}\sigma}{\mathrm{d}\theta} + \alpha_2\sigma\frac{\mathrm{d}\sigma}{\mathrm{d}\theta} + \alpha_3\sigma = \beta\frac{\partial^2\sigma}{\partial\theta^2} , \tag{4.78}$$

which is a transport equation that is a slight generalization of Burgers' equation (Burgers 1948) for purely dissipative effects (without dispersive effects 'à la KdV'). In Germain (1971), the reader can find various pertinent references up to 1970 concerning a unified theory of progressive waves. We should add that our initial assumption (4.55b) of a relation between a low Mach number and large Strouhal and Reynolds numbers was not fortuitous; the choice was motivated by our wish to derive a model equation which retains all of the main features of the starting NSF problem. More precisely, the relative orders of the two large parameters must be chosen in order to obtain the least degenerate system of three equations (4.70).

4.3 The Blasius Problem for a Slightly Compressible Flow

A basic problem in the theory of fluids with a small viscosity (high Reynolds number) is that of steady 2D flow past a solid, flat plate placed in a uniform stream. More precisely, the plate is understood to be a half-plane, say $y > 0$, with $0 < x < \infty$, and the flow is taken to be uniform far upstream, with a velocity parallel to the plate and normal to its edge, so that

$$\boldsymbol{u} = U_0\boldsymbol{i} \quad \text{for } x \to -\infty \text{ and all } x \text{ and } z, \tag{4.79a}$$

where $U_0 = \text{const} > 0$, and $\boldsymbol{i}$ denotes the unit vector in the direction of increasing $x > 0$. For $x \to -\infty$ we assume also that the pressure p, the density ρ and the temperature T have constant prescribed values p_0, ρ_0 and T_0. In a compressible, viscous, thermally conducting flow, the temperature T has a prescribed value on the flat plate also. If this value is a function of x only, then the symmetry of the Blasius (1908) (steady) problem indicates that there should be a 2D flow with $\boldsymbol{u} = (u, v, 0)$ such that u, v, p, ρ and T are functions of x and y only, and our attention will be restricted to such a flow. The flat plate is understood to be impermeable, but with a variable temperature, so that we assume the following boundary conditions:

$$u = v = 0 \quad \text{and } T = T_0 + \Delta T_0\,\Theta(x) \quad \text{on } y = 0, 0 < x < \infty , \tag{4.79b}$$

where $\Delta T_0 > 0$ is a constant reference wall temperature and $\Theta(x)$ is a given function of x. In view of the symmetry of the flow, moreover, there is no loss of generality in restricting our attention to the half-space $y > 0$.

4.3.1 Basic Dimensionless Equations and Conditions

We start from the NSF equations (1.28) for a compressible, viscous, thermally conducting fluid and we suppose that the first and second coefficients of viscosity, μ and λ, and the heat conductivity coefficient k are functions of T only. First, the boundary conditions (4.79) make it possible to define a Blasius reference length, namely

$$l_0 = \frac{\nu_0}{U_0}\left(\frac{\Delta T_0}{T_0}\right)^2 , \tag{4.80}$$

where $\nu_0 = \mu(1)/\rho_0 = \text{const}$, and $\mu = \mu(T/T_0)$. In this case, with a proper choice of dimensionless quantities, the steady NSF equations for a thermally conducting perfect gas depend, in particular, on the Reynolds number (Re) and the upstream Mach number (M), where

$$Re = \frac{U_0 l_0}{\nu_0}, \quad M = \frac{U_0}{[\gamma R T_0]^{1/2}} . \tag{4.81}$$

The boundary condition for the temperature, according to (4.79b) in dimensionless form, includes the parameter

$$\kappa = \frac{\Delta T_0}{T_0} , \tag{4.82}$$

and from the above relations we have necessarily the following two similarity rules:

$$\kappa = M^2 \quad \text{and } M^2 Re^{1/2} = 1 . \tag{4.83}$$

These two similarity relations (4.83) are consistent with the asymptotic analysis carried out below. With (4.83), the dimensionless form of the NSF steady equations for a two-dimensional flow is

$$\frac{\partial(\rho u)}{\partial x} + \frac{\partial(\rho v)}{\partial y} = 0 , \tag{4.84a}$$

$$\begin{aligned} \rho\left(u\frac{\partial u}{\partial x} + v\frac{\partial u}{\partial y}\right) + \frac{1}{\gamma M^2}\frac{\partial p}{\partial x} = M^4\Bigg\{&\mu(T)\left(\frac{\partial^2 u}{\partial x^2} + \frac{\partial^2 u}{\partial y^2}\right) \\ &+ \frac{\mathrm{d}\mu}{\mathrm{d}T}\left[2\frac{\mathrm{d}T}{\mathrm{d}x}\frac{\partial u}{\partial x} + \frac{\mathrm{d}T}{\mathrm{d}y}\left(\frac{\partial u}{\partial y} + \frac{\partial v}{\partial x}\right)\right] \\ &+ \frac{\lambda_0}{\mu_0}\frac{\mathrm{d}\lambda}{\mathrm{d}T}\frac{\partial T}{\partial x}\left(\frac{\partial u}{\partial x} + \frac{\partial v}{\partial y}\right) \\ &+ \left[\frac{\lambda_0}{\mu_0}\lambda(T) + \mu(T)\right]\frac{\partial}{\partial x}\left(\frac{\partial u}{\partial x} + \frac{\partial v}{\partial y}\right)\Bigg\} , \end{aligned} \tag{4.84b}$$

$$\rho\left(u\frac{\partial v}{\partial x}+v\frac{\partial v}{\partial y}\right)+\frac{1}{\gamma M^2}\frac{\partial p}{\partial y}=M^4\left\{\mu(T)\left(\frac{\partial^2 v}{\partial x^2}+\frac{\partial^2 v}{\partial y^2}\right)\right.$$
$$+\frac{\mathrm{d}\mu}{\mathrm{d}T}\left[2\frac{\mathrm{d}T}{\mathrm{d}y}\frac{\partial v}{\partial y}+\frac{\mathrm{d}T}{\mathrm{d}x}\left(\frac{\partial u}{\partial y}+\frac{\partial v}{\partial x}\right)\right]$$
$$+\frac{\lambda_0}{\mu_0}\frac{\mathrm{d}\lambda}{\mathrm{d}T}\frac{\partial T}{\partial y}\left(\frac{\partial u}{\partial x}+\frac{\partial v}{\partial y}\right)$$
$$\left.+\left[\frac{\lambda_0}{\mu_0}\lambda(T)+\mu(T)\right]\frac{\partial}{\partial y}\left(\frac{\partial u}{\partial x}+\frac{\partial v}{\partial y}\right)\right\}, \tag{4.84c}$$

$$\rho\left(u\frac{\partial T}{\partial x}+v\frac{\partial T}{\partial y}\right)-\frac{\gamma-1}{\gamma}\left(u\frac{\partial p}{\partial x}+v\frac{\partial p}{\partial y}\right)$$
$$=M^4\left\{k(T)\left(\frac{\partial^2 T}{\partial x^2}+\frac{\partial^2 T}{\partial y^2}\right)\right.$$
$$\left.+\frac{\mathrm{d}k}{\mathrm{d}T}\left[\left(\frac{\mathrm{d}T}{dx}\right)^2+\left(\frac{\partial T}{\partial y}\right)^2\right]\right\}$$
$$+(\gamma-1)M^6\left\{\mu(T)\left[2\left(\frac{\partial u}{\partial x}\right)^2+2\left(\frac{\partial v}{\partial y}\right)^2\right.\right.$$
$$\left.\left.+\left(\frac{\partial u}{\partial y}+\frac{\partial v}{\partial x}\right)^2\right]+\frac{\lambda_0}{\mu_0}\lambda(T)\left(\frac{\partial u}{\partial x}+\frac{\partial v}{\partial y}\right)^2\right\}, \tag{4.84d}$$

where $p=\rho T, Pr=\mu_0 C_p/k_0\equiv 1, \mu_0=\mu(1), \lambda_0=\lambda(1)$ and $k_0=k(1)$. All the variables are now understood to be in units of their respective scales (we have not changed the notation). With the Stokes relation, we have $\lambda_0/\mu_0=-2/3$, and the non-dimensional form of the boundary conditions (4.79b) is

$$u=v=0 \text{ and } T=1+M^2\Theta(x) \quad \text{on } y=0, 0<x<\infty\,. \tag{4.85a}$$

We have also the following conditions far upstream:

$$u\to 1, v\to 0, (p,\rho,T)\to 1, \quad \text{as } x\to-\infty \text{ for all } y\,. \tag{4.85b}$$

4.3.2 The Limiting Euler Equations for $M^2\to 0$

We consider now the solution of the problem (4.84a–c) and (4.85) in the outer limit as

$$M^2\to 0 \quad \text{with } x \text{ and } y \text{ fixed}\,, \tag{4.86}$$

and we associate with the limiting process (4.86) the following outer asymptotic representation:

$$\begin{aligned} u &= 1 + M^2 u_1(x,y) +, \dots \\ v &= M^2 v_1(x,y) + \dots, \quad p = 1 + M^4 p_1(x,y) + \dots, \\ (\rho, T) &= (1.1) + M^2(\rho_1(x,y), T_1(x,y)) + \dots . \end{aligned} \tag{4.87}$$

The far-upstream conditions (4.85b) imply that $\rho_1 = T_1$ and, as a consequence, we derive the following Laplace equation from (4.84a–c):

$$\frac{\partial^2 \psi_1}{\partial x^2} + \frac{\partial^2 \psi_1}{\partial y^2} = 0, \quad \text{where } u_1 = \frac{\partial \psi_1}{\partial y}, v_1 = -\frac{\partial \psi_1}{\partial x}, \tag{4.88}$$

and ψ_1 is the associated 'incompressible' stream function satisfying the far-upstream condition $\psi_1(-\infty, y) = 0$. Next, we can determine the function $p_1(x,y)$ from the outer limit equation $\partial u_1/\partial x + (1/\gamma)\,\partial p_1/\partial x = 0$, with $p_1(-\infty, y) = 0$, and as a consequence we obtain

$$p_1(x,y) = -\gamma u_1(x,y) = -\gamma \frac{\partial \psi_1}{\partial y} . \tag{4.89}$$

4.3.3 The Limiting Prandtl Equations for $M^2 \to 0$

Now we consider the solution of our problem (4.84a–c) and (4.85) in the inner limit as

$$M^2 \to 0 \quad \text{with } x \text{ and } \eta = \frac{y}{M^2} \text{ fixed} . \tag{4.90}$$

We associate with (4.90) the following asymptotic inner representation:

$$\begin{aligned} u &= U_0(x,\eta) + M^2 U_1(x,\eta) + \cdots, \\ v &= M^2[V_0(x,\eta) + M^2 V_1(x,\eta) + \cdots], \\ p &= 1 + M^4 P_1(x,\eta) + \dots, \\ (\rho, T) &= (1,1) + M^2(R_1(x,\eta), T_1(x,\eta)) + \cdots . \end{aligned} \tag{4.91}$$

The substitution of (4.91) into the NSF equations (4.84a–c) gives, for the functions $U_0(x,\eta)$ and $V_0(x,\eta)$, the classical Prandtl boundary-layer equations for the associated incompressible Blasius problem:

$$\frac{\partial U_0}{\partial x} + \frac{\partial V_0}{\partial \eta} = 0 , \tag{4.92a}$$

$$U_0 \frac{\partial U_0}{\partial x} + V_0 \frac{\partial U_0}{\partial \eta} = \frac{\partial^2 U_0}{\partial \eta^2} . \tag{4.92b}$$

From (4.85a), we derive the following boundary conditions for (4.92):

$$U_0 = V_0 = 0 \quad \text{on } \eta = 0, 0 < x < \infty . \tag{4.93}$$

Matching between the outer and inner representations (4.87) and (4.91) gives also the conditions

$$v_1(x,0) = V_0(x,\infty), \quad U_0(x,\infty) = 1 \, . \tag{4.94}$$

It is important to note that the matching conditions (4.94) are a direct consequence of the second of the similarity relations (4.83) and give, for the incompressible outer first-order stream function $\psi_1(x,y)$, a non-degenerate solution different from zero (see (4.97) below).

4.3.4 Flow Due to Displacement Thickness

Now it is necessary to obtain the behaviour of the Blasius problem (4.92) when $\eta \to \infty$. For the present purpose, the essential result of this incompressible Blasius problem is contained in the expansion of its solution for large η,

$$V_0(x,\infty) = \frac{\beta}{2} x^{-1/2} \, , \tag{4.95}$$

where $\beta = 1.7208$ (Meyer 1971, p. 106). As a consequence, we obtain in place of (4.94) for $v_1(x,0)$ the following boundary condition for the outer stream function $\psi_1(x,y)$:

$$\psi_1(x,0) = -\beta x^{1/2}, \quad 0 < x < \infty \, . \tag{4.96}$$

For this outer Euler stream function $\psi_1(x,y)$, we derive the following classical problem for the flow due to the displacement thickness:

$$\frac{\partial^2 \psi_1}{\partial x^2} + \frac{\partial^2 \psi_1}{\partial y^2} = 0 \, , \tag{4.97a}$$

$$\psi_1 = 0, x < 0 \text{ and } \psi_1(x,0) = -\beta x^{1/2}, 0 < x < \infty \, . \tag{4.97b}$$

The full solution of the outer problem (4.97) is obvious from the viewpoint of complex-variable theory; namely, we can write the solution as

$$\psi_1(x,y) = -\beta \text{ Real } \left[(x + \mathrm{i}y)^{1/2} \right] , \tag{4.98}$$

and because the displacement speed vanishes at $y = 0$, we have

$$\frac{\partial \psi_1}{\partial y} = 0 \quad \text{on } y = 0 \, .$$

4.3.5 Boundary-Layer Equations with Weak Compressibility

Substituting the inner expansions (4.91) into the full NSF equations (4.84a–c) yields, for the functions $U_1(x,\eta), V_1(x,\eta), P_1(x,\eta), T_1(x,\eta)$ and $R_1(x,\eta)$, the following boundary-layer equations, which take the effects of weak compressibility into account:

$$\frac{\partial U_1}{\partial x} + \frac{\partial V_1}{\partial \eta} + U_0 \frac{\partial R_1}{\partial x} + V_0 \frac{\partial R_1}{\partial \eta} = 0 \, ; \tag{4.99a}$$

$$\begin{aligned} &U_0 \frac{\partial U_1}{\partial x} + V_0 \frac{\partial U_1}{\partial \eta} + U_1 \frac{\partial U_0}{\partial x} + V_1 \frac{\partial U_0}{\partial \eta} \\ &+ R_1 \left[U_0 \frac{\partial U_0}{\partial x} + V_0 \frac{\partial U_0}{\partial \eta} \right] + \frac{1}{\gamma} \frac{\partial P_1}{\partial x} \\ &= \frac{\partial^2 U_1}{\partial \eta^2} + \left(\frac{\mathrm{d}\mu}{\mathrm{d}T} \right)_{T=1} \left[\frac{\partial^2 U_0}{\partial \eta^2} + \frac{\partial T_1}{\partial \eta} \frac{\partial U_0}{\partial \eta} \right] ; \end{aligned} \tag{4.99b}$$

$$\frac{\partial P_1}{\partial \eta} = 0 \, ; \tag{4.99c}$$

$$U_0 \frac{\partial T_1}{\partial x} + V_0 \frac{\partial T_1}{\partial \eta} = \frac{\partial^2 T_1}{\partial \eta^2} + (\gamma - 1) \left(\frac{\partial U_0}{\partial \eta} \right)^2 . \tag{4.99d}$$

Here, $R_1 = -T_1$. From (4.99c), we have $P_1 \equiv P_1(x)$, and matching with the outer flow gives

$$P_1(x) = p_1(x, 0) = -\gamma \left(\frac{\partial \psi_1}{\partial y} \right)_{y=0} = 0 \quad \Rightarrow P_1(x) = 0 \, , \tag{4.100}$$

according to (4.89) and (4.98). For the boundary-layer equations (4.99a,b,d), together with (4.100), we have the following conditions for the inner functions U_1, V_1, and T_1, from (4.85a) and matching with the outer expansion:

$$U_1 = V_1 = 0, \quad T_1 = \Theta(x), \quad \text{on } \eta = 0, \; 0 < x < \infty \, , \tag{4.101a}$$

$$U_1 = 0, \quad T_1 = 0, \text{when } \eta \to \infty \, . \tag{4.101b}$$

Now, the continuity equation (4.99a) makes it possible to introduce the boundary-layer stream function $\Psi_1(x, \eta)$, such that

$$U_1 - T_1 U_0 = \frac{\partial \Psi_1}{\partial \eta}, \quad \text{and } V_1 - T_1 V_0 = -\frac{\partial \Psi_1}{\partial x} \, , \tag{4.102}$$

since, because $\partial U_0 / \partial x + \partial V_0 / \partial \eta = 0$, we can also introduce an incompressible (Blasius) boundary-layer stream function $\Psi_0(x, \eta)$, where

$$U_0 = \frac{\partial \Psi_0}{\partial \eta} \quad \text{and } V_0 = -\frac{\partial \Psi_0}{\partial x} \, . \tag{4.103}$$

4.3.6 Self-Similar Solution

If we assume, in the boundary condition (4.85a), that

$$\Theta(x) \equiv 1 \quad \text{for } 0 < x < \infty \,, \tag{4.103}$$

then a self-similar solution exists for the flat-plate Blasius problem because the problem has a group property that permits its reduction to an ordinary differential equation. We choose the following form:

$$\Psi_0(x,\eta) = x^{1/2} f(Y) \quad \text{and} \quad \Psi_1(x,\eta) = x^{1/2} F(Y) \,, \tag{4.104}$$

where $Y = \eta / x^{1/2}$. Substituting (4.104) for $\Psi_0(x,\eta)$ into the boundary-layer problem (4.92a,b), (4.93) and (4.94) gives the following classical problem:

$$2\frac{\mathrm{d}^3 f}{\mathrm{d}Y^3} + f\frac{\mathrm{d}^2 f}{\mathrm{d}Y^2} = 0, \quad \mathrm{f}(0) = \left(\frac{\mathrm{d}f}{\mathrm{d}Y}\right)_{Y=0} = 0, \quad \left(\frac{\mathrm{d}f}{\mathrm{d}Y}\right)_{Y=\infty} = 1\,. \tag{4.105}$$

This is the problem considered by Blasius (1908). Reintroducing the original dimensional variables at this point shows that the length l_0 (see (4.80)) disappears from f, F and Y in accordance with the fact that it is irrelevant for a boundary layer on a flat plate with $\Theta(x) \equiv 1$ for $0 < x < \infty$, and this is an another way of justifying the group transformation (4.104). For our purposes, the essential results of the numerical solution are contained in the expansion for small Y,

$$f(Y) = \frac{1}{2}\alpha\, Y^2 + O(Y^5), \quad \text{where } \alpha = \left(\frac{\mathrm{d}^2 f}{\mathrm{d}Y^2}\right)_{Y=0} = 0.33206\,. \tag{4.106}$$

If we now introduce the function $G(Y) = -T_1(x,\eta)$, then we obtain the following classical thermal problem for $G(Y)$ from the equation (4.99d), the condition (4.101a) for T_1 with $\Theta(x) \equiv 1$ when $0 < x < \infty$, and (4.101b) for T_1, namely

$$2\frac{\mathrm{d}^2 G}{\mathrm{d}Y^2} + f\frac{\mathrm{d}G}{\mathrm{d}Y} = 2(\gamma - 1)\left(\frac{\mathrm{d}^2 f}{\mathrm{d}Y^2}\right)^2, \quad G(0) = -1, \quad G(\infty) = 0\,, \tag{4.107}$$

and the full solution of (4.107)) is obvious:

$$G(Y) = -1 + \frac{1}{2}(3-\gamma)\frac{\mathrm{d}f}{\mathrm{d}Y} + \frac{1}{2}(\gamma - 1)\left(\frac{\mathrm{d}f}{\mathrm{d}Y}\right)^2. \tag{4.108}$$

Finally, substituting (4.102), (4.104) and the function $G(Y) = -T_1 = R_1$ into the equation (4.99b) and conditions (4.101) for U_1 and V_1, we obtain the following linear, inhomogeneous equation for the function $F(Y)$:

$$\begin{aligned} 2\frac{\mathrm{d}^3 F}{\mathrm{d}Y^3} + f\frac{\mathrm{d}^2 F}{\mathrm{d}Y^2} + \frac{\mathrm{d}^2 f}{\mathrm{d}Y^2}F &= 2\left[\frac{\mathrm{d}f}{\mathrm{d}Y}\frac{\mathrm{d}^2 G}{\mathrm{d}Y^2} + 2\frac{\mathrm{d}^2 f}{\mathrm{d}Y^2}\frac{\mathrm{d}G}{\mathrm{d}Y}\right] + f\frac{\mathrm{d}f}{\mathrm{d}Y}\frac{\mathrm{d}G}{\mathrm{d}Y} \\ &\quad - 2\left(\frac{\mathrm{d}\mu}{\mathrm{d}T}\right)_{T=1}\left[\frac{\mathrm{d}^3 f}{\mathrm{d}Y^3} - \frac{\mathrm{d}G}{\mathrm{d}Y}\frac{\mathrm{d}^2 f}{\mathrm{d}Y^2}\right], \end{aligned} \tag{4.109}$$

with the boundary conditions

$$F(0) = 0, \quad \left(\frac{\mathrm{d}F}{\mathrm{d}Y}\right)_{Y=0} = 0 \quad \text{and} \quad \left(\frac{\mathrm{d}F}{\mathrm{d}Y}\right)_{Y=\infty} = 0\,. \tag{4.110}$$

Here $f(Y)$ and $G(Y)$ are known functions, from the classical Blasius problem (4.105) and the solution (4.108) for $G(Y)$. At this stage, the two-term inner expansion for the longitudinal velocity component u is

$$u = \frac{\mathrm{d}f}{\mathrm{d}Y} + M^2\left[\frac{\mathrm{d}F}{\mathrm{d}Y} + \left(\frac{\mathrm{d}f}{\mathrm{d}Y}\right)G\right] + O\left(M^4\right)\,, \tag{4.111}$$

and the skin friction coefficient is

$$C_{\mathrm{f}} = 0.6641\,(Re_x)^{-1/2} + \frac{M^2}{(Re_x)^{1/2}}\left\{0.6641 + 2\left(\frac{\mathrm{d}^2F}{\mathrm{d}Y^2}\right)_{Y=0} + 0.6641\left(\frac{\mathrm{d}\mu}{\mathrm{d}T}\right)_{T=1}\right\} + \ldots\,, \tag{4.112}$$

where $Re_x = (U_0/\nu_0)x$ is the local Reynolds number, based on the dimensional distance x from the leading edge of the flat plate. The second term in the skin friction coefficient (4.112) is a direct result of the slight effect of compressibility. According to Godts and Zeytounian (1990), for $\mu(T) = T^\omega$, when $0 < \omega < 1$, we obtain a value of -0.403 for $(\mathrm{d}^2F/\mathrm{d}Y^2)_{Y=0}$ if $\omega = 0$. For other values of $(\mathrm{d}^2F/\mathrm{d}Y^2)_{Y=0}$ as a function of ω, see Godts and Zeytounian (1990, p. 69).

In conclusion, we note two fundamental points of our analysis: the first concerns the hyposonic viscous similarity relation $1/Re = M^4$ between the small Mach number and the large Reynolds number, and the second is connected with the skin friction coefficient C_{f}, in which we have a new term as a direct consequence of the slight effect of compressibility coupled with a small effect of viscosity. Naturally, it would be possible to generalize the above asymptotic analysis to an arbitrary body, but then a full numerical computation would be required to estimate the influence of weak compressibility on the coefficient C_{f}. Such a generalization would be interesting for various low-speed, viscous, hyposonic fluid flow phenomena (in particular, in "road vehicle aerodynamics").

4.4 Compressible Flow at Low Reynolds Number in Low-Mach-Number Asymptotics

Two limiting processes play a fundamental role (see the discussion in Lagerstrom 1964) in the asymptotic study of low-Reynolds-number flow, namely the Stokes limit (inner limit) and the Oseen limit (outer limit). However, for a compressible flow, when $Re \to 0$ it is necessary to specify also the role of the Mach number M. In fact, it is necessary to pose a problem concerning the behaviour of solutions of the full dimensionless NSF equations (1.29), when simultaneously

$$Re \to 0 \quad \text{and } M \to 0 \,. \tag{4.113}$$

Naturally, for the validity of the NSF equations, it is obvious that it is assumed that the limiting compressible flow at low Reynolds and Mach numbers remains that of a continuous medium, and this implies that the (dimensionless) Knudsen number is also a small parameter:

$$Kn = \frac{M}{Re} \ll 1 \,. \tag{4.114}$$

As a consequence, the double limiting process (4.113) must be performed with the following similarity rule:

$$Re = M^{1-a}, \text{where } 1 > a > 0, \text{when } M \to 0 \,, \tag{4.115a}$$

such that

$$Kn = M^{a} \ll 1, \text{where } 1 > a > 0 \,. \tag{4.115b}$$

For an external flow of a weakly compressible but strongly viscous fluid around a bounded body in an unbounded atmosphere, it is necessary to consider two low-Mach-number expansions: (i) the *Stokes expansion*, which is compressible and proximal, with $\boldsymbol{x}$ and t fixed when $M \to 0$, and (ii) the *Oseen expansion*, which is compressible and distal, with $\xi = (M^{1-a})\,\boldsymbol{x}$ and t fixed when $M \to 0$. In general, the distal, Oseen limit is applied only at points which are not on the boundary of the body – in fact, at points far away from the boundary of the body. The proximal, Stokes limit may be thought of as the limit where the 'viscosity tends to infinity' while the other parameters (except the Mach number, which satisfies the similarity rule (4.115a)) are kept fixed. From this it is evident that in the Stokes limit ($M \to 0$ with $\boldsymbol{x}$ and t fixed) the influence of the boundary is strong, and for this reason the Stokes limit is called a proximal limit. The value of the velocity vector $\boldsymbol{u}$ at the wall, i.e. zero, will have an increasingly large influence on any fixed point not on the boundary as the viscosity tends to infinity. However, for any given viscosity one can find points sufficiently far away from the boundary of the body that $\boldsymbol{u}$ there is arbitrarily close to $U_\infty \boldsymbol{i}$, if we assume that $\boldsymbol{u} \to U_\infty \boldsymbol{i}$ at infinity ($\boldsymbol{u}$ is a dimensionless velocity vector). Here, the free-stream direction is taken to coincide with the positive horizontal direction x_1 so that $\boldsymbol{u}(\infty)$ may be written as $U_\infty \boldsymbol{i}$, with $U_\infty = \text{const}$. The Oseen limit ($M \to 0$ with ξ and t fixed) may be thought of as the limit where the characteristic length L_c (a measure of a finite body) tends to zero while the other parameters (except M) and the point under consideration (in 3D space) are fixed. The body then tends to a conical body, i.e. a body for which no length can be defined. The limiting conical body may, in particular, be a point, a finite line or a semi-infinite line. Such objects have no arresting power, i.e. they cannot cause a finite disturbance in the fluid, and as a consequence, in the Oseen limit, the conditions at infinity determine the value of the limit.

4.4.1 The Stokes Limiting Case and the Steady, Compressible Stokes Equations

We now consider the Stokes limit of the *steady* NSF equations (1.29). If we take into account the similarity rule (4.115a), then, when $M \to 0$, it is necessary to study the approximate solutions of the NSF equations in the following asymptotic form (Stokes proximal expansion):

$$\boldsymbol{u} = \boldsymbol{u}_{\mathrm{s}} + \cdots, \quad p = 1 + M^{a+1}[p_{\mathrm{s}} + \cdots], \quad \rho = \rho_{\mathrm{s}} + \cdots, \quad T = T_{\mathrm{s}} + \cdots, \tag{4.116}$$

where the 'Stokes' limiting functions (with subscript 's') depend only on the position variable $\boldsymbol{x} = (x_i), i = 1, 2, 3$. In this case we can derive for $\boldsymbol{u}_{\mathrm{s}}, p_{\mathrm{s}}, \rho_{\mathrm{s}}$ and T_{s} an 'à la Stokes' steady, compressible system, which is written in three parts:

$$\nabla \cdot [k(T_{\mathrm{s}})\, \nabla T_{\mathrm{s}}] = 0, \tag{4.117a}$$

$$T_{\mathrm{s}} = T_{\mathrm{w}}(\boldsymbol{P}, \tau) \equiv 1 + \tau \Xi(\boldsymbol{P}), \quad \boldsymbol{P} \in \partial\Omega, \quad \text{on } \partial\Omega, \tag{4.117b}$$

$$\rho_{\mathrm{s}} = \frac{1}{T_{\mathrm{s}}}, \tag{4.118}$$

$$\nabla \cdot \mathbf{u}_{\mathrm{s}} = \mathbf{u}_{\mathrm{s}} \cdot \nabla \log T_{\mathrm{s}}, \tag{4.119a}$$

$$\nabla p_{\mathrm{s}} = \gamma\, \nabla \cdot [2\mu(T_{\mathrm{s}})\mathbf{D}_{\mathrm{s}}\boldsymbol{u}_{\mathrm{s}} + \lambda(T_{\mathrm{s}})(\nabla \cdot \boldsymbol{u}_{\mathrm{s}})\mathbf{I}],$$
$$\mathbf{D}_{\mathrm{s}}\boldsymbol{u}_{\mathrm{s}} = \frac{1}{2}[(\nabla \boldsymbol{u}_{\mathrm{s}}) + (\nabla \boldsymbol{u}_{\mathrm{s}})^T], \tag{4.119b}$$

$$\boldsymbol{u}_{\mathrm{s}} = 0 \quad \text{on } \partial\Omega. \tag{4.119c}$$

Equation (4.117a) with the associated boundary condition on $\partial\Omega$ (4.117b) determines T_{s}, once the behaviour of the temperature at infinity is specified. Equation (4.118) is a relation (a limiting form of the equation of state in (1.29)) between T_{s} and ρ_{s}, and determines ρ_{s} when T_{s} is known. Finally, the three equations (4.119a–c) give a closed system for the determination of the velocity vector $\boldsymbol{u}_{\mathrm{s}}$ and the pressure perturbation p_{s}, if we assume that T_{s} is known. If the rate of fluctuation of the temperature τ in (4.117b) tends to zero as $M \to 0$, then we can obtain a particularly simple (incompressible) solution for (4.117a,b) and (4.118), namely

$$T_{\mathrm{s}} = 1, \quad \rho_{\mathrm{s}} = 1, \tag{4.120}$$

and the equations (4.119a,b) together with (4.120) reduce to the classical Stokes equations for a fluid with a divergenceless velocity:

$$\nabla \cdot \boldsymbol{u}_{\mathrm{s}} = 0, \quad \nabla p_{\mathrm{s}} = \gamma\mu(1)\, \nabla^2 \boldsymbol{u}_{\mathrm{s}}. \tag{4.121}$$

We note that the Stokes equations (4.121) for an incompressible flow may be obtained either by linearization or by letting Re tend to zero. That these two procedures give the same result in the incompressible case is fortuitous; for compressible fluids, the low-Mach-number Stokes equations, with the similarity rule (4.115a) and the compressible Stokes proximal expansion (4.116), are nonlinear, as shown by the above system of equations (4.117a)–(4.119c). In fact, when solving these equations, one first finds T_s from the energy equation and then ρ_s from the equation of state. The continuity and momentum equations then become linear equations, whose variable coefficients involve the known function T_s. Finally, we observe that the positive scalar $1 > a > 0$ in the similarity rule (4.115a)/(4.115b) and in the Stokes proximal expansion (4.116) must be consistently determined during the matching process with the Oseen distal expansion, considered in the next subsection.

4.4.2 The Oseen Limiting Case and the Steady, Compressible Oseen Equations

An alternative form of the steady dimensionless NSF equations, using the Oseen (outer) space variable? If so, please ensure that all instances of ξ used as a vector are bold italic (but not, of course, in the case of components ξ_i). $\boldsymbol{\xi} = M^{1-a}\boldsymbol{x}$, is

$$\boldsymbol{u} \,.\, \nabla^{\mathrm{Os}}\rho + \rho\, \nabla^{\mathrm{Os}} \,.\, \boldsymbol{u} = 0\,, \tag{4.122a}$$

$$\rho(\boldsymbol{u} \,.\, \nabla^{\mathrm{Os}})\boldsymbol{u} + \frac{1}{\gamma M^2}\nabla^{\mathrm{Os}} p = \nabla^{\mathrm{Os}} \,.\, [2\mu \mathbf{D}^{\mathrm{Os}}(\boldsymbol{u}) + \lambda(\nabla^{\mathrm{Os}} \,.\, \boldsymbol{u})\boldsymbol{I}]\,, \tag{4.122b}$$

$$\begin{aligned}\rho\boldsymbol{u} \,.\, \nabla^{\mathrm{Os}} T + (\gamma-1)p\nabla^{\mathrm{Os}} \,.\, \boldsymbol{u} &= \frac{\gamma}{Pr}\nabla^{\mathrm{Os}} \,.\, [k\, \nabla^{\mathrm{Os}} T] \\ &+\gamma(\gamma-1)M^2\left[\left(\frac{\mu}{2}\right)(\mathbf{D}^{\mathrm{OS}}(\boldsymbol{u}) : \mathbf{D}^{\mathrm{OS}}(\boldsymbol{u})) + \lambda(\boldsymbol{\nabla}^{\mathrm{OS}}.\boldsymbol{u})^2\right]\end{aligned} \tag{4.122c}$$

where $p = \rho T$. The gradient vector ∇^{Os} is formed with respect to the Oseen space variable $\boldsymbol{\xi} = (\xi_i), i = 1, 2, 3$, and we have the following relations for the gradient operator and the rate-of-strain tensor: $\nabla = M^{1-a}\nabla^{\mathrm{Os}}$ and $\boldsymbol{D} = M^{1-a}\boldsymbol{D}^{\mathrm{Os}}$. The unknown functions $\boldsymbol{u}, p, \rho$ and T and the rate-of-strain tensor $\boldsymbol{D}^{\mathrm{Os}}$ depend on all of the Oseen space variables ξ_i (we consider a steady flow). In the steady NSF equations (4.122), the Reynolds number $Re = M^{1-a}$ has been eliminated! However, Re will reappear in the boundary conditions. For example, if the finite solid body is a sphere of diameter L_c, then the boundary of the body is given in Oseen coordinates ξ_i by $(\xi_1)^2 + (\xi_2)^2 + (\xi_3)^2 = Re/2$, and when $Re = M^{1-a} \to 0$ with $\xi_i (i = 1, 2, 3)$ fixed, the body shrinks to a point! For a steady flow past a solid finite body at rest, when the domain is infinite, the following uniform conditions have to be added:

$$\boldsymbol{u} = U_\infty \boldsymbol{i},\ p = \rho = T = 1, \quad \text{at infinity}\,, \tag{4.123}$$

Now, we assume the following asymptotic expansions for $\boldsymbol{u}$, p, ρ and T for small M (Oseen distal expansion):

$$\boldsymbol{u} = U_\infty \boldsymbol{i} + \mu_1(M)\boldsymbol{u}^{\mathrm{Os}}(\boldsymbol{\xi}) + \ldots, \quad p = 1 + M^2[\mu_1(M)p^{\mathrm{Os}}(\boldsymbol{\xi}) + \ldots] \tag{4.124a}$$

$$\rho = 1 + \mu_1(M)\rho^{\mathrm{Os}}(\boldsymbol{\xi}) + \ldots, \quad T = 1 + \mu_1(M)T^{\mathrm{Os}}(\boldsymbol{\xi}) + \ldots, \tag{4.124b}$$

where $\mu_1(M)$ is some suitable gauge function of M which tends to zero as $M \downarrow 0$. In the Oseen distal, outer region, a finite 3D body shrinks to a point, which cannot cause a finite disturbance in the Oseen fluid flow, and hence the values of u, p, ρ and T, at any distant fixed point, in the Oseen region, will tend to the free-stream values (4.123). The expansions (4.124) take this property into account. If we then insert these expansions (4.124) into the NSF equations written in Oseen variables ((4.122) with $p = \rho T$) and retain only terms of order $\mu_1(Re)$, it is found that $\boldsymbol{u}^{\mathrm{Os}}(\boldsymbol{\xi})$, $p^{\mathrm{Os}}(\boldsymbol{\xi})$, $\rho^{\mathrm{Os}}(\boldsymbol{\xi})$ and $T^{\mathrm{Os}}(\boldsymbol{\xi})$ satisfy the following Oseen (linear) limiting equations:

$$\nabla^{\mathrm{Os}} . \boldsymbol{u}^{\mathrm{Os}} = U_\infty \frac{\partial T^{\mathrm{Os}}}{\partial \xi_1}, \tag{4.125a}$$

$$U_\infty \frac{\partial \boldsymbol{u}^{\mathrm{Os}}}{\partial \xi_1} + \frac{1}{\gamma}\nabla^{\mathrm{Os}} p^{\mathrm{Os}} = \mu(1)\left[\nabla^{\mathrm{Os}2}\, T^{\mathrm{Os}} + \frac{U_\infty}{3}\nabla^{\mathrm{Os}}\left(\frac{\partial T^{\mathrm{Os}}}{\partial \xi_1}\right)\right], \tag{4.125b}$$

$$U_\infty \frac{\partial T^{\mathrm{Os}}}{\partial \xi_1} = \frac{k(1)}{Pr}\nabla^{\mathrm{Os}2}\, T^{\mathrm{Os}}, \tag{4.125c}$$

$$\rho^{\mathrm{Os}} = -T^{\mathrm{Os}}. \tag{4.125d}$$

Here we assume that $\lambda(1) = -(2/3)\mu(1)$ and take into account the similarity rule (4.115a). For the above Oseen equations (4.125a,b,c), together with (4.125d), we can write the following boundary conditions:

$$u^{\mathrm{Os}}, p^{\mathrm{Os}}, T^{\mathrm{Os}} \text{ and } \rho^{\mathrm{Os}} \to 0 \quad \text{at infinity}. \tag{4.126}$$

When T^{Os} is identically zero, we recover the classical steady, incompressible Oseen equations for $\boldsymbol{u}^{\mathrm{Os}}$ and p^{Os}:

$$\nabla^{\mathrm{Os}} . \boldsymbol{u}^{\mathrm{Os}} = 0, \tag{4.127a}$$

$$U_\infty \frac{\partial \boldsymbol{u}^{\mathrm{Os}}}{\partial \xi_1} + \frac{1}{\gamma}\nabla^{\mathrm{Os}} p^{\mathrm{Os}} = \mu(1)\nabla^{\mathrm{Os}2}\, \boldsymbol{u}^{\mathrm{Os}}. \tag{4.127b}$$

Therefore, the fundamental problem for the compressible Oseen equations (4.125a,b,c), together with (4.125d), is the study of the behaviour of the temperature fluctuation $T^{\mathrm{Os}}(\boldsymbol{\xi})$, which is a solution of the linear equation

$$\left[\frac{\partial^2}{\partial \xi_1^2} + \frac{\partial^2}{\partial \xi_2^2} + \frac{\partial^2}{\partial \xi_3^2}\right] T^{\mathrm{Os}} - \beta \frac{\partial T^{\mathrm{Os}}}{\partial \xi_1} = 0; \quad \beta = \frac{Pr\, U_\infty}{k(1)}, \tag{4.128}$$

when

$$(\xi_1)^2 + (\xi_2)^2 + (\xi_3)^2 \to 0 ,$$

in accordance with the condition for matching with the Stokes equations (4.117a).

A consequence of the incompressible Oseen equations (4.127) is that p^{Os} is a harmonic function of $\boldsymbol{\xi}$:

$$\nabla^{\mathrm{Os}2} p^{\mathrm{Os}} = 0 . \tag{4.129}$$

Surprisingly, this property (4.129) is also true for the compressible Oseen equations (4.125a,b,c,), together with (4.125d), if we assume for the Prandtl number that $Pr = (3/4)((k(1)/\mu(1))$. In the case of incompressible flow, the Oseen equations (4.127) are actually uniformly valid, since the inner (Stokes) limit of these equations gives the classical incompressible Stokes equations (4.121). This is a fortuitous coincidence, closely linked to the fact that the Stokes (proximal) equations for incompressible flow are linear. The Oseen equations for compressible flow are not uniformly valid near the body. In solving a problem of low-Mach-number compressible flow, with the similarity rule (4.115a), around a solid finite body, it is not sufficient to use the Oseen solution only; the outer (distal, Oseen) solution must be matched to an inner solution, i.e. a solution of the compressible Stokes equations. The method of matching is in principle the same as in the case of incompressible flow, although the computational difficulties are of course considerably greater in the case of compressible flow.

In order to obtain a 'crude estimate' of the solution, one may solve the compressible Oseen equations as if they were uniformly valid. The matching in the compressible steady case considered above is in fact an open problem, and the *unsteady* compressible case with the similarity rule (4.115a) has not yet been considered. In a paper by Kassoy et al. (1966), the reader can find an extension of the classical Stokes problem for the Navier incompressible flow past a sphere to a compressible low-Reynolds-number flow around a sphere, while in a more recent paper by Boris and Fridlender (1981), the slow motion of a gas near a strongly heated or cooled sphere is considered. The authors of the latter paper take into account, in the expression of the stress tensor $\mathbf{T}$ (see the Sect. 1.1), the Burnett (1936) stress temperature, which arises as a consequence of the slow, non-isothermal motion.

4.4.3 The Rarefied-Gas Point of View for Small Knudsen Number

In a recent paper by Sone et al. (2000), an asymptotic theory of a Boltzmann system, devoted to a steady flow of a slightly rarefied gas with a finite Mach number (of the order of unity) around a body, is described. The authors of this paper observe judiciously that, 'incidentally, when a finite Mach number is mentioned, one may think that this is a general case that covers a small Mach number case'. This is not so in the case of a small Knudsen number, because a finite Mach number means a very large Reynolds number, owing to the relation among these three parameters (see (4.114)). It is interesting to note that Chapman and Enskog (see Chapman and Cowling 1952 and also Grad 1958, where the structure of the Chapman–Enskog expansion is clearly but briefly explained) developed a 'skilful' expansion, where the velocity distribution function is expanded, but the macroscopic variables are not, and a series of fluid-dynamic-type equations is obtained. The leading set of equations is the Euler set of inviscid equations, the second is the Navier–Stokes set of compressible equations (our NSF equations), and the third is the Burnett set of equations. Owing to the derivation of the NSF set of equations, the Chapman–Enskog expansion is usually referred to when the relation between fluid dynamics and kinetic theory is mentioned; however, disadvantages of this expansion have also been reported (see, for instance, Sone 1984, 1998, Sone and Aoki 1994, and Cercignani 1988). In the introduction of Sone et al. (2000), the reader can find various pieces of pertinent information concerning the limiting case $Kn \downarrow 0$.

Here, we make some remarks related to the Mach number and, in particular, to the case $M \ll 1$. First, in Sone (1991), the case where the state of the gas is close to a uniform state of rest is considered, and its asymptotic behaviour for small Knudsen numbers is analysed on the basis of the linearized Boltzmann equation. The overall behaviour of the gas is described by the Stokes set of equations at any order of approximation. The boundary condition for the Stokes set, and the Knudsen-layer correction in the neighbourhood of the boundary are obtained up to the second order in the Knudsen number. At the second order in the Knudsen number, a thin layer with a thickness of the order of the mean free path squared divided by the radius of curvature of the boundary appears at the bottom of the Knudsen layer over a convex boundary (see, for instance, Sone 1973 and Sone and Takada 1992). In Sone and Aoki (1987), the linearized theory was extended to the case where the Reynolds number of the system is very small ($Re \ll 1$). The case when $Re = O(1)$, for which the linearized Boltzmann equation is no longer valid, corresponds in fact to the case when the Mach number is of the same order as the Knudsen number in the Boltzmann dimensionless equation, i.e. $M \approx Kn \ll 1$. In such a case, the leading fluid-dynamic-type equations are the Navier set of equations (the Navier–Stokes equations for an incompressible fluid). The next-order equations are the second set of equations of the Mach number expansion of the Navier–Stokes set, with an additional thermal-stress term in the momentum equation owing to the effect of rarefaction of the gas (see, for instance, Sone et al. 1996, and also Galkin and Kogan 1970, 1979 and Galkin 1974); this shows the incompleteness of continuum gas dynamics (classical NSF theory) for small Mach numbers when we consider the Boltzmann equation from the start. In view of the above, it seems that the fundamental (open) problem is to derive a Stokes compressible set of equations from the (non-linearized) Boltzmann equation under the similarity rule (4.115b) when the Reynolds number Re satisfies the constraint (4.115a).

References

H. Blasius. Z. Math. Phys., **56** (1908), 1–37.

A. Yu. Boris and O. G. Fridlender, Small flow of a gas near a strongly heated or cooled sphere. Izv. Akad. Nauk SSSR, Mech. Liquid Gas No. 6, (1981), 170–175 (in Russian).

J. M. Burgers, A mathematical model illustrating the theory of turbulence. Adv. Appl. Mech., **1** (1948), 171–199.

D. Burnett. Proc. Lond. Math. Soc., **40**(2)(1936), 382.

C. Cercignani, *The Boltzmann Equation and its Applications*. Springer, Berlin, 1988.

S. Chapman and T. G. Cowling, *The Mathematical Theory of Nonuniform Gases*. Cambridge University Press, Cambridge, 1952.

Y. Choquet-Bruhat, Ondes asymptotiques et approchées pour des systèmes d'équations aux dérivées partielles non linéaires. J. Math. Pures Appl., **134** (1969), 117–158.

S. C. Crow, Aerodynamic sound emission as a singular perturbation problem. UCRL 70189 (Preprint), 1966.

S. C. Crow, Aerodynamic sound emission as a singular perturbation problem. Stud. Appl. Math., **49** (1970), 21–44.

N. Curle, The influence of solid boundaries upon aerodynamic sound. Proc. Roy. Soc. A, **231** (1955), 505–514.

J. E. Ffowcs Williams, Acoustic analogy – Thirty years on. IMA J. Appl. Math., **32** (1984), 113–124.

V. S. Galkin. Stud. TSAGUI, **5**(4) (1974), 40–47 (in Russian).

V. S. Galkin and M. N. Kogan. Izv. Akad. Nauk SSSR, Mech. Liquid Gas, No. 3, (1970), 13–21 (in Russian).

V. S. Galkin and M. N. Kogan. Izv. Akad. Nauk SSSR, Mech. Liq. Gas, No. 6 (1979), 77–84 (in Russian).

P. Germain, *Progressive Waves*. 14th Ludwig Prandtl Memorial Lecture (Mannheim, 15 April 1971). Sonderdruck aus dem Jahrbuch 1971 der DGLR, 1971.

S. Godts and R. Kh. Zeytounian, A note on the Blasius problem for a slightly compressible flow. ZAMM, **70**(1)(1990), 67–69.

H. Grad, *Principles of the Kinetic Theory of Gases*. In: Handbuch der Physik, Vol. 12, edited by S. Flüge. Springer, Berlin, 1958, pp. 205–294.

J.-P. Guiraud, La théorie du bang supersonique d'après la théorie nonlinéaire des ondes courtes. *Colloque sur le bang des avions supersoniques*, ONERA TP 521, 1967.

J.-P. Guiraud, *Low Mach Number Asymptotics of the Steichen Equation*. Unpublished notes, 1983.

D. R. Kassoy, T. S. Adamson and A. F. Messiter, Compressible low Reynolds number flow around a sphere. Phys. Fluids, **9**(4) (1966), 671–681.

P. A. Lagerstrom, Laminar flow theory. In: *Theory of Laminar Flows*, edited by F. K. Moore. Princeton University Press Princeton, NJ, 1964, pp. 61–83.

V. R. Lauvstad, On non-uniform Mach number expansion of the Navier–Stokes equations and its relation to aerodynamically generated sound. J. Sound Vib., **7** (1968), 90–105.

F. G. Leppington and H. Levine, The sound field of a pulsating sphere in unsteady rectilinear motion. Proc. Roy. Soc. A, **412** (1987), 199–221.

M. J. Lighthill, On sound generated aerodynamically. I. General Theory. Proc. Roy. Soc. Lon. A, **211** (1952), 564–514. See also M. J. Lighthill, Proc. Roy. Soc. Lond. A, **267** (1962), 147–182.

M. J. Lighthill, Viscosity in waves of finite amplitude. In: *Surveys in Mechanics*, edited by G. Batchelor and R. M. Davis. Cambridge University Press, 1956, pp. 250–351.

R. E. Meyer, *Introduction to Mathematical Fluid Dynamics.* Wiley Interscience, New York, 1971.

F. Obermeier, *The Application of Singular Perturbation Methods to Aerodynamic Sound Generation.* Lecture Notes in Mathematics, No. 594, Springer, Berlin, 1976, pp. 400–421.

R. Sery Baye, *Contribution à l'étude de certains écoulements à faibles nombres de Mach.* Thèse de Doctorat de Mécanique, Université de Paris VI, 1994.

Y. Sone, New kind of boundary layer over a convex solid boundary in a rarefied gas. Phys. Fluids, **16**, (1973), 1422–1424.

Y. Sone, Analytical studies in rarefied gas dynamics. In: *Rarefied Gas Dynamics*, edited by H. Oguchi. University of Tokyo Press, Tokyo, 1984, pp. 71–84.

Y. Sone, Asymptotic theory of a steady flow of a rarefied gas past bodies for small Knudsen numbers. In: *Advances in Kinetic Theory and Continuum Mechanics*, edited by R. Gatignol and Soubbaramayer. Springer, Berlin, 1991, pp. 19–31.

Y. Sone, *Theoretical and Numerical Analyses of the Boltzmann Equation – Theory and Analysis of Rarefied Gas Flows – Part I.* Lecture Notes, Department of Aero-Astronautics, Graduate School of Engineering, Kyoto University, Kyoto, 1998.

Y. Sone and K. Aoki, Steady gas flows past bodies as small Knudsen numbers. Transp. Theory Stat. Phys., **16** (1987), 189–199.

Y. Sone and K. Aoki, *Molecular Gas Dynamics.* Asakura, Tokyo, 1994 (in Japanese).

Y. Sone and S. Takada, Discontinuity of the velocity distribution function in a rarefied gas around a convex body and the S layer atthe bottom of the Knudsen layer. Transp. Theory Stat. Phys., **21**, (1992), 501–530.

Y. Sone, K. Aoki, S. Takada, H. Sugimoto and A.V. Bobylev, Inappropriateness of the heat-conduction equation for description of a temperature field of a stationary gas in the continuum limit: examination by asymptotic analysis and numerical computation of the Boltzmann equation. Phys. Fluids, **8** (1996), 628–638 (erratum in Phys. Fluids, **8**, (1996), 841).

Y. Sone, C. Bardos, F. Golse and H. Sugimoto, Asymptotic theory of the Boltzmann system, for a steady flow of a slightly rarefied gas with a finite Mach number: general theory. Eur. J. Mech. B, **19** (2000), 325–360.

H. Viviand, Etude des écoulements instationnaires à faibles nombre de Mach avec application au bruit aérodynamique. J. de Mécanique, **9** (1970), 573–599.

C. Wilcox, *Scattering Theory for the d'Alembert Equation in Exterior Domain.* Springer, Berlin, 1975.

R. Kh. Zeytounian, Analyse asymptotique des écoulements de fluides visqueux compressibles à faible nombre de Mach, I: Cas des fluides non pesants. USSR Comput. Math. Math. Phys., **17**(1) (1977), 175–182.

R. Kh. Zeytounian, *Theory and Applications of Nonviscous Fluid Flows.* Springer, Heidelberg, 2002a.

R. Kh. Zeytounian, *Asymptotic Modelling of Fluid Flow Phenomena.* Fluid Mechanics and Its Applications, vol. 64, series Ed. R. Moreau, Kluwer Academic, Dordrecht, 2002b.

5

Some Aspects of Low-Mach-Number Internal Flows

In the present chapter, the main problem concerns low-Mach-number flow affected by acoustic effects in a confined gas over a long time. In some working notes of Zeytounian and Guiraud (1980), an asymptotic multiple-timescale approach to the flow of a perfect gas within a bounded domain $D(t)$, the walls $\partial D(t) = \Sigma(t)$ of which deform very slowly in comparison with the speed of sound, was investigated. The result of a 'naive' asymptotic analysis with single time and space scales (see Sect. 3.3.4) is that the appropriate (quasi-incompressible Euler) model is one for which the velocity is divergence-free, while the density is not a constant but a function of time. Unfortunately, this 'naive' main model is valid only if the component of the wall displacement velocity, normal to $\Sigma(t), W_{\Sigma}$, at the initial time $t = 0$ is zero! Obviously, this is not the case when the motion of $\Sigma(t)$ is started impulsively from rest, and the same holds if the motion of $\Sigma(t)$ is accelerated from rest to a finite velocity in a short (dimensionless) time $O(M)$. As a consequence, it is necessary to elucidate how the 'naive' main (quasi-incompressible Euler) model fits with the initial conditions (at $t = 0$) associated with the exact unsteady, compressible, inviscid equations. We see (now!) that for this purpose, it is necessary to consider an initial layer in the vicinity of $t = 0$ and to introduce a short time $\tau = t/M$. However, when we consider the adjustment process $\tau \to \infty$, we see again (according to the relations (3.47)) that only the case where the motion of the wall from rest is *progressive*, during a time $O(1)$, allows a matching of the local acoustic model with the 'naive' main (quasi-incompressible Euler) model when $t \to 0$. As a consequence, if the wall $\Sigma(t)$ is started impulsively from rest, then the acoustic oscillations remain undamped at the end of the stage $t = O(M)$ and enter into the longer-time stage $t = O(1)$. A multiple-timescale technique is necessary for the investigation of the long-time evolution of these rapid oscillations. Unfortunately, a double-timescale technique (with t and $\tau = t/M$) is not sufficient to eliminate all the secular terms. Indeed, with the unit of speed chosen, the speed of sound is $O(1/M)$ and, consequently, the periods of the natural (free) vibrations of the bounded domain $D(t)$ are $O(M)$. Therefore, we expect that the solution will oscillate on a timescale

R. Kh. Zeytounian: *Topics in Hyposonic Flow Theory*, Lect. Notes Phys. **672**, 115–162 (2006)
www.springerlink.com

$O(M)$, while, on the other hand, it will also evolve on a timescale $O(1)$ owing to the slow evolution of $\Sigma(t)$. As a consequence, we must introduce into the structure of the solution a multiplicity (an 'enumerable' infinity!) of fast timescales. First we must use the time t, a slow time, and then we must bring into the solution an infinity of fast times designed to cope with the infinity of periods of the free vibrations of $D(t)$.

In Sect. 5.1, we give a brief review of the acoustic effects by looking the acoustic waves inside a time-independent cavity D, the boundary of which is a rigid wall $\Sigma(0) = \Gamma$. Section 5.2 is devoted to the investigation of a simplified dissipative acoustic system of equations when we have a small parameter, the inverse of the acoustic Reynolds number $1/Re_{\text{ac}} \equiv \delta^2 \ll 1$, in front of the dissipative terms. This investigation is only a test problem but gives us the possibility to detect some interesting features of the flow inside a cavity when we take dissipative effects into account. In Sect. 5.3, we return to our main problem for a bounded domain $D(t)$, the walls $\partial D(t) = \Sigma(t)$ of which deform very slowly in comparison with the speed of sound, and consider as the starting equations the Euler nonviscous equations for an inviscid perfect gas. The case where the perfect gas is viscous and heat-conducting is considered in Sect. 5.4 when the Reynolds number $Re = O(1)$, but we also give some results concerning the (more difficult) case when M and $1/Re$ are both much less than 1. Namely, in Sect. 5.4.1, the Navier–Fourier averaged model problem is derived, and in Sect. 5.4.2, the problem of damping of acoustic oscillations by viscosity is investigated via a Stokes layer of thickness $\chi = \sqrt{M/Re}$, when the Reynolds number $Re \gg 1$.

5.1 Acoustic Waves Inside a Cavity with a Rigid Wall

We now see that close to the initial time $t = 0$, where it is necessary to assign initial conditions for the unsteady NSF equations (3.48), the low-Mach-number limiting case leads to the classical acoustic equations (3.68) derived in Sect. 3.4.1 when we consider the short (acoustic) time $\tau = t/M$ as the significant fixed time and assume that the Reynolds number is $O(1)$. Here, as a first simple example, we look at these acoustic waves inside a *time-independent* cavity D, the boundary of the cavity being a rigid wall Γ. For the perturbation of the density $r(\tau, x) = (\rho - 1)/M$ and the velocity $\boldsymbol{v}(\tau, x)$, when the Mach number $M \downarrow 0$ with τ and x fixed, we write

$$\frac{\partial r}{\partial \tau} + \nabla \cdot \boldsymbol{v} = 0, \quad \frac{\partial \boldsymbol{v}}{\partial \tau} + \nabla r = 0\,, \tag{5.1}$$

with the condition ($\boldsymbol{n}$ is the unit vector normal to Γ pointing towards the exterior of the cavity D)

$$\boldsymbol{v} \,.\, \boldsymbol{n}_{|\Gamma} = 0\,. \tag{5.2}$$

If we start with the normal modes

$$\boldsymbol{v} = \boldsymbol{V}(\boldsymbol{x})\cos(\omega\tau + \phi), \quad r = R(\boldsymbol{x})\sin(\omega\tau + \phi) \tag{5.3}$$

(where ϕ is a constant), we derive the following eigenvalue problem for $\boldsymbol{V}(\boldsymbol{x})$ and $R(\boldsymbol{x})$:

$$\omega R + \nabla . \boldsymbol{V} = 0, \quad \omega \boldsymbol{V} - \nabla R = 0, \quad \boldsymbol{V} . \boldsymbol{n}_{|\Gamma} = 0 \,. \tag{5.4a}$$

First, it may be proven that there exists a sequence of eigenfrequencies

$$\omega_1 \leq \omega_2 \leq \omega_3 \leq \ldots \omega_n \leq \ldots , \tag{5.4b}$$

arranged with non-decreasing values, such that the eigenfunctions corresponding to different indices are orthogonal to each other. More precisely, if we set

$$\langle R'; R''\rangle = \int_D R'R''\mathrm{d}v, \quad \langle \boldsymbol{V}'; \boldsymbol{V}''\rangle = \int_D \boldsymbol{V}' . \boldsymbol{V}''\mathrm{d}v \,, \tag{5.4c}$$

we may prove (through a proper normalization) that

$$\langle R_n; R_m\rangle = \langle \boldsymbol{V}_n; \boldsymbol{V}_m\rangle = \delta_{n,m} \,, \tag{5.4d}$$

where $\delta_{n,m}$ is zero if $n \neq m (\omega_n \neq \omega_m)$ and is equal to 1 if $n = m$. If it happens that distinct eigenmodes correspond to the same eigenfreqency, one may always consider linear combinations of the modes which are orthogonal to each other, and the reason why the two terms in (5.4d) are equal for $n = m$ and may both be set equal to one is readily seen to be a consequence of the previous relation. One should notice that zero is an eigenfrequency but a peculiar one, and in this case there is an infinity (more precisely, a continuum) of eigenmodes with zero eigenfrequency, namely

$$\omega = 0 \quad \Rightarrow R = \text{const} \quad \text{and} \quad \nabla . \boldsymbol{V} = 0 \text{ with } \boldsymbol{V} . \boldsymbol{n}_{|\Gamma} = 0 \,, \tag{5.5}$$

$\boldsymbol{V}$ being otherwise arbitrary. The existence of eigenfrequencies and eigenmodes is a consequence of some deep mathematical results on the spectral theory of elliptic operators. We make the obvious remark, from (5.4), that all the eigenmodes with non-zero frequency are irrotational, so that they are eigenmodes of the wave equation also. On the other hand, the zero-frequency eigenmodes are rotational. Indeed, if one adds to (5.5) the condition that $\boldsymbol{V}$ is irrotational one finds that $\boldsymbol{V}$ must be zero. This fact is a consequence of

$$\boldsymbol{V} = \nabla\psi, R = \omega\psi \Rightarrow \Delta\psi + \omega^2\psi = 0, \frac{\mathrm{d}\psi}{\mathrm{d}n}_{|\Gamma} = 0 \,, \tag{5.6}$$

and in fact $R = 0$ in (5.5) when $\omega = 0$, and, from (5.6), $\psi = 0$ if $\omega = 0$. The reason why these zero-frequency modes are parasitic with respect to the wave equation is that they correspond, as we shall soon see, to modes that do not start from rest. They are eliminated during the process of deriving the wave equation.

An interesting by-product of spectral theory is that if ω is different from the whole set of eigenfrequencies, then the inhomogeneous problem

$$\omega R + \nabla \cdot \boldsymbol{v} = F, \quad \omega \boldsymbol{V} - \nabla R = \boldsymbol{G}, \quad \boldsymbol{V} \cdot \boldsymbol{n}_{|\Gamma} = H \tag{5.7}$$

has a unique solution, but if $\omega = \omega_n$, one of the eigenfreqencies, the above problem (5.7) has no solution except when the following compatibility condition is satisfied:

$$\langle F; R_n \rangle + \langle \boldsymbol{G}; \boldsymbol{V}_n \rangle + \int_\Gamma H R_n \, \mathrm{d}s = 0 \,. \tag{5.8}$$

In such a case, there is a one-parameter family (at least) of solutions, any two of which differ by a multiple of a (normalized) eigenmode. The relation (5.4d) for $m = n$ is related to a well-known property of acoustic waves, namely, that the dimensionless acoustic energy in the whole of the cavity D is $(1/2)[\langle r; r \rangle + \langle \boldsymbol{v}; \boldsymbol{v} \rangle]$. Now, if we specialize this to one of the eigenmodes, we obtain $(1/2)(\langle R_n; R_n \rangle Cn^2 + \langle \boldsymbol{V}_n; \boldsymbol{V}_n \rangle Sn^2)$, where Cn and Sn stand for $\cos(\omega_n t)$ and $\sin(\omega_n t)$ respectively. Averaging over a period, we obtain $(\pi/2\omega_n)[\langle R_n; R_n \rangle + \langle \boldsymbol{V}_n; \boldsymbol{V}_n \rangle]$, and the equality (5.4d) for $n = m$ means that both parts of the acoustic energy, the kinetic and the internal parts, are equal, on average, for an eigenmode. On the other hand, it is obvious that there is a two-parameter family of eigenmodes corresponding to an eigenfrequency, because the phase lag ϕ in (5.3) may be changed arbitrarily. As a matter of fact, we see that if $(R_n, \boldsymbol{V}_n)$ is an eigenmode corresponding to ω_n, then $(-R_n, \boldsymbol{V}_n)$ corresponds to $-\omega_n$, that is, to the same eigenfrequency $|\omega_n|$. We may either consider that the eigenfrequencies are positive and associate two orthogonal eigenmodes with each eigenfrequency, or consider that the eigenfrequencies may be either negative and that there is, in general, only one eigennnmode for one eigenfrequency. Under some circumstances, it may happen that several eigenmodes correspond to the same eigenfrequency. In such circumstances we use the index n in order to label the eigenmodes (we may use negative indices for modes with negative eigenfrequencies if use is made of the property mentioned above) in such a way that different indices correspond to different modes; we then need only to order the modes according to non-increasing eigenfrequency. We may assume that all the modes corresponding to the same eigenfrequency are orthogonal to each other. It is sufficient to use a standard orthogonalization process in the subspace spanned by the eigenmodes concerned.

A quite difficult mathematical issue is related with the proof that the eigenmodes, including the ones with zero eigenfrequency, are complete. This is the result of self-adjointness of the acoustic operator, a property which is reflected in (5.4d), and some regularity results for elliptic operators. The proper mathematical result is generally stated in terms of the completeness of the generalized eigenmodes, and the argument must be performed by using the fact that, under the condition of self-adjointness, generalized eigenmodes are, in fact, simply eigenmodes. The distinction between eigenmodes and generalized eigenmodes is best understood by reference to matrices. It is well known that symmetric matrices are diagonalizable but that non-symmetric matrices may not be, because it may happen that there are not sufficiently many eigenvectors and that, in order to reduce the matrix (to Jordan form), one needs to add generalized eigenvectors to the eigenvectors proper.

5.1.1 The Solution of the Wave Equation for 'Free' Acoustic Oscillations

We see that the eigenmodes with non-zero frequency are irrotational and may be written as

$$\boldsymbol{V}_n(\boldsymbol{x}) = \nabla\psi_n(\boldsymbol{x}), \quad R_n(\boldsymbol{x}) = \omega_n\psi_n(\boldsymbol{x}) \, , \tag{5.9}$$

where $\psi_n(\boldsymbol{x})$ is a solution of the eigenvalue problem

$$\Delta\psi_n + \omega_n^2\psi_n = 0, \quad \frac{d\psi_n}{dn}|_\Gamma = 0 \, , \tag{5.10}$$

and $\psi_n(\boldsymbol{x})\cos(\omega_n\tau + \phi_n)$ is a solution of the wave equation

$$\frac{\partial^2\psi}{\partial\tau^2} - \Delta\psi = 0 \, . \tag{5.11}$$

The initial-value problem for (5.11), with the initial data

$$\psi(0, \boldsymbol{x}) = \psi^0(\boldsymbol{x}) \quad \text{and} \quad \frac{\partial\psi}{\partial\tau}\bigg|_{\tau=0} = \psi^1(\boldsymbol{x}) \tag{5.12}$$

and a homogeneous boundary condition, is solved through the formula

$$\psi(\tau, \boldsymbol{x}) = \sum\nolimits_n [A_n\cos(\omega_n\tau) - B_n\sin(\omega_n\tau)]\psi_n(\boldsymbol{x}) \, , \tag{5.13}$$

where the coefficients A_n and B_n are related to the initial data by using (5.4d), namely

$$A_n = \langle\psi_n; \psi^0\rangle, \quad Bn = -\frac{1}{\omega_n}\langle\psi_n; \psi^1\rangle \, . \tag{5.14}$$

On the other hand, the solution to the initial-value problem

$$\frac{\partial r}{\partial\tau} + \nabla\,.\,\boldsymbol{v} = 0, \quad \frac{\partial\boldsymbol{v}}{\partial\tau} + \nabla r = 0, \quad \boldsymbol{v}\,.\,\boldsymbol{n}_{|\Gamma} = 0, \quad \boldsymbol{v}_{|\tau=0} = \boldsymbol{v}^0, \quad r_{|\tau=0} = r^0 \tag{5.15}$$

is found from (5.13) by properly taking into account the zero-frequency eigenmodes, that is;

$$\boldsymbol{v} = \boldsymbol{U}(\boldsymbol{x}) + \sum\nolimits_n [A_n\cos(\omega_n\tau) - B_n\sin(\omega_n\tau)]\,\boldsymbol{V}_n(\boldsymbol{x}) \, , \tag{5.16a}$$

$$r = \sum\nolimits_n [A_n\sin(\omega_n\tau) + B_n\cos(\omega_n\tau)]R_n(\boldsymbol{x}) + q_0 \, . \tag{5.16b}$$

Here A_n and B_n are found to be

$$A_n = \langle \boldsymbol{V}_n; \boldsymbol{v}^0 \rangle, \quad Bn = -\frac{1}{\omega_n} \langle R_n; r^0 \rangle \ . \tag{5.16c}$$

The constant q_0 is found from the compatibility relation

$$q_0 \int_D \mathrm{d}v = \int_D r^0 \,\mathrm{d}v \ , \tag{5.16d}$$

and $\boldsymbol{U}(\boldsymbol{x})$ is the rotational part of the initial velocity, such that $\boldsymbol{v}^0 = \boldsymbol{U}(\boldsymbol{x}) + \nabla \psi^0(\boldsymbol{x})$. We can find a number of equivalent ways to define the function $\boldsymbol{U}(\boldsymbol{x})$; one of them is the solution of the following problem:

$$\nabla \wedge \boldsymbol{U}(\boldsymbol{x}) = \nabla \wedge \boldsymbol{v}^0, \quad \langle \boldsymbol{U}; \boldsymbol{V}_n \rangle = 0, \quad \text{for all } n \ . \tag{5.16e}$$

5.1.2 The Solution of the Wave Equation for 'Forced' Acoustic Oscillations

For forced acoustic oscillations, solution of the wave equation amounts mathematically to solving the inhomogeneous acoustic system

$$\frac{\partial r}{\partial \tau} + \nabla \,.\, \boldsymbol{v} = F \cos(\omega\tau + \phi), \quad \frac{\partial \boldsymbol{v}}{\partial \tau} + \nabla r = \boldsymbol{G} \sin(\omega\tau + \phi) \ , \tag{5.17a}$$

with the (inhomogeneous) boundary condition

$$\boldsymbol{v} \,.\, \boldsymbol{n}_{|\Gamma} = H \cos(\omega\tau + \phi) \ . \tag{5.17b}$$

As this problem is linear, it is sufficient to solve the above problem (5.17) with the same phase lag ϕ for the two equations (5.17a) and (5.17b). We set

$$\boldsymbol{v} = \boldsymbol{V}(\boldsymbol{x}) \cos(\omega\tau + \phi) + \boldsymbol{v}^*(\tau, \boldsymbol{x}), \quad r = R(\boldsymbol{x}) \sin(\omega\tau + \phi) + r^*(\tau, \boldsymbol{x}) \ , \tag{5.18}$$

such that $R(\boldsymbol{x}) \sin(\omega\tau + \phi)$ and $\boldsymbol{V}(\boldsymbol{x}) \cos(\omega\tau + \phi)$ gives the part of the acoustic field which is tuned to the frequency of the forcing, while $\boldsymbol{v}^*(\tau, \boldsymbol{x})$ and $r^*(\tau, \boldsymbol{x})$ give the transient part. This transient part is easily obtained from the above solution (5.16a,b) together with (5.16c–e), where we need only substitute $-R \sin\phi$ for r^0 and $-\boldsymbol{V} \cos\phi$ for $\boldsymbol{v}^0$. We leave aside this transient solution and focus our attention on the tuned response to the amplitudes of the forcing terms $F, \boldsymbol{G}$ and H in (5.17). We have to solve

$$\omega R + \nabla \,.\, \boldsymbol{V} = F, \quad -\omega \boldsymbol{V} + \nabla R = \boldsymbol{G}, \quad \boldsymbol{V} \,.\, \boldsymbol{n}_{|\Gamma} = H \ . \tag{5.19}$$

A mathematical argument shows that a solution exists and is unique provided that the frequency ω is different from all the eigenfrequencies ω_n in the solution (5.16a,b). We assume here that this is the case, and we now present the solution. We first compute a constant q^* and a rotational vector $\boldsymbol{U}^*$, orthogonal to all the $\boldsymbol{V}_n$, by use of

$$F = \sum_n \langle R_n; F\rangle R_n + q^*; \quad \boldsymbol{G} = \sum_n \langle \boldsymbol{V}_n; \boldsymbol{G}\rangle \boldsymbol{V}_n + \boldsymbol{U}^* , \tag{5.20}$$

then we find a scalar field $Q(\boldsymbol{x})$ by solving

$$\Delta Q + \omega^2 Q = \omega q^* + \nabla . \boldsymbol{U}^*; \quad \frac{\mathrm{d}Q}{\mathrm{d}n}\bigg|_{\Gamma} = \omega H + \boldsymbol{U}^* . \boldsymbol{n}|_{\Gamma} , \tag{5.21}$$

which is uniquely solvable provided that, as was assumed, the frequency ω is different from all the eigenfrequencies, and we finish the process by using the formulae

$$R = \sum_n C_n R_n + Q, \quad \boldsymbol{V} = \sum_n D_n \boldsymbol{V}_n + \boldsymbol{W} , \tag{5.22}$$

where

$$C_n = \frac{1}{(\omega^2 - \omega_n{}^2)} \{\omega\langle R_n; F\rangle - \omega_n\langle \boldsymbol{V}_n; \boldsymbol{G}\rangle\} , \tag{5.23a}$$

$$D_n = \frac{1}{(\omega^2 - \omega_n{}^2)} \{\omega_n\langle R_n; F\rangle - \omega\langle \boldsymbol{V}_n; \boldsymbol{G}\rangle\} , \tag{5.23b}$$

and

$$\boldsymbol{W} = \frac{1}{\omega}[\nabla Q - \boldsymbol{U}^*] . \tag{5.24}$$

The solution for $Q(\boldsymbol{x})$ may be stated in closed form. We first look for any scalar field $\sigma(\boldsymbol{x})$ that satisfies the relation

$$\frac{\mathrm{d}\sigma}{\mathrm{d}n}\bigg|_{\Gamma} = \omega H + \boldsymbol{U}^* . \boldsymbol{n}_{|\Gamma} , \tag{5.25a}$$

and then, by adding some constant value to it, we choose this field in such a way that the following relation holds:

$$\int_D [\omega q^* + \nabla . \boldsymbol{U}^* - \Delta\sigma - \omega^2\sigma]\mathrm{d}v = 0 . \tag{5.25b}$$

Then we obtain the following solution for $Q(\boldsymbol{x})$:

$$Q(\boldsymbol{x}) = \sigma(\boldsymbol{x}) + \sum_n \frac{1}{(\omega^2 - \omega_n{}^2)} \langle \omega F + \nabla . \boldsymbol{U}^* - \Delta\sigma - \omega^2\sigma; R_n\rangle R_n . \tag{5.26}$$

Of course, our discussion of forced acoustic waves has been a little more technical than we might have desired, but we must keep this degree of generality if we want to be able to deal with the situations encountered below. Anyway, we emphasize that the above formulae (5.23) and (5.26) exibit the phenomenon of resonance. Namely, when ω approaches one of the eigenfrequencies (i.e. $\omega \to \omega_n$), it is readily seen that the eigenmode corresponding to this eigenfrequency is considerably enhanced over all the others, the amplitude growing as $(\omega - \omega_n)^{-1}$. It may be checked that the condition under which resonance is avoided is that the forcing functions $F, \boldsymbol{G}$ and H in the right-hand sides of the system (5.17a), with the condition (5.17b), are related through (5.8).

5.2 Damping of Acoustic Waves by Viscosity and Heat Conduction Inside a Cavity With a Rigid Wall

We return to the system of nonlinear acoustic equations (1.37) (Sect. 1.5), with

$$\frac{1}{Re_{\mathrm{ac}}} \equiv \delta^2 \quad \text{and} \quad \frac{1}{3} + \frac{\mu_{\mathrm{vc}}}{\mu_{\mathrm{c}}} \equiv \lambda \,. \tag{5.27}$$

When $M \downarrow 0$ with t and $\boldsymbol{x}$ fixed, the parameters δ^2 and λ being also fixed, we can write in place of (1.37) the following set of approximate equations:

$$\frac{\partial \omega}{\partial t} + \nabla \cdot \boldsymbol{v} = 0 \,, \tag{5.28a}$$

$$\frac{\partial \boldsymbol{v}}{\partial t} + \frac{1}{\gamma}\nabla \pi = \delta^2 \{\nabla^2 \boldsymbol{v} + \lambda \nabla(\nabla \cdot \boldsymbol{v})\} \,, \tag{5.28b}$$

$$\frac{\partial \theta}{\partial t} + (\gamma - 1)\nabla \cdot \boldsymbol{v} = \delta^2 \sigma \nabla^2 \theta \,, \tag{5.28c}$$

with $\pi = \theta + \omega$ and $\sigma = \gamma/Pr$. As the boundary conditions on the rigid heated wall Γ (with a constant dimensional characteristic temperature T_{c}) of the cavity, we write

$$\boldsymbol{v} = 0 \quad \text{and } \theta = 0 \quad \text{on } \Gamma \,. \tag{5.29}$$

At the initial time $t = 0$, we assume (for generality) that

$$t = 0 : \quad \boldsymbol{v} = \boldsymbol{v}^0, \quad \theta = \theta^0, \quad \omega = \omega^0 \,. \tag{5.30}$$

In fact, we want to investigate the behaviour of the fluid inside the cavity D when the time t tends to infinity under the hypothesis that $\delta \ll 1$. Since t is large but δ is small, it is judicious to introduce a *slow* time (using a double-time technique)

$$\tau = \delta t \quad \Rightarrow \frac{\partial}{\partial t} = \frac{\partial}{\partial t} + \delta \frac{\partial}{\partial \tau} \,, \tag{5.31}$$

together with the expansion

$$(\boldsymbol{v}, \theta, \omega, \pi) \equiv U = U_0 + \delta U_1 + \delta^2 U_2 + \dots \,. \tag{5.32}$$

5.2.1 Solution for U_0

At the zeroth order, δ^0, we obtain (for U_0) the following acoustic (inviscid) problem for the functions $\boldsymbol{v}_0(t, \boldsymbol{x}), \theta_0(t, \boldsymbol{x}), \omega_0(t, \boldsymbol{x})$ and $p_0(t, \boldsymbol{x}) \equiv (1/\gamma)\pi_0(t, \boldsymbol{x})$;

$$\frac{\partial \boldsymbol{v}_0}{\partial t} + \nabla p_0 = 0, \quad \frac{\partial p_0}{\partial t} + \nabla \cdot \boldsymbol{v}_0 = 0 \,, \tag{5.33a}$$

$$\boldsymbol{v}_0 \cdot \boldsymbol{n} = 0 \text{ on } \Gamma \tag{5.33b}$$

$$\text{at } t = 0\colon \boldsymbol{v}_0 = \boldsymbol{v}^0, p_0 = \frac{1}{\gamma}[\omega^0 + \theta^0] \,, \tag{5.33c}$$

$$\omega_0 = p_0 + \frac{1}{\gamma}[(\gamma - 1)\omega^0 - \theta^0], \quad \theta_0 = (\gamma - 1)p_0 - \frac{1}{\gamma}[(\gamma - 1)\omega^0 - \theta^0] \,. \tag{5.33d}$$

We observe that the slip condition (5.33b) is, as a matter of fact, a matching condition with a boundary-layer solution valid near the wall Γ. We see that the two equations (5.33a), for $\boldsymbol{v}_0$ and p_0, admit an infinity of acoustic solutions, such that

$$\{Cn(t)\ \boldsymbol{V}_n(\boldsymbol{x}), Sn(t)\ P_n(\boldsymbol{x})\}, \quad \{-Sn(t)\ \boldsymbol{V}_n(\boldsymbol{x}), Cn(t)\ P_n(\boldsymbol{x})\} \,, \tag{5.34a}$$

where

$$\nabla \cdot \boldsymbol{V}_n + \omega_n P_n = 0, \quad \nabla P_n - \omega_n \boldsymbol{V}_n = 0, \ \boldsymbol{V}_n \cdot \boldsymbol{v} = 0 \text{ on } \Gamma \,, \tag{5.34b}$$

and $Cn(t) = \cos \omega_n t$ and $Sn(t) = \sin \omega_n t$. The solutions (5.34a) are the oscillatory eigenmodes of the cavity D containing the fluid, and we see that the eigenmodes are orthogonal:

$$\omega_n \neq \omega_m \quad \Rightarrow \langle P_n; P_m \rangle = \langle \boldsymbol{V}_n; \boldsymbol{V}_m \rangle = 0 \,. \tag{5.35}$$

Below, we assume that all $\omega_n, n = 1, 2, \ldots$, are different, and we want to solve the problem (5.33). First, we write for the initial velocity field

$$\boldsymbol{v}^0 = \boldsymbol{U}^0(\boldsymbol{x}) + \nabla \phi^0, \quad \text{with } \Delta \phi^0 = \nabla \cdot \boldsymbol{v}^0, \quad \frac{\mathrm{d}\phi^0}{\mathrm{d}n} = 0 \text{ on } \Gamma \,, \tag{5.36}$$

assuming that we have the following (strong) compatibility condition:

$$\int_D (\nabla \cdot \boldsymbol{v}^0)\, \mathrm{d}v = 0 \,. \tag{5.37a}$$

It is then necessary to determine the constant q_0 such that

$$\int_D (p^0 - q_0)\, \mathrm{d}v = 0, \quad \text{where } p^0 \equiv \frac{1}{\gamma}[\omega^0 + \theta^0] \,. \tag{5.37b}$$

The solution of the problem (5.33) is sought in the following form (for $n > 1$):

$$\boldsymbol{v}_0 = \boldsymbol{U}^0(\boldsymbol{x}) + \sum_n [A_n(\tau)\ Cn(t) - B_n(\tau)\ Sn(t)]\, \boldsymbol{V}_n \,, \tag{5.38a}$$

$$p_0 = q_0 + \sum_n [A_n(\tau)\ Sn(t) + B_n(\tau)\ Cn(t)] P_n \,. \tag{5.38b}$$

The initial conditions for $\boldsymbol{v}_0$ and p_0 are satisfied when

$$A_n(0) = \frac{\langle \nabla \phi^0; \boldsymbol{V}_n \rangle}{\langle \boldsymbol{V}_n; \boldsymbol{V}_n \rangle}, \quad B_n(0) = \frac{\langle p^0; P_n \rangle}{\langle P_n; P_n \rangle} \,, \tag{5.39}$$

and we note that $\langle q_0; P_n \rangle = 0$. Finally, we observe that it is inconsistent to assume that q_0 and $\boldsymbol{U}^0(\boldsymbol{x})$ depend on the slow time $\tau = \delta t$.

5.2.2 Solution for U_1

Now, it is necessary to elucidate how the amplitudes $A_n(\tau)$ and $B_n(\tau)$ depend on the slow time τ. For this purpose it is necessary to consider the first-order, δ^1, system of equations for U_1 in the expansion (5.32). For $\boldsymbol{v}_1(t, \boldsymbol{x}; \tau)$ and $p_1(t, \boldsymbol{x}; \tau)$, we derive the following two acoustic, inviscid inhomogeneous equations:

$$\frac{\partial \boldsymbol{v}_1}{\partial t} + \nabla p_1 + \sum\nolimits_{n \geq 1} \left[Cn(t) \frac{\mathrm{d}A_n(\tau)}{\mathrm{d}\tau} - Sn(t) \frac{\mathrm{d}B_n(\tau)}{\mathrm{d}\tau} \right] \boldsymbol{V}_n = 0 \,, \quad (5.40a)$$

$$\frac{\partial p_1}{\partial t} + \nabla \cdot \boldsymbol{v}_1 + \sum\nolimits_{n \geq 1} \left[Sn(t) \frac{\mathrm{d}A_n(\tau)}{\mathrm{d}\tau} + Cn(t) \frac{\mathrm{d}B_n(\tau)}{\mathrm{d}\tau} \right] P_n = 0 \,. \quad (5.40b)$$

The value of $\boldsymbol{u}_1 \cdot \boldsymbol{n} \equiv w_1$ on Γ should be given by a matching condition with a boundary-layer solution valid near the wall Γ.

Boundary-Layer Equations

If we want to find the corresponding boundary-layer solution, which must match with the solution of the above first-order equations (5.40), then it is necessary to derive the boundary-layer equations significant near the wall Γ associated with (5.33) from the starting equations (5.28), with the small parameter $\delta^2 \ll 1$. For the position vector near the wall, we write

$$\boldsymbol{M} = \boldsymbol{P}(\boldsymbol{s}) + \delta \zeta \boldsymbol{n}, \quad \boldsymbol{s} \in \Gamma \,, \quad (5.41a)$$

where $\zeta = z/\delta$ is the boundary-layer coordinate normal to Γ, $\boldsymbol{s} = (s_1, s_1)$ are the horizontal coordinates on Γ, and z is the dimensionless distance from the wall (directed towards the inside of the cavity) by the unit normal $\boldsymbol{n}$ to the wall. Together with (5.41a), we write

$$\nabla = \delta^{-1} \frac{\partial}{\partial \zeta} \boldsymbol{n} + \boldsymbol{D} + \mathbf{O}(\delta) \,, \quad (5.41b)$$

where $\boldsymbol{D} = (\partial/\partial s_1, \partial/\partial s_2)$ is the gradient operator on the wall Γ. Now, if we take into account the usual slip condition for the inviscid zero-order problem (5.33), then we can assume in the boundary layer that

$$\boldsymbol{v} = \boldsymbol{v}_0|_\Gamma + \boldsymbol{v}_{\mathrm{BL}0} + \delta \left[w_{\mathrm{BL}0} + \zeta \left(\frac{\partial \boldsymbol{v}_0}{\partial z} \cdot \boldsymbol{n} \right)_{|\Gamma} \right] \boldsymbol{n} + \ldots,$$
$$\text{with } \boldsymbol{v}_{\mathrm{BL}0} \cdot \boldsymbol{n} = 0 \,, \quad (5.42a)$$
$$\theta = \theta_{0|\Gamma} + \theta_{\mathrm{BL}0} + \ldots,$$
$$p = p_{0|\Gamma} + \ldots, \quad \omega = \omega_{0|\Gamma} - \theta_{\mathrm{BL}0} + \ldots \,, \quad (5.42b)$$

in place of the expansion (5.32). We observe that if we use (5.42), the matching conditions with the acoustic solution U_0 given in Sect. 5.2.1 are automatically satisfied, and as a consequence we write

$$\boldsymbol{v}_{\mathrm{BL0}} \text{ and } \theta_{\mathrm{BL0}} \text{ both tend to zero when } \zeta \to +\infty\,. \tag{5.43}$$

we then observe that the slip condition (5.33b), $\boldsymbol{v}_0 \cdot \boldsymbol{n} = 0$ on Γ, for the equations (5.33a), is consistent in the framework of the leading-order problem (5.33), the boundary layer being coupled to the first-order inviscid solution U_1. With the conditions (5.41) and (5.42), we can derive the following leading-order (Prandtl) boundary-layer equations from the starting equations (5.28) by a classical boundary-layer analysis:

$$\frac{\partial \boldsymbol{v}_{\mathrm{BL0}}}{\partial t} - \frac{\partial^2 \boldsymbol{v}_{\mathrm{BL0}}}{\partial \zeta^2} = 0\,, \tag{5.44a}$$

$$\frac{\partial \theta_{\mathrm{BL0}}}{\partial t} - \frac{\sigma}{\gamma}\frac{\partial^2 \theta_{\mathrm{BL0}}}{\partial \zeta^2} = 0\,, \tag{5.44b}$$

with the boundary conditions

$$\text{at } \zeta = 0: \quad \boldsymbol{v}_{\mathrm{BL0}} = -\boldsymbol{v}_{0|\Gamma}, \quad \theta_{\mathrm{BL0}} = -\theta_{0|\Gamma}\,, \tag{5.44c}$$

and the initial conditions

$$\text{at } t = 0: \quad \boldsymbol{v}_{\mathrm{BL0}} = 0, \quad \theta_{\mathrm{BL0}} = 0\,. \tag{5.44d}$$

In (5.44c), from the solution (5.38) for $\boldsymbol{v}_0$ and p_0 and also the second relation of (5.33d) for θ_0, we have the following values on the wall Γ:

$$\boldsymbol{v}_{0|\Gamma} = U^0_{|\Gamma} + \sum\nolimits_{n>1} [A_n(\tau)\, Cn(t) - B_n(\tau)\, Sn(t)]\, \boldsymbol{V}_{n|\Gamma}\,, \tag{5.45a}$$

$$\begin{aligned}\theta_{0|\Gamma} &= (\gamma-1) q_{0|_\Gamma} - \frac{1}{\gamma}[(\gamma-1)\omega^0 - \theta^0]_{|\Gamma} \\ &\quad - (\gamma-1)\sum\nolimits_{n\geq 1} [A_n(\tau)\, Sn(t) + B_n(\tau)\, Cn(t)] P_{n|\Gamma}\,. \end{aligned}\tag{5.45b}$$

The solution of the above equations (5.44a,b), with the conditions (5.44c,d), (5.45) and (5.43), is obtained via a Laplace transform with respect to the time t. If we assume, for simplicity, that the parameter σ/γ is equal to 1, then we obtain the following solution:

$$\boldsymbol{v}_{\mathrm{BL0}} = -\left[1 - \operatorname{erf}\left(\frac{\zeta}{2\sqrt{t}}\right)\right] \boldsymbol{U}^0_{|\Gamma} - \sum_{n\geq 1} \mathrm{Real}\{[A_n(\tau) + \mathrm{i} B_n(\tau)] I_n\}\, \boldsymbol{V}_{n|\Gamma}\,, \tag{5.46a}$$

$$\begin{aligned}\theta_{\mathrm{BL0}} &= -\left[1 - \operatorname{erf}\left(\frac{\zeta}{2\sqrt{t}}\right)\right] \left[(\gamma-1) q_{0|_\Gamma} - \frac{1}{\gamma}[(\gamma-1) r^0 - \theta^0]_{|\Gamma}\right] \\ &\quad -(\gamma-1)\sum_{n\geq 1} \mathrm{Real}\{[B_n(\tau) - \mathrm{i} A_n(\tau)] I_n\} P_{n|\Gamma}\,, \end{aligned}\tag{5.46b}$$

where

$$I_n = \frac{1}{2\sqrt{\pi}} \int_0^t \frac{\zeta}{t'^{3/2}} \exp[\mathrm{i}\omega_n(t-t') - \frac{\zeta}{4t'}]\mathrm{d}t' \tag{5.46c}$$

We observe that, for large times t, an asymptotic evaluation of the solution (5.46a,b), with (5.46c), shows that local Rayleigh boundary-layer solutions are associated with the steady part of the inviscid solution. Boundary-layer oscillations are associated with the oscillatory part of this inviscid solution; these oscillations tend, when t tends to infinity, to a local Stokes layer with a residual term of the order $O(1/t)$, which occurs in a layer of thickness $O(\sqrt{t})$. This thickness $O(\sqrt{t})$ is that of the increasing local Rayleigh layer, while the Stokes layer is independent of the time t. Now, it is necessary to calculate the vertical velocity w_{BL0}. To do this, we take into account the continuity equation associated with the (Prandtl) boundary-layer system of equations (5.44a,b), namely, $\partial\omega_{\mathrm{BL0}}/\partial t + \boldsymbol{D} \cdot \boldsymbol{v}_{\mathrm{BL0}} + \partial w_{\mathrm{BL0}}/\partial\zeta = 0$, and we consider the limiting value

$$(w_{\mathrm{BL0}})^\infty = \int_0^\infty \left[\frac{\partial\theta_{\mathrm{BL0}}}{\partial t} - \boldsymbol{D} \cdot \boldsymbol{v}_{\mathrm{BL0}}\right] \mathrm{d}\zeta \,, \tag{5.47}$$

since here only $(w_{\mathrm{BL0}})^\infty$ – the behaviour of w_{BL0} when ζ tends to infinity – is of interest. This gives us the possibility to write, by matching, the following boundary condition on the wall for the first-order inviscid system (5.40).

$$\boldsymbol{v}_1 \cdot \boldsymbol{n} \equiv w_1 = (w_{\mathrm{BL0}})^\infty \quad \text{on } \varGamma \,, \tag{5.48}$$

with $w_{\mathrm{BL0}} = 0$ on $\zeta = 0$.

The Resonance Phenomenon

The expression for $(w_{\mathrm{BL0}})^\infty$ is very awkward-looking, but for the purpose of our discussion below we observe that in this expression we have a collection of terms containing $\cos(\omega_n t)$ and $\sin(\omega_n t)$. As a consequence, we expect a resonance phenomenon of oscillatory eigenmodes excited by these terms. It is necessary to consider the following test problem (as our problem is linear):

$$\begin{aligned}
&\frac{\partial \boldsymbol{u}_1}{\partial t} + \nabla p_1 + [\lambda_n \cos(\omega_n t) + \mu_n \sin(\omega_n t)]\, \boldsymbol{V}_n = 0 \,, \\
&\frac{\partial p_1}{\partial t} + \nabla \cdot \boldsymbol{u}_1 + [\nu_n \cos(\omega_n t) + \rho_n \sin(\omega_n t)] P_n = 0 \,, \\
&\quad (\boldsymbol{u}_1 \cdot \boldsymbol{n})_{|\varGamma} = [\xi_n \cos(\omega_n t) + \eta_n \sin(\omega_n t)] F_n \,,
\end{aligned} \tag{5.49}$$

where $\lambda_n, \mu_n, \nu_n, \rho_n, \xi_n$ and η_n are constant coefficients. According to the compatibility conditions discussed at end of Sect. 5.1.2, a resonance phenomenon is avoided if we assume the following compatibility conditions:

$$\lambda_n \langle P_n; P_n \rangle - \rho_n \langle \boldsymbol{V}_n;\, \boldsymbol{V}_n \rangle = \xi_n \int F_n P_n \,\mathrm{d}s \,, \tag{5.50a}$$

$$\mu_n \langle P_n; P_n \rangle + \nu_n \langle \boldsymbol{V}_n;\, \boldsymbol{V}_n \rangle = \eta_n \int F_n P_n \mathrm{d}s \,. \tag{5.50b}$$

If we introduce

$$\varphi_n = \frac{1}{\chi_n^2}\left[(\gamma-1)\sqrt{\omega_{n/2}}\int_\Gamma P_{n|\Gamma}^2\,\mathrm{d}s - \frac{1}{\sqrt{2\omega_n}}\int_\Gamma P_{n|\Gamma}(\boldsymbol{D}\,.\,\boldsymbol{V}_n|\Gamma)\,\mathrm{d}s\right], \tag{5.51a}$$

where

$$\chi_n^2 = \int_\Gamma [P_n^2 + |\boldsymbol{V}_n|^2]\,\mathrm{d}s\,, \tag{5.51b}$$

then the compatibility conditions (5.50) can be written as

$$\frac{\mathrm{d}A_n}{\mathrm{d}\tau} + \varphi_n(A_n - B_n) = 0\,, \tag{5.52a}$$

$$\frac{\mathrm{d}B_n}{\mathrm{d}\tau} + \varphi_n(A_n + B_n) = 0\,, \tag{5.52b}$$

by comparison with the system of equations (5.40) and the expression for $(w_{\mathrm{BL0}})^\infty$. On the other hand, when we take (5.34b) into account, a straightforward calculation shows that

$$\varphi_n > 0\,, \tag{5.53}$$

since $-P_{n|\Gamma}(\boldsymbol{D}\,.\,\boldsymbol{V}_{n|\Gamma}) \equiv \omega_n|\boldsymbol{V}_{n|\Gamma}|^2$, and in this case we obtain the solution of the equations (5.52) as

$$A_n = \alpha_n \exp(-\varphi_n\tau)\cos(\varphi_n\tau + \beta_n), \quad B_n = -\alpha_n \exp(-\varphi_n\tau)\sin(\varphi_n\tau+\beta_n)\,. \tag{5.54}$$

From (5.39), we write the initial conditions as

$$\alpha_n \cos\beta_n = \frac{\langle\nabla\phi^0;\boldsymbol{V}_n\rangle}{\langle\boldsymbol{V}_n;\boldsymbol{V}_n\rangle}, \quad -\alpha_n \sin\beta_n = \frac{1}{\gamma}\frac{\langle T^0+\omega^0;P_n\rangle}{\langle P_n;P_n\rangle}\,, \tag{5.55}$$

and we see that the damping time for the acoustic waves is of the order of $O(1/\delta) = O[\sqrt{Re/M}]$.

A Simple Solution After the Transition Stage

After the transition stage (relative to the slow time τ), we must consider the following simplified system of two equations in place of (5.40):

$$\frac{\partial p_1}{\partial t} + \nabla\,.\,\boldsymbol{v}_1 = 0, \quad \frac{\partial \boldsymbol{v}_1}{\partial t} + \nabla p_1 = 0\,, \tag{5.56a}$$

with the slip condition

$$(\boldsymbol{v}_1\,.\,\boldsymbol{n})_{|\Gamma} = 2\sqrt{\frac{t}{\pi}}(\boldsymbol{D}\,.\,\boldsymbol{U}_{|\Gamma}^0) + O\left(\frac{1}{\sqrt{t}}\right)\,, \tag{5.56b}$$

where we take into account in $(w_{\mathrm{BL0}})^\infty$ only the term that is important for large times t, after the transition stage in which $\tau = O(1)$. When $t \to \infty$, we have

$$\boldsymbol{v}_1 = \sqrt{t}\boldsymbol{v}_{1,1} + O\left(\frac{1}{\sqrt{t}}\right) \quad \text{and } p_1 = \left(\frac{1}{\sqrt{t}}\right) p_{1,1} + O\left(\frac{1}{t^{3/2}}\right) , \tag{5.56c}$$

$$\boldsymbol{v}_{1,1} = -2\,\nabla p_{1,1} , \tag{5.56d}$$

$$\Delta p_{1,1} = 0 \quad \text{and} \quad \left(\frac{\mathrm{d}p_{1,1}}{\mathrm{d}n}\right)_{|\Gamma} = \left(\frac{1}{\sqrt{\pi}}\right) (\boldsymbol{D}\,.\,\boldsymbol{U}^0_{|\Gamma}) , \tag{5.56e}$$

and the Neumann problem (5.56e) for $p_{1,1}$ is well posed, since

$$\int_\Gamma \boldsymbol{D}\,.\,\boldsymbol{U}^0_{|\Gamma}\,\mathrm{d}s = 0 . \tag{5.56f}$$

For large times, when $t = O(1/\delta)$, we obtain from the solution (5.46a,b) the result that the term containing $\boldsymbol{U}^0_{|\Gamma}$ is $O(1)$, but that the term proportional to the integral I_n is $O(\delta)$ near the wall within a layer with a thickness $O(\delta^{1/2})$, in place of a thickness $O(\delta)$. On the other hand, when $t = O(1/\delta^2)$, these two terms are, $O(1)$ and $O(\delta^2)$, respectively, at a distance $O(1)$ from the wall. If we now assume that $\tau = O(1)$ then we see that in the expansion (5.32) it is necessary to insert a term

$$\delta^{1/2} U_{1/2}, \quad \text{where } U_{1/2} = (\sqrt{\tau}\boldsymbol{v}_{1,1}, 0, 0, 0) , \tag{5.57}$$

since $\sqrt{t} = \sqrt{\tau}/\sqrt{\delta}$ and $\delta(1/\sqrt{\delta}) = \delta^{1/2}$. As a matter of fact, a term U_1 proportional to δ does not exist in the expansion (5.32).

5.2.3 Solution for U_2

Below, we consider, for a time $t = O(1/\delta^2)$, a new slow (large) time

$$t^* = \delta^2 t , \tag{5.58}$$

and in this case, obviously, when $t^* = O(1)$, the term $\delta^{1/2}U_{1/2}$ is $O(1)$. We return to the expansion (5.32) and consider the term $\delta^2 U_2$ – we observe that, according to (5.38) and (5.57), $\boldsymbol{v}_0$ and p_0 are functions of t^* and $\boldsymbol{x}$ only, such that, for $t^* = 0$, we have $\boldsymbol{v}_0 = \boldsymbol{U}^0(\boldsymbol{x})$ and $p_0 = q_0 = \text{const}$, given by (5.37b); θ_0 and ω_0 are given by (5.33d), where in place of p_0 we have q_0. If we add a constant to p_0 then we can assume that $q_0 = 0$.

According to our discussion above (Sect. 5.2.2), it seems that, for a slow time $t^* = O(1)$, the term δU_1 is zero in the expansion (5.32). From the starting system (5.28), taking into account the fact that U_0 is a function of t^* and $\boldsymbol{x}$ only, we obtain the following equations for the term $U_2 = (\boldsymbol{v}_2, \theta_2, \omega_2, p_2)$:

$$\frac{\partial \omega_2}{\partial t} + \nabla \cdot \boldsymbol{v}_2 + \frac{\partial \omega_0}{\partial t^*} = 0 \,,$$
$$\frac{\partial \boldsymbol{v}_2}{\partial t} + \nabla p_2 + \frac{\partial \boldsymbol{v}_0}{\partial t^*} - \Delta \boldsymbol{v}_0 - \lambda \nabla(\nabla \cdot \boldsymbol{v}_0) = 0 \,,$$
$$\frac{\partial \theta_2}{\partial t} + (\gamma - 1)\nabla \cdot \boldsymbol{v}_2 + \frac{\partial \theta_0}{\partial t^*} - \sigma \Delta \theta_0 = 0 \,. \tag{5.59a}$$

We observe that we do not have an associated boundary layer for the functions $U_0 = (\boldsymbol{v}_0, \theta_0, \omega_0, p_0)$, and we write the boundary conditions on the wall simply as

$$\boldsymbol{v}_0 = 0, \quad \theta_0 = 0 \quad \text{on } \Gamma \,, \tag{5.59b}$$

with the following conditions when $t^* = 0$:

$$\boldsymbol{v}_{0\,|t^*=0} = \boldsymbol{U}^0(\boldsymbol{x}), \quad \theta_{0\,|t^*=0} = \frac{1}{\gamma}[\theta^0 - (\gamma - 1)\omega^0] = -\omega_{0\,|t^*=0} \,. \tag{5.59c}$$

Obviously, we have the possibility to consider p_0 as a function of t^* only, and the condition $\nabla \cdot \boldsymbol{U}^0 = 0$ gives us the possibility to seek a divergenceless solution for $\boldsymbol{v}_0$, i.e.

$$\nabla \cdot \boldsymbol{v}_0 = 0 \,. \tag{5.60}$$

From the first and last equations of (5.59a), with $\gamma p_2 = \omega_2 + \theta_2$, by the elimination of the term $(\gamma - 1)\nabla \cdot \boldsymbol{v}_2$, we derive the following equation (again with $\sigma/\gamma = 1$):

$$\frac{\partial}{\partial t}[(\gamma - 1)\omega_2 - \theta_2] = \gamma \left[\frac{\partial \theta_0}{\partial t^*} - \Delta \theta_0 - (\gamma - 1)\frac{\mathrm{d}p_0}{\mathrm{d}t^*}\right] \,. \tag{5.61}$$

This shows that it is necessary to consider the right-hand-side of (5.61) equal to zero because this right-hand side is not a function of the time t and, for very large t, the term $(\gamma - 1)\omega_2 - \theta_2$ becomes very large. As a consequence, from (5.61) we obtain the following equation:

$$\frac{\partial \theta_0}{\partial t^*} = \Delta \theta_0 + (\gamma - 1)\frac{\mathrm{d}p_0}{\mathrm{d}t^*} \,. \tag{5.62}$$

We obtain the following solution for θ_0 from (5.62):

$$\theta_0 = (\gamma - 1)p_0(t^*) + \Theta_0(t^*, \boldsymbol{x}), \quad \text{with } \frac{\partial \Theta_0}{\partial t^*} = \Delta \Theta_0 \,. \tag{5.63}$$

Now, if we take into account the divergenceless condition (5.60) for $\boldsymbol{v}_0$, we obtain the following equation from the second equation of (5.59a):

$$\frac{\partial \boldsymbol{v}_2}{\partial t} = -\left(\frac{\partial \boldsymbol{v}_0}{\partial t^*} + \nabla p_2\right) + \Delta \boldsymbol{v}_0 \,. \tag{5.64}$$

On the other hand, in place of the first equation of the system (5.59a), we write the equation

$$\frac{\partial \omega_2}{\partial t} = \frac{\partial \Theta_0}{\partial t^*} - \frac{\partial p_0}{\partial t^*} - \nabla \cdot \boldsymbol{v}_2 , \tag{5.65}$$

and also, using (5.65) and the last equation of the system (5.59a), the equation

$$\gamma \frac{\partial p_2}{\partial t} = \frac{\partial [\omega_2 + \theta_2]}{\partial t} = -\gamma \left[\nabla \cdot \boldsymbol{v}_2 + \frac{\partial p_0}{\partial t^*} - \Delta \Theta_0 \right] . \tag{5.66}$$

Now, when in (5.65)

$$\nabla \cdot \boldsymbol{v}_2 = -\frac{\partial p_0}{\partial t^*} + \Delta \Theta_0 , \tag{5.67}$$

which is, in fact, a consequence of (5.66), we recover the second equation for Θ_0 in (5.63) and then, since $\partial p_2 / \partial t = 0$ according to (5.67), we obtain the following equation for $\boldsymbol{v}_0$:

$$\frac{\partial \boldsymbol{v}_0}{\partial t^*} + \nabla p_2 = \Delta \boldsymbol{v}_0 . \tag{5.68}$$

The two equations (5.68) and (5.60) are a closed system for the functions $\boldsymbol{v}_0(t^*, \boldsymbol{x})$ and $p_2(\boldsymbol{x})$. The solution of (5.67) is

$$\boldsymbol{v}_2 = \nabla \Theta_0 - \frac{\partial p_0}{\partial t^*} \boldsymbol{v}_2^*, \quad \text{with } \nabla \cdot \boldsymbol{v}_2^* = 1 . \tag{5.69}$$

We observe that the equation $\partial \Theta_0 / \partial t^* = \Delta \Theta_0$ for $\Theta_0(t^*, \boldsymbol{x})$ is a consequence of the compatibility condition (non-secularity) (5.62), with (5.63) for $\theta_0(t^*, \boldsymbol{x})$. As a consequence, the two equations (5.67) and (5.68), which lead to $\partial \omega_2 / \partial t = 0$ and $\partial p_2 / \partial t = 0$, are also, in fact non-secularity conditions with which to back up our non-secularity condition (5.62), $\partial \theta_2 / \partial t$ being also equal to 0. Finally, the slip condition for the velocity $\boldsymbol{v}_2$ (U_2 is an inviscid approximation, because the dissipative terms in the system (5.59a) are linked to $U_0(t^*, \boldsymbol{x})$),

$$(\boldsymbol{v}_2 \cdot \boldsymbol{n})_{|\Gamma} = 0 , \tag{5.70}$$

leads to the following compatibility relation (from (5.67)), since $\Delta \Theta_0 = \nabla \cdot (\nabla \Theta_0)$:

$$\int_\Gamma \left(\frac{\mathrm{d}\Theta_0}{\mathrm{d}n} \right)_{|\Gamma} \mathrm{d}s - \frac{\partial p_0}{\partial t^*} |D| = 0 , \tag{5.71}$$

where $|D|$ is the volume of the cavity D. With the condition $p_0(0) = 0$, this relation (5.71) determines the function $p_0(t^*)$, when the function Θ_0, the solution of the (parabolic) heat conduction equation with the conditions

$$t^* = 0 : \quad \Theta_0 = \frac{1}{\gamma}[\theta^0 - (\gamma - 1)\omega^0], \quad \text{and } \Theta_0 = (1 - \gamma) p_0(t^*), \quad \text{on } \Gamma , \tag{5.72}$$

is known. For $\boldsymbol{v}_0$ and p_2 to be solutions of the two equations (5.68) and (5.60), we have as conditions

$$t^* = 0 : \quad \boldsymbol{v}_0 = \boldsymbol{U}^0(\boldsymbol{x}), \quad \text{and } \boldsymbol{v}_0 = 0 \quad \text{on } \Gamma . \tag{5.73}$$

We shall not go further, because our main purpose in this section was to show that the solution $U_0(t^*, \boldsymbol{x})$ is only completely determined when, in fact, we consider the system of equations (5.59a) for U_2 and write some compatibility conditions (non-secularity conditions that cancel the terms which do not remain bounded when the time t increases indefinitely). This characterizes the multiple-timescale approach.

5.2.4 Further Considerations

Obviously, the problem considered above in this section deserves further, more careful investigations. It has been considered by Vasilieva and Boutousov (1990, Sect. 17) in the case where the time dependence of the functions is represented by $\exp(\mathrm{i}\omega t)$ and $\partial/\partial t \to \mathrm{i}\omega$:

$$(\boldsymbol{v}, \omega, \theta, \pi) \equiv U(t, \boldsymbol{x}) = U(\boldsymbol{x}) \exp(\mathrm{i}\omega t) , \tag{5.74}$$

where $U(\boldsymbol{x}) \equiv (\boldsymbol{V}, R, \Theta, \Pi)$. Obviously, under the assumption (5.74), we have to deal with an oscillating cavity and, in fact, we have as the boundary condition on the wall of the cavity

$$\boldsymbol{v} = \boldsymbol{v}^*(\boldsymbol{P}) \exp(\mathrm{i}\omega t) , \tag{5.75}$$

which creates harmonic oscillations. In this case we do not have the possibility to investigate the initial transition stage close to the initial time and we do not write initial conditions, the motion being periodic in time. In (5.75), $\boldsymbol{v}^*(\boldsymbol{P})$ is a given vectorial function (amplitude) of the two-dimensional position vector $\boldsymbol{P}$ on the wall Γ which is maintained at $\theta = 0$. In this case, we write the following system of equations in place of (5.28):

$$\begin{aligned}
\mathrm{i}\omega \boldsymbol{V} &= -\frac{1}{\gamma} \nabla(\Pi) + \delta^2 \{\Delta \boldsymbol{V} + \lambda \nabla(\nabla \cdot \boldsymbol{V})\} , \\
\mathrm{i}\omega R &+ \nabla \cdot \boldsymbol{V} = 0 , \\
\mathrm{i}\omega \Theta &= (\gamma - 1)\mathrm{i}\omega R + \delta^2 \sigma \Delta \Theta , \\
\Pi &= R + \Theta .
\end{aligned} \tag{5.76}$$

This is the system of equations considered by Vasilieva and Boutousov (1990, Sect. 17). These authors investigated a boundary-value problem in a cavity D with a sufficiently smooth boundary $\partial D = \Gamma$. In accordance with (5.75), the following two boundary conditions on Γ are written

$$\boldsymbol{V}|_\Gamma = \boldsymbol{v}^*(\boldsymbol{P}), \boldsymbol{P} \in \Gamma, \quad \text{and } \Theta|_\Gamma = 0 . \tag{5.77}$$

When we write for the velocity vector $\boldsymbol{V}$ the classical representation

$$\boldsymbol{V} = \nabla Q + \boldsymbol{W}, \quad \text{with } \nabla \cdot \boldsymbol{W} = 0 , \tag{5.78}$$

we know that this representation (5.78)) is not unique, but that it is possible to show that among all such representations (5.78) there exists only one such that the pair ($\boldsymbol{Q}$, $\boldsymbol{W}$) is a solution of the following system of three compatible equations:

$$\mathrm{i}\omega \boldsymbol{Q} = -\frac{1}{\gamma}\Pi + \delta^2 \Delta Q, \quad \nabla \cdot \boldsymbol{W} = 0, \quad \delta^2 \Delta \boldsymbol{W} - \mathrm{i}\omega \boldsymbol{W} = 0 . \tag{5.79}$$

5.3 Long-Time Evolution of Acoustic Waves and the Concept of Incompressibility: Inviscid Perfect Gas in a Cavity with a Boundary that is Deformable as a Function of Time

In Zeytounian and Guiraud (1980), an initial approach to the problem of the unsteady flow of an inviscid perfect gas within a bounded domain $\Omega(t)$, the wall $\partial\Omega(t) = \Sigma(t)$ of which deforms very slowly in comparison with the speed of sound, was investigated. This problem has been considered in Sect. 3.3.2 of the present book (see (3.19) and (3.20)) and, for the case when the wall is started impulsively from rest, we have derived (in Sect. 3.3.4) a Neumann problem for the limiting case $M \to 0$ with t fixed, with the Bernoulli integral (3.31d)) for $\Pi = \pi/\gamma\rho_0(t)$, where π is the pseudo-pressure term proportional to M^2 in the expansion (3.13)). However, when the wall is started impulsively from rest, unfortunately, this limiting case (with t fixed) is singular close to the initial time, and it is necessary to consider a local-in-time limiting case $M \to 0$ with $\tau = t/M$ fixed. In such a case, close to the initial time $t = 0$, we derive the classical acoustic equations (3.37) and obtain the solution (3.43). For an impulsive motion we have also (3.47), which shows that when (in accordance with matching) $\tau \to \infty$, the solution (3.43) together with (3.47) does not tend to the initial value (at $t = 0$) of the solution of the Neumann problem (3.31a,b) corresponding to the limiting case $M \to 0$ with t fixed. Indeed, when the motion of the wall of the bounded cavity is started impulsively (or even only rapidly during a time $t = O(M)$), then, at the end of the initial-time layer transition stage, when $t = O(M) \Leftrightarrow \tau = O(1)$, the acoustic oscillations (generated at the initial moment by the motion of the deformable wall) remain undamped and continue into an evolution stage $t = O(1)$. A multiple-scale technique, in place of the MAE method, is necessary for the investigation of the long-time evolution of these rapid oscillations. Unfortunately, a two-time 'naive' technique, with t and $\tau = t/M$, is not adequate, because such a technique does not give us the possibility to eliminate all secular terms. The main reason is that the acoustic eigenfrequencies of the cavity appear in our internal aerodynamic problem and, because the cavity is a function of the slow time t (the time of the boundary velocity), these eigenfrequencies are also functions of the (slow) time t.

Below, in this section, we consider as the starting equations the Euler dimensionless equations (where the Strouhal number is unity) valid within the cavity $\Omega(t)$. We write for the dimensionless density ρ, the velocity vector $\boldsymbol{u}$, the pressure p and the specific entropy S, the following system:

$$\frac{\partial \rho}{\partial t} + \boldsymbol{u} \,.\, \nabla \rho \,.\, + \nabla \,.\, \boldsymbol{u} = 0 \,,$$
$$\rho \left[\frac{\partial \boldsymbol{u}}{\partial t} + (\boldsymbol{u} \,.\, \nabla) \boldsymbol{u} \right] + \frac{1}{\gamma M^2} \nabla p = 0 \,,$$
$$\frac{\partial S}{\partial t} + \boldsymbol{u} \,.\, \nabla S = 0 \,, \tag{5.80}$$

where

$$p = \rho^{\gamma} \exp(S) \,. \tag{5.81}$$

As the initial condition, we write

$$t = 0 : \quad \boldsymbol{u} = 0, S = 0 \quad \text{and } p = \rho = 1 \,, \tag{5.82}$$

and we apply a slip condition on the wall $\partial\Omega(t) = \Sigma(t)$,

$$(\boldsymbol{u} \,.\, \boldsymbol{n})_{\Sigma(t)} = w_{\Sigma}(t, \boldsymbol{P}) \,, \tag{5.83}$$

where $w_{\Sigma}(t, \boldsymbol{P})$ is the given value (the positive normal component when the cavity $\Omega(t)$ is contracting) of the displacement velocity of the wall $\Sigma(t)$; $\boldsymbol{n}$ is the unit inward normal to the boundary $\Sigma(t)$, directed towards the inside of the domain of flow $\Omega(t)$; and $\boldsymbol{P}$ is the position vector on $\Sigma(t)$. Our purpose in this section is to derive a formal expansion of the solution $U = (\boldsymbol{u}, \rho, p, S)$ of (5.80)–(5.83) with respect to the small parameter M, and to investigate (in the inviscid case) how the approximately incompressible motion is affected by the acoustic oscillations generated in the initial transition stage of the motion (close to $t = 0$), when the wall is started impulsively at rest.

5.3.1 Persistence of the Oscillations

With the unit of speed chosen, the speed of sound is $O(1/M)$ and, consequently, the periods of the natural (free) vibrations of the domain $\Omega(t)$ are $O(M)$. We expect that the solution will oscillate on a timescale $O(M)$, while, on the other hand, it will evolve on a timescale $O(1)$ owing to the evolution of boundary $\partial\Omega(t)$. Therefore, we must build into the structure of the solution U of (5.80)–(5.85) a multiplicity of times (in particular, a family of fast times, in contrast to Müller (1998), Meister (1999) and Ali (2003)).

The Multiple-Scale Technique

First we use the time t, a *slow* time, and then we bring into the solution an infinity of *fast* times designed to cope with the infinity of periods of free vibration of the cavity $\Omega(t)$. We use the symbol U^* for the solution U expressed through this variety of timescales, and write

$$\frac{\partial U}{\partial t} = \frac{\partial U^*}{\partial t} + \frac{1}{M} D U^* , \tag{5.84}$$

where $\partial U^*/\partial t$ stands for the time derivative computed when all fast times are maintained constant and $(1/M)DU^*$ is the time derivative (where D is a differential operator) that occurs as a result of all the fast times. We carry such a change into the starting Euler equations (3.80) and then expand according to

$$U^* = U_0^* + MU_1^* + M^2 U_2^* + \ldots, \quad \text{with } U^* = \langle U^* \rangle + U^{*\prime} , \tag{5.85}$$

where $\langle U^* \rangle$ is the average of U^* (over all rapid oscillations; this average depends only on the slow time t and the spatial position $\boldsymbol{x}$), and $U^{*\prime}$ is the fluctuating (oscillating) part of U^* (which depends, also, on all fast times). More precisely, the operation $U^* \Rightarrow \langle U^* \rangle$ erases all the oscillations associated with the fast times, and obviously $D\langle U^* \rangle = 0$. We observe also that for two functions f and g, we must write

$$\langle fg \rangle = \langle f \rangle \langle g \rangle + \langle f' g' \rangle, (fg)' = \langle f \rangle g' + f' \langle g \rangle + (f'g')' . \tag{5.86}$$

For instance, we see (according to (5.102)) that, for the fluctuating parts of $\boldsymbol{u}_0^{*\prime}$ and $\rho_1^{*\prime}$, we can write the solution as

$$\boldsymbol{u}_0^{*\prime} = \sum_{n \geq 1} [A_n(t)\, Cn - B_n(t) Sn]\, \boldsymbol{U}_n , \tag{5.87a}$$

$$\rho_1^{*\prime} = \rho_0^*(t) \left[\frac{\rho_0^*(t)}{p_0^*(t)} \right]^{1/2} \sum_{n \geq 1} [A_n(t)\, Sn + B_n(t)\, Cn] R_n , \tag{5.87b}$$

respectively, where

$$Cn = \cos \left[\left(\frac{1}{M} \right) \phi_n(t) \right] \quad \text{and } Sn = \sin \left[\left(\frac{1}{M} \right) \phi_n(t) \right] ,$$

and

$$\langle Cp\, Cq \rangle = \langle Sp\, Sq \rangle = \frac{1}{2} \delta_{pq}, \quad \delta_{pq} = 0 \text{ if } p \neq q, \delta_{pq} = 1 \text{ if } p \equiv q ,$$

$$\langle Cn \rangle = 0, \langle Sn \rangle = 0, \langle Cp\, Sq \rangle = \langle Cq\, Sp \rangle \equiv 0 , \tag{5.88a}$$

$$D\, Cn = -\frac{\mathrm{d}\phi_n(t)}{\mathrm{d}t} Sn, \quad D\, Sn = \frac{\mathrm{d}\phi_n(t)}{\mathrm{d}t} Cn , \tag{5.88b}$$

where

$$\frac{\mathrm{d}\phi_n(t)}{\mathrm{d}t} = \sqrt{p_0^*(t)/\rho_0^*(t)}\, \omega_n, \quad \phi_n(0) = 0 . \tag{5.88c}$$

In (5.87), U_n and R_n are the normal modes of vibration of $\Omega(t)$ with eigenfrequencies ω_n, namely

$$\omega_n R_n + \nabla . \boldsymbol{U}_n = 0, \quad -\omega_n \boldsymbol{U}_n + \nabla R_n = 0, (\boldsymbol{U}_n \cdot \boldsymbol{n})_{\Sigma(t)} = 0 . \tag{5.89}$$

The above relation (5.88c) defines the scales of the fast times in relation to the speed of sound in the cavity (at time t) and to the eigenfrequencies of the cavity at that time.

Derivation of an Average Continuity Equation

With (5.84), we obtain for the functions $\boldsymbol{u}^*, \rho^*, p^*$ and S^* from the Euler equations (5.80) the following equations:

$$D\rho^* + M\left(\frac{\partial \rho^*}{\partial t} + \boldsymbol{u}^* \,.\, \nabla \rho^* + \rho^* \nabla \,.\, \boldsymbol{u}^*\right) = 0\,, \tag{5.90a}$$

$$\frac{1}{\gamma}\nabla p^* + M\rho^* D\boldsymbol{u}^* + M^2 \rho * \left[\frac{\partial \boldsymbol{u}^*}{\partial t} + (\boldsymbol{u}^* \,.\, \nabla)\boldsymbol{u}^*\right] = 0\,; \tag{5.90b}$$

$$DS^* + M\left(\frac{\partial S^*}{\partial t} + \boldsymbol{u}^* \,.\, \nabla S^*\right) = 0\,, \tag{5.90c}$$

$$p^* = \rho^{*\gamma} \exp(S^*)\,; \tag{5.90d}$$

with the slip condition

$$(\boldsymbol{u}^* \,.\, \boldsymbol{n})_{\Sigma(t)} = w_\Sigma(t, \boldsymbol{P})\,. \tag{5.91}$$

From the expansion (5.85), at the zeroth order, and from the above system (5.90), we derive

$$\nabla p_0^* = 0, \quad D\rho_0^* = 0, \quad DS_0^* = 0\,, \tag{5.92}$$

and this shows that ρ_0^* and S_0^* are independent of the fast times and, as a consequence of the equation of state (5.90d), that this is also the case for p_0^*, which is, in fact, a function of only the slow time t; namely:

$$p_0^* = p_0^*(t), \quad \rho_0^* = \rho_0^*(t, \boldsymbol{x}), \quad S_0^* = S_0^*(t, \boldsymbol{x})\,. \tag{5.93}$$

Now, at the first-order, from (5.90c), we derive the equation

$$DS_1^* + \frac{\partial S_0^*}{\partial t} + \boldsymbol{u}_0^* \,.\, \nabla S_0^* = 0 \tag{5.94}$$

for S, and, since $S_0^* = S_0^*(t, \boldsymbol{x})$ is independent of the fast time and $D\langle S_1^*\rangle = 0$, we have the following average equation for $S_0^*(t, \boldsymbol{x})$:

$$\frac{\partial S_0^*}{\partial t} + \langle \boldsymbol{u}_0^* \rangle \,.\, \nabla S_0^* = 0\,. \tag{5.95}$$

However, close to the initial time, when we consider the Euler equations (5.80) with (5.81) written with the short time $\tau = t/M$ in place of the slow time t, and use the local-in-time asymptotic expansion

$$S = S_{\mathrm{a}0} + MS_{\mathrm{a}1} + M^2 S_{\mathrm{a}2} + \dots\,, \tag{5.96a}$$

where $S_{\mathrm{ak}} = S_{\mathrm{ak}}(\tau, \boldsymbol{x}), k = 0, 1, 2, \dots$, and the initial condition $S = 0$ at $\tau = 0$, we derive

$$\frac{\partial S_{\mathrm{a}0}}{\partial \tau} = 0, \quad \frac{\partial S_{\mathrm{a}1}}{\partial \tau} = 0 \quad \Rightarrow S_{\mathrm{a}0} = 0, \quad S_{\mathrm{a}1} = 0\,, \tag{5.96b}$$

and, as a consequence, from (5.95), by continuity,

$$S_0^* = 0 . \tag{5.97}$$

On the other hand, with (5.97), we obtain from (5.81)

$$p_0^*(t) = [\rho_0^*(t)]^\gamma, \tag{5.98a}$$

the function $\rho_0^*(t)$ being determined by the relation

$$\int_{\Omega(t)} \rho_0^*(t)\, \mathrm{d}v = \int_{\Omega(t)} \mathrm{d}v \Rightarrow \rho_0^*(t)|\Omega|(t) = M_0 , \tag{5.98b}$$

where $|\Omega|(t)$ is the volume of the cavity, such that

$$\frac{\mathrm{d}|\Omega|(t)}{\mathrm{d}t} = -\int_{\Sigma(t)} w_\Sigma(t, \boldsymbol{P})\, \mathrm{d}s , \tag{5.98c}$$

and M_0 (= const) is the whole mass of the cavity. From the initial condition for the density, we have $|\Omega|(0) \equiv M_0$. We observe that if, in particular, we assume that $\rho_0^*(t) \equiv 1$ (and, as a consequence, $p_0^*(t) \equiv 1$ also) then the volume $|\Omega|(t)$ is not a function of time, and for all t in the motion we have $|\Omega|(t) \equiv M_0 =$ const. Obviously, this is not the case in the various applications.

At the first order, from the first two equations of (5.80) and with the above results, we derive the following two equations:

$$D\rho_1^* + \frac{\mathrm{d}\rho_0^*}{\mathrm{d}t} + \rho_0^* \nabla \cdot \boldsymbol{u}_0^* = 0 , \tag{5.99a}$$

$$\frac{1}{\gamma} \nabla p_1^* + \rho_0^* D\boldsymbol{u}_0^* = 0 , \tag{5.99b}$$

together with, from (5.83),

$$(\boldsymbol{u}_0^* \cdot \boldsymbol{n})_{\Sigma(t)} = w_\Sigma(t, \boldsymbol{P}) . \tag{5.99c}$$

Since $D\langle \rho_1^* \rangle = 0$, we derive from (5.99a) an average (zero-order) continuity equation

$$\frac{1}{\rho_0} \frac{\mathrm{d}\rho_0^*}{\mathrm{d}t} + \nabla \cdot \langle \boldsymbol{u}_0^* \rangle = 0 , \tag{5.100a}$$

together with (from (5.99c))

$$(\langle \boldsymbol{u}_0^* \rangle \cdot \boldsymbol{n})_{\Sigma(t)} = w_\Sigma(t, \boldsymbol{P}) . \tag{5.100b}$$

Equation (5.100a) can be identified with (3.20a) in Sect. 3.3.2, and obviously the model system (3.29) must be interpreted as a model for the averaged flow after erasure of the acoustic oscillations (see, for instance, Sect. 5.3.2 below). From (5.99b), since $D\langle \boldsymbol{u}_0^* \rangle = 0$, we have also

$$\nabla \langle p_1^* \rangle = 0 \quad \Rightarrow \langle p_1^* \rangle = 0, \quad p_1^* \equiv {p^*}'_1 . \tag{5.100c}$$

As a consequence, we obtain the equation (5.100a) for the average part, together with (5.100b,c), from the two equations (5.99a,b) together with (5.99c).

Solution for $u_0^{*\prime}$ and $\rho_1^{*\prime}$

For the fluctuations, we derive the following acoustic-type equations with slip from the two equations (5.99a,b) together with (5.99c):

$$D\rho_1^{*\prime} + \rho_0^* \nabla \cdot \boldsymbol{u}_0^{*\prime} = 0 \,, \tag{5.101a}$$

$$\frac{1}{\gamma}\nabla p_1^{*\prime} + \rho_0^* D\boldsymbol{u}_0^{*\prime} = 0 \,, \tag{5.101b}$$

$$(\boldsymbol{u}_0^{*\prime} \cdot \boldsymbol{n})_{\Sigma(t)} = 0 \,. \tag{5.101c}$$

Concerning the equation for the specific entropy, we have (because $S_0^* = 0$)

$$DS_1^{*\prime} = 0 \quad \Rightarrow S_1^{*\prime} = 0 \,, \tag{5.101d}$$

and, as a consequence, from the equation of state, for the fluctuation of the pressure we derive

$$p_1^{*\prime} = \gamma \left(\frac{p_0^*}{\rho_0^*}\right) \rho_1^{*\prime} \,. \tag{5.101e}$$

As a consequence of (5.100c), we have also

$$\langle \rho_1^* \rangle = 0 \quad \Rightarrow \rho_1^* \equiv \rho_1^{*\prime} \,. \tag{5.101f}$$

The solution of the two equations for $\boldsymbol{u}_0^{*\prime}$ and $(\rho_1^{*\prime}/\rho_0^*)$ obtained from (5.101a,b) when we use (5.101e) is given by (5.87a,b). Indeed, if we use the solutions (5.87a,b) in (5.101a,b), then

$$D\left(\frac{\rho_1^{*\prime}}{\rho_0^*}\right) + \nabla \cdot \boldsymbol{u}^{*\prime}{}_0$$
$$= \sum_{n\geq 1} [A_n(t)\, Cn - B_n(t)\, Sn] \left\{ \left[\frac{\rho_0^*(t)}{p_0^*(t)}\right]^{1/2} \frac{\mathrm{d}\phi_n(t)}{\mathrm{d}t} R_n + \nabla \cdot \boldsymbol{U}_n \right\} \,, \tag{5.102a}$$

and

$$\frac{p_0^*}{\rho_0^*} \nabla \left(\frac{\rho_1^{*\prime}}{\rho_0^*}\right) + D\boldsymbol{u}_0^{*\prime}$$
$$= \sum_{n\geq 1} [A_n(t)\, Sn + B_n(t)\, Cn] \left\{ \left[\frac{p_0^*(t)}{\rho_0^*(t)}\right]^{1/2} \nabla R_n - \frac{\mathrm{d}\phi_n(t)}{\mathrm{d}t} \boldsymbol{U}_n \right\} \,, \tag{5.102b}$$

and, with (5.88b,c), we obtain the result that the right-hand sides of the above equations (5.102a,b) are zero. We observe also that the eigenfunctions (the normal modes of vibration of $\Omega(t)$, with eigenfrequencies ω_n) $\boldsymbol{U}_n$ and R_n are normalized according to

$$\int_{\Omega(t)} \left[(\boldsymbol{U}_n)^2 + (R_n)^2\right] \mathrm{d}v = 1 \,. \tag{5.103}$$

Now, we observe that the matching condition between the above solution (5.87) and the solution (3.43) obtained in Sect. 3.3.4 is satisfied if we assume that

$$\langle \boldsymbol{u}_0^* \rangle|_{t=0} = \nabla \varphi^\infty \,, \tag{5.104}$$

and, in the case of an impulsive motion,

$$A_n(0) = \alpha_n \quad \text{and } B_n(0) = 0 \,, \tag{5.105}$$

where α_n is given by (3.47). We have also $p_0^*(0) = 1$ and $\rho_0^*(0) = 1$. However, it is now necessary to determine, from the system (5.90) together with (5.91), the equations for the second-order approximation and then to derive, first, a system of two equations for the amplitudes $A_n(t)$ and $B_n(t)$ which give the possibility to consider the long-time evolution of the rapid oscillations, and, second, an equation for the average value of $\boldsymbol{u}_0^*$ which gives, with the average continuity equation (5.100a), a system of two average equations for $\langle \boldsymbol{u}_0^* \rangle$ and $\langle p_2^* \rangle$.

5.3.2 The Second-Order Approximation

We return to the system of equations (5.90a–d) with (5.91), and consider the second-order approximation for S_2^*, p_2^* and ρ_2^*. First, from (5.90c), we obtain

$$DS_2^* + \frac{\partial \langle S_1^* \rangle}{\partial t} + [\langle \boldsymbol{u}_0^* \rangle \,.\, \nabla] \langle S_1^* \rangle = 0 \,, \tag{5.106a}$$

but, according to (5.101d), $S_1^{*\prime} = 0$ and also, $\partial\langle S_1^*\rangle/\partial t + [\langle \boldsymbol{u}_0^* \rangle \,.\, \nabla]\langle S_1^* \rangle = 0$. With a zero initial condition at $t = 0$, we have

$$S_1^* \equiv 0, \quad \text{and then} \quad S_2^{*\prime} = 0 \,. \tag{5.106b}$$

For the third-order approximation, we have $DS_3^* + \partial\langle S_2^*\rangle/\partial t + [\langle \boldsymbol{u}_0^* \rangle \,.\, \nabla]\langle S_2^* \rangle = 0$, and again (according to the second relation in (5.106b) we obtain

$$S_2^* = 0 \quad \text{and } S_3^{*\prime} = 0 \,. \tag{5.106c}$$

Finally, from the equation of state (5.90d), when we take into account that

$$S_0^* = S_1^* = S_2^* \equiv 0 \,, \tag{5.107}$$

we derive the following relation between p_2^* and ρ_2^*:

$$p_2^* = \gamma \left(\frac{p_0^*}{\rho_0^*} \right) \left[\rho_2^* + \frac{1}{2\rho_0^*} (\gamma - 1)(\rho_1^*)^2 \right] \,. \tag{5.108}$$

Now, from (5.90a,b), we derive two second-order equations,

$$D\rho_2^* + \rho_0^* \nabla \,.\, \boldsymbol{u}_1^* + \frac{\partial \rho_1^*}{\partial t} + \boldsymbol{u}_0^* .\, \nabla \rho_1^* + \rho_1^* \, \nabla \boldsymbol{u}_0^* = 0 \,, \tag{5.109a}$$

$$\nabla\left(\frac{p_2^*}{\gamma\rho_0^*}\right)+D\boldsymbol{u}_1^*+\frac{\rho_1^*}{\rho_0^*}D\boldsymbol{u}_0^*+\frac{\partial\boldsymbol{u}_0^*}{\partial t}+(\boldsymbol{u}_0^*\cdot\nabla)\boldsymbol{u}_0^*=0\,, \tag{5.109b}$$

together with

$$(\boldsymbol{u}_1^*\cdot\boldsymbol{n})_{\Sigma(t)}=0\,. \tag{5.109c}$$

From (5.109b), we have, first, the possibility to derive the following average equation for $\langle\boldsymbol{u}_0^*\rangle$:

$$\frac{\partial\langle\boldsymbol{u}_0^*\rangle}{\partial t}+\langle(\boldsymbol{u}_0^*\cdot\nabla)\boldsymbol{u}_0^*\rangle+\nabla\left(\frac{\langle p_2^*\rangle}{\gamma\rho_0^*}\right)+\frac{1}{\rho_0^*}\langle\rho_1^*D\boldsymbol{u}_0^*\rangle=0\,. \tag{5.110}$$

we write this average equation explicitly below. One the other hand, we write the above two equations (5.109a,b) as an inhomogeneous, acoustic-type system for ρ_2^*/ρ_0^* and $\boldsymbol{u}_1^*$. Namely,

$$D\left(\frac{\rho_2^*}{\rho_0^*}\right)+\nabla\cdot\boldsymbol{u}_1^*+G=0\,, \tag{5.111a}$$

$$\frac{p_0^*}{\rho_0^*}\nabla\left(\frac{\rho_2^*}{\rho_0^*}\right)+D\boldsymbol{u}_1^*+\boldsymbol{F}=0\,, \tag{5.111b}$$

$$(\boldsymbol{u}_1^*\cdot\boldsymbol{n})_{\Sigma(t)}=0\,, \tag{5.111c}$$

where

$$G=\frac{\partial}{\partial t}\left(\frac{\rho_1^*}{\rho_0^*}\right)+\nabla\cdot\left[\frac{\rho_1^*}{\rho_0^*}\boldsymbol{u}_0^*\right]+\frac{1}{\rho_0^*}\frac{\mathrm{d}\rho_0^*}{\mathrm{d}t}\frac{\rho_1^*}{\rho_0^*}\,, \tag{5.112a}$$

$$\boldsymbol{F}=\frac{\partial\boldsymbol{u}_0^*}{\partial t}+(\boldsymbol{u}_0^*\cdot\nabla)\boldsymbol{u}_0^*+(\gamma-2)\frac{p_0^*}{\rho_0^*}\frac{\rho_1^*}{\rho_0^*}\nabla\left(\frac{\rho_1^*}{\rho_0^*}\right)\,. \tag{5.112b}$$

In G and $\boldsymbol{F}$ above, according to (5.112), we have three categories of terms: (i) the average terms ($\langle G\rangle$ and $\langle\boldsymbol{F}\rangle$), independent of the scale of the fast times; (ii) the terms (G_L and $\boldsymbol{F}_L$), which depend linearly on the Cn and Sn; and (iii) the terms which depend quadratically (G_Q and $\boldsymbol{F}_Q$) on these Cn and Sn, and are proportional to $\cos[(1/M)(\phi_p(t)\pm\phi_q(t))]$ or $\sin[(1/M)(\phi_p(t)\pm\phi_q(t))]$. As a consequence, we can write, for the system (5.111a,b) for the inhomogeneous terms G and $\boldsymbol{F}$, the following formal representation:

$$G=\langle G\rangle+\left[\frac{p_0^*(t)}{\rho_0^*(t)}\right]^{1/2}\sum_{n\geq1}[G_{n\mathrm{C}}\,Cn+G_{n\mathrm{S}}\,Sn]+G_Q\,, \tag{5.113a}$$

$$\boldsymbol{F}=\langle\boldsymbol{F}\rangle+\frac{p_0^*(t)}{\rho_0^*(t)}\sum_{n\geq1}[\boldsymbol{F}_{n\mathrm{S}}\,Sn+\boldsymbol{F}_{n\mathrm{C}}\,Cn]+\boldsymbol{F}_Q\,, \tag{5.113b}$$

where $\langle G\rangle$ and $\langle\boldsymbol{F}\rangle$ and also the coefficients $G_{n\mathrm{C}}$, $G_{n\mathrm{S}}$, $\boldsymbol{F}_{n\mathrm{C}}, F_{n\mathrm{S}}$, are determined from (5.112) according to (5.113). Below, we assume that the quadratic terms G_Q and $\boldsymbol{F}_Q$ are not resonant triads satisfying the relation

$$|\phi_p(t)| \pm |\phi_q(t)| = \phi_r(t), \quad \forall p, q, r , \tag{5.114}$$

and thus none of the quadratic terms can interfere with any of the terms depending linearly on the Cn and Sn. As a consequence of the linearity of our system (5.111a,b), we can, in particular, write the solution for the *fluctuations* $(\rho_2^{*\prime}/\rho_0^*)$ and $\boldsymbol{u}_1^{*\prime}$, corresponding only to the terms linearly dependent on the C_n and S_n in (5.113), in the following form:

$$\frac{\rho_2^{*\prime}}{\rho_0^*} = \sum_{n\geq 1}[R_{n\mathrm{C}}\, Cn + R_{n\mathrm{S}}\, Sn] , \tag{5.115a}$$

$$\boldsymbol{u}_1^{*\prime} = \left(\frac{p_0^*}{\rho_0^*}\right)^{1/2} \sum_{n\geq 1}[\boldsymbol{U}_{n\mathrm{C}}\, Cn - \boldsymbol{U}_{n\mathrm{S}}\, Sn] . \tag{5.115b}$$

For example, the amplitudes $R_{n\mathrm{S}}$ and $\boldsymbol{U}_{n\mathrm{C}}$ satisfy the system

$$\begin{aligned} &\omega_n R_{n\mathrm{S}} + \nabla \cdot \boldsymbol{U}_{n\mathrm{C}} + G_{n\mathrm{C}} = 0 , \\ &-\omega_n \boldsymbol{U}_{n\mathrm{C}} + \nabla R_{n\mathrm{S}} + \boldsymbol{F}_{n\mathrm{S}} = 0 , \\ &\boldsymbol{U}_{n\mathrm{C}} \cdot \boldsymbol{n} = 0 \text{ on } \Sigma(t) . \end{aligned} \tag{5.116}$$

Obviously, we obtain a similar system for $R_{n\mathrm{C}}$ and $\boldsymbol{U}_{n\mathrm{S}}$ when, in place of $R_{n\mathrm{S}}$, $\boldsymbol{U}_{n\mathrm{C}}$, $G_{n\mathrm{C}}$ and $\boldsymbol{F}_{n\mathrm{S}}$, we write $R_{n\mathrm{C}}$, $\boldsymbol{U}_{n\mathrm{S}}$, $G_{n\mathrm{S}}$ and $\boldsymbol{F}_{n\mathrm{C}}$. For the existence of a solution of these two inhomogeneous systems, it is necessary that two compatibility relations (which are, in fact, a consequence of the Fredholm alternative), related to $(G_{n\mathrm{C}}, \boldsymbol{F}_{n\mathrm{S}})$ and $(G_{n\mathrm{S}}, \boldsymbol{F}_{n\mathrm{C}})$, respectively, are satisfied. For this, the system (5.89), for the normal modes $(R_n, \boldsymbol{U}_n)$ of vibration of the cavity $\Omega(t)$ with eigenfrequencies ω_n, must be taken into account. From (5.89), after an integration by parts, it follows that

$$\begin{aligned} 0 &= \int_{\Omega(t)} \{[\omega_n R_n + \nabla \cdot \boldsymbol{U}_n]R_{n\mathrm{C}} - [\omega_n \boldsymbol{U}_n - \nabla R_n]\,\boldsymbol{U}_{n\mathrm{C}}\}\mathrm{d}v \\ &= \int_{\Omega(t)} \{[\omega_n R_{n\mathrm{C}} + \nabla \cdot \boldsymbol{U}_{n\mathrm{C}}]R_{n\mathrm{C}} - [\omega_n \boldsymbol{U}_{n\mathrm{C}} - \nabla R_{n\mathrm{C}}]\,\boldsymbol{U}_n\}\mathrm{d}v , \end{aligned} \tag{5.117}$$

where we take into account also the boundary conditions (on $\partial\Omega(t) = \Sigma(t)$)

$$\boldsymbol{U}_{n\mathrm{C}} \cdot \boldsymbol{n} = 0 \text{ and } \boldsymbol{U}_{n\mathrm{S}} \cdot \boldsymbol{n} = 0 \quad \text{on } \Sigma(t) ,$$

As a a consequence, we derive the following compatibility condition for the solvability of the inhomogeneous system (5.116):

$$\int_{\Omega(t)} [G_{n\mathrm{C}} R_n - \boldsymbol{F}_{n\mathrm{S}} \cdot \boldsymbol{U}_n] \, \mathrm{d}v = 0 , \tag{5.118}$$

Of course, compatibility relation a similar to (5.118) is obtained if we write $G_{n\mathrm{S}}$ and $\boldsymbol{F}_{n\mathrm{C}}$ in place of $G_{n\mathrm{C}}$ and $\boldsymbol{F}_{n\mathrm{S}}$, respectively, after using a system similar to (5.116) for Rn_{C} and $\boldsymbol{U}n_{\mathrm{S}}$, with $G_{n\mathrm{S}}$ and $\boldsymbol{F}_{n\mathrm{C}}$.

The Average System of Equations for the Slow Variation

With only the average continuity equation (5.100a) and the slip condition (5.105b) for $\langle \boldsymbol{u}_0^* \rangle$, we lack sufficient information for the determination of the slow (nearly incompressible) variation. Such information is provided by the average equation (5.110). Namely, according to the solution (5.87), we obtain first

$$\langle (\boldsymbol{u}_0^* \cdot \nabla)\boldsymbol{u}_0^* \rangle = (\langle \boldsymbol{u}_0^* \rangle \cdot \nabla)\langle \boldsymbol{u}_0^* \rangle + \frac{1}{2}\sum_{n\geq 1}(A_n^2 + B_n^2)[\boldsymbol{U}_n \cdot \nabla]\boldsymbol{U}_n \,, \tag{5.119a}$$

$$\frac{1}{\rho_0^*}\langle \rho_1^* D\boldsymbol{u}_0^* \rangle = -\frac{1}{2}\sum_{n\geq 1}(A_n^2 + B_n^2)[R_n \cdot \nabla]R_n \,, \tag{5.119b}$$

where we have also used (5.89). From (5.89), we derive also the relation $[\boldsymbol{U}_n \cdot \nabla]\boldsymbol{U}_n = 1/2)|\nabla \boldsymbol{U}_n|^2$. Finally, for $\langle \boldsymbol{u}_0^* \rangle$, we derive the following average equation of motion:

$$\frac{\partial \langle \boldsymbol{u}_0^* \rangle}{\partial t} + (\langle \boldsymbol{u}_0^* \rangle \cdot \nabla)\langle \boldsymbol{u}_0^* \rangle + \nabla \Pi = 0 \,, \tag{5.120}$$

where

$$\Pi = \frac{\langle p_2^* \rangle}{\gamma \rho_0^*} + \frac{1}{4}\sum_{n\geq 1}(A_n^2 + B_n^2)\{|\boldsymbol{U}_n|^2 - |R_n|^2\} \,. \tag{5.121}$$

However in Sect. 5.3.1 we observed (see (5.104)) that $\langle \boldsymbol{u}_0^* \rangle|_{t=0} = \nabla \varphi^\infty$, and because $\langle \boldsymbol{u}_0^* \rangle$ is irrotational according to the average equation (5.120), it remains irrotational and we can write $\langle \boldsymbol{u}_0^* \rangle = \nabla \varphi$. In such a case, the average continuity equation (5.100a) for $\langle \boldsymbol{u}_0^* \rangle$, with the slip condition (5.100b) on $\Sigma(t)$, gives the possibility to determine $\langle \boldsymbol{u}_0^* \rangle$ thanks to the following Neumann problem:

$$\Delta \varphi + \frac{\mathrm{d} \log\, \rho_0^*(t)}{\mathrm{d}t} = 0 \,, \tag{5.122a}$$

with the condition

$$\frac{\mathrm{d}\varphi}{\mathrm{d}n}\Big|_{\Sigma(t)} = w_\Sigma(t, \boldsymbol{P}) \,. \tag{5.122b}$$

A comparison with the results derived in Sect. 3.3.4 (see (3.31a,b,d) shows that the model problem (3.31a,b) is formally correct but that (3.31d) must be replaced by the relation

$$\frac{\langle p_2^* \rangle}{\gamma \rho_0^*} = -[\partial\varphi/\partial t + \frac{1}{2}|\nabla\varphi|^2] - \frac{1}{4}\sum_{n\geq 1}(A_n^2 + B_n^2)\{|\boldsymbol{U}_n|^2 - |R_n|^2\} \,, \tag{5.123}$$

which takes into account the influence of the acoustics on the pressure. We observe also that for the initial condition for $\langle \boldsymbol{u}_0^* \rangle$ at $t = 0$, the solution of the average equation (5.120), we can write, according to the solution (5.87a) and the initial condition (5.82),

$$\langle \boldsymbol{u}_0^* \rangle + \sum_{n\geq 1} A_n(0)\,\boldsymbol{U}_n(0, \boldsymbol{x}) = 0 \quad \text{for } t = 0 \,. \tag{5.124}$$

The Long-Time Evolution of the Fast Oscillations

With the above derivation of the average system of equations for the slow variation, we have only eliminated part of the secular terms in $\boldsymbol{u}_1^*$ and ρ_2^*. As a consequence, it is necessary to consider in detail the system of compatibility conditions (5.118) for $G_{n\mathrm{C}}$ and $\boldsymbol{F}_{n\mathrm{S}}$, and the similar system for $G_{n\mathrm{S}}$ and $\boldsymbol{F}_{n\mathrm{C}}$. First, we write $G_{n\mathrm{C}}$, $\boldsymbol{F}_{n\mathrm{S}}$, $G_{n\mathrm{S}}$ and $\boldsymbol{F}_{n\mathrm{C}}$ explicitly, where we take into account the relations (5.112) and (5.113) and the solution (5.87) for $\boldsymbol{u}_0^{*\prime}$ and $\rho_1^{*\prime}$, where $\boldsymbol{u}_0^* = \langle \boldsymbol{u}_0^* \rangle + \boldsymbol{u}_0^{*\prime}$ and $\rho_1^* \equiv \rho_1^{*\prime}$. A straightforward calculation gives the following formulae:

$$G_{n\mathrm{C}} = \frac{\rho_0^*}{p_0^*}\left\{\frac{\mathrm{d}B_n}{\mathrm{d}t}R_n + B_n\left[\frac{\partial R_n}{\partial t} + \frac{\mathrm{d}\log\,\rho_0^*}{\mathrm{d}t}R_n + \nabla\cdot(\langle \boldsymbol{u}_0^*\rangle R_n)\right]\right\}, \tag{5.125a}$$

$$\boldsymbol{F}_{n\mathrm{S}} = -\frac{\rho_0^*}{p_0^*}\left\{\frac{\mathrm{d}B_n}{\mathrm{d}t}\boldsymbol{U}_n + \boldsymbol{B}_n\left[\frac{\partial \boldsymbol{U}_n}{\partial t} + (\boldsymbol{U}_n\cdot\nabla)\langle \boldsymbol{u}_0^*\rangle + (\langle \boldsymbol{u}_0^*\rangle\cdot\nabla)\boldsymbol{U}_n\right]\right\}, \tag{5.125b}$$

$$G_{n\mathrm{S}} = -\frac{\rho_0^*}{p_0^*}\left\{\frac{\mathrm{d}A_n}{\mathrm{d}t}R_n + A_n\left[\frac{\partial R_n}{\partial t} + \frac{\mathrm{d}\log\,\rho_0^*}{\mathrm{d}t}R_n + \nabla\cdot(\langle \boldsymbol{u}_0^*\rangle R_n)\right]\right\}, \tag{5.125c}$$

$$\boldsymbol{F}_{n\mathrm{S}} = \frac{\rho_0^*}{p_0^*}\left\{\frac{\mathrm{d}A_n}{\mathrm{d}t}\boldsymbol{U}_n + A_n\left[\frac{\partial \boldsymbol{U}_n}{\partial t} + (\boldsymbol{U}_n\cdot\nabla)\langle \boldsymbol{u}_0^*\rangle + (\langle \boldsymbol{u}_0^*\rangle\cdot\nabla)\boldsymbol{U}_n\right]\right\}. \tag{5.125d}$$

We observe that in (5.125a,c), according to (5.100a),

$$\nabla\cdot(\langle \boldsymbol{u}_0^*\rangle R_n) = R_n(\nabla\cdot\langle \boldsymbol{u}_0^*\rangle) + \langle \boldsymbol{u}_0^*\rangle.\nabla R_n \equiv \langle \boldsymbol{u}_0^*\rangle\cdot\nabla R_n - \frac{\mathrm{d}\log\,\rho_0^*}{\mathrm{d}t}R_n\,.$$

Then, from the compatibility relation (5.118), together with (5.125a,b), and from a similar compatibility relation, together with (5.125c,d), we derive the two ordinary differential equations for the amplitudes $A_n(t)$ and $B_n(t)$, taking into account the normalization condition (5.103), namely

$$\frac{\mathrm{d}A_n}{\mathrm{d}t} + \gamma_n(t)A_n = 0, \quad \frac{\mathrm{d}B_n}{\mathrm{d}t} + \gamma_n(t)B_n = 0\,, \tag{5.126}$$

where

$$\gamma_n(t) = \frac{1}{2}\int_{D(t)}\frac{\partial}{\partial t}\left[|\boldsymbol{U}_n|^2 + |R_n|^2\right]\mathrm{d}v + \frac{1}{2}\int_{D(t)}\{\langle \boldsymbol{u}_0^*\rangle\cdot\nabla[|\boldsymbol{U}_n|^2 + |R_n|^2]\}\,\mathrm{d}v$$
$$+ \int_{D(t)}[(\boldsymbol{U}_n\cdot\nabla)\langle \boldsymbol{u}_0^*\rangle]\cdot\boldsymbol{U}_n\,\mathrm{d}v\,.$$

Alternatively,

$$\gamma_n(t) = \frac{1}{2}\frac{\mathrm{d}\log\,\rho_0^*}{\mathrm{d}t} + \int_{D(t)}[(\boldsymbol{U}_n\cdot\nabla)\langle \boldsymbol{u}_0^*\rangle]\cdot\boldsymbol{U}_n\,\mathrm{d}v\,, \tag{5.127}$$

where we take into account that

$$\frac{1}{2}\int_{D(t)} \frac{\partial}{\partial t}\left[|\boldsymbol{U}_n|^2 + |R_n|^2\right] \mathrm{d}v = -\frac{1}{2}\int_{\Sigma(t)} \left[|\boldsymbol{U}_n|^2 + |R_n|^2\right] w_\Sigma(t, \boldsymbol{P})\, \mathrm{d}s\,,$$

thanks to (5.103) and (5.98c), and also that

$$\frac{1}{2}\int_{D(t)} \left\{\langle \boldsymbol{u}_0^*\rangle \cdot \nabla\left[|\boldsymbol{U}_n|^2 + |R_n|^2\right]\right\} \mathrm{d}v = \frac{1}{2}\int_{D(t)} \nabla\left\{\left[|\boldsymbol{U}_n|^2 + |R_n|^2\right]\langle u_0^*\rangle\right\} \mathrm{d}v$$
$$-\frac{1}{2}\int_{D(t)} \left[|\boldsymbol{U}_n|^2 + |R_n|^2\right](\nabla \cdot \langle \boldsymbol{u}_0^*\rangle)\, \mathrm{d}v\,.$$

However,

$$\frac{1}{2}\int_{D(t)} \nabla\left\{\left[|\boldsymbol{U}_n|^2 + |R_n|^2\right]\langle \boldsymbol{u}_0^*\rangle\right\} \mathrm{d}v = \frac{1}{2}\int_{\Sigma(t)} \left[|\boldsymbol{U}_n|^2 + |R_n|^2\right] w_\Sigma(t, \boldsymbol{P})\, \mathrm{d}s\,,$$

thanks to the slip condition (5.100b), and

$$-\frac{1}{2}\int_{D(t)} \left[|\boldsymbol{U}_n|^2 + |R_n|^2\right](\nabla \cdot \langle \boldsymbol{u}_0^*\rangle)\, \mathrm{d}v = \frac{1}{2}\frac{\mathrm{d}\log\, \rho_0^*}{\mathrm{d}t}$$

according to (5.100a) and (5.103). At $t = 0$ we have as initial conditions, from (5.124),

$$t = 0: \quad \sum_{n\geq 1} A_n(0)\, \boldsymbol{U}_n(0, \boldsymbol{x}) = -\langle \boldsymbol{u}_0^*\rangle \quad \text{and } B_n = 0, n = 1, 2, \ldots\,. \tag{5.128}$$

We observe that we derive initial conditions for $A_n(t)$ and $B_n(t)$ by applying the starting initial conditions (5.82) for $\boldsymbol{u}$ and ρ; this gives the condition (5.124) for A_n because $\boldsymbol{u} = 0$ at $t = 0$, when we take into account the solution (5.87a) for $\boldsymbol{u}_0^{*\prime}$ and the decomposition given in (5.85). The value of B_n is related to the initial condition $\rho = 1$ at $t = 0$, which is compatible with the leading-order solution $\rho_0^*(t = 0) = 1$, and $\rho_1^{*\prime}(0, x) = 0$ gives $B_n(0) = 0$. Thanks to the equation (5.126) for B_n, obviously $B_n(t) \equiv 0$ for all t. Concerning $A_n(0)$, its values must be derived from (5.124) and (5.128), and depend on the value of $\langle \boldsymbol{u}_0^*\rangle$ at $t = 0$. On the other hand, obviously, if $w_\Sigma(0, P) = 0$ in the condition (5.83), then $\langle u_0^*\rangle$ is also zero at $t = 0$, and then $A_n(0) = 0$, which implies also that $A_n(t) \equiv 0$ (i.e. it is zero for all t), and then the oscillations are absent! However, if the motion of the wall of the deformable cavity is started impulsively from rest (or accelerated from rest to a finite velocity in a time $O(M)$), then we have $w_\Sigma(0^-, \boldsymbol{P}) = 0$, but $w_\Sigma(0^+, \boldsymbol{P}) \neq 0$, and the same holds for the averaged velocity, $\langle \boldsymbol{u}_0^*\rangle$. In this case $A_n(0^+) \neq 0$ and, as a consequence, $A_n(t)$ is also non-zero, when $t \geq 0^+$.

5.3.3 Some Comments

As a most important result, we obtain the following: *If the motion of the wall of a deformable (as a function of time) cavity, in which an inviscid gas is confined is started impulsively from rest, then acoustic oscillations remain present, have an effect on the pressure which would be felt by a gauge, and are not related to the mean (averaged) motion. The same holds if the wall is accelerated from rest to a finite velocity in a time* $O(M)$. If we are dealing with a slightly viscous flow when the Mach number is much less than 1, we must start from the full unsteady NSF equations (see Sect. 5.4) in place of (5.80) and (5.81). In a such dissipative (viscous and heat-conducting) case, we can bring into the analysis a second small parameter Re^{-1}, the inverse of a (large) Reynolds number, and then we must expect that the acoustic oscillations will be damped out. Unfortunately, a precise analysis (when a similarity rule between M and Re^{-1} is assumed in a multiple-timescale asymptotic investigation) of this damping phenomenon appears to be a very difficult problem and raises many questions. In Sect. 5.4, this damping problem is considered mainly in the framework of the hypothesis $Re \gg 1/M$. Müller (1998) gives an insight into the compressible Navier–Stokes equations at low Mach number when a slow flow is affected by acoustic effects in a bounded domain over a long time. As an application, Müller mentions a closed piston–cylinder system in which isentropic compression due to slow motion is modified by acoustic waves. Müller uses only a two-timescale analysis, which obviously is insufficient for the elimination of the secular terms in derived approximate systems, as has been made clear in Sect. 5.3.2 above.

The Results of Ali (2003)

The results obtained recently by Ali (2003) are more interesting than those formally derived by Müller (1998), in spite of the fact that in Ali (2003), a two-timescale analysis is again used. In Ali (2003), the Euler equations for a compressible, perfect fluid are considered in a bounded time-dependent domain $\Omega_t \in \Re^n$, and Ω_0 denotes the domain at the initial time $t = 0$. The evolution of the bounded time-dependent domain is described by a family of invertible maps $\Phi_t : \Re^n \to \Re^n$, depending continuously on the time t such that $\Omega_t = \Phi_t(\Omega_0)$ for all t. This severe assumption on the domain Ω_t is, nevertheless, general enough to include a moving rigid domain, a cylinder cut by a fixed surface and a moving surface (piston problem), or a contacting and expanding sphere (star). In the particular case of a moving rigid domain, Ali (2003, p. 2023) writes $\Phi_t(\boldsymbol{x}) = \boldsymbol{x} + \boldsymbol{c}(t)$. The map Φ_t has a geometric meaning and is related neither to the fluid motion nor to the Lagrangian variables. Moreover, Φ_t does not need to be globally unique, since only its restriction to a neighbourhood of the boundary $\partial\Omega_0$ characterizes the motion of the domain's boundary $\partial\Omega_t$. In particular, Ali (2003) uses the map Φ_t to define the velocity $\boldsymbol{u}_\Omega$ of the boundary by

$$\boldsymbol{u}_\Omega(\boldsymbol{x}) = \frac{\partial \Phi_t}{\partial t}\left(\Phi_t^{-1}(\boldsymbol{x})\right) \quad \text{for all } \boldsymbol{x} \in \partial\Omega_t \,, \tag{5.129}$$

the aim being to describe the fluid motion when the velocity of the boundary is small compared with the speed of sound. The starting dimensionless Euler equations in Ω_t are

$$\partial_t \rho + \nabla \cdot \boldsymbol{m} = 0, \quad \partial_t \boldsymbol{m} + \nabla \cdot (\boldsymbol{m} \otimes \boldsymbol{m}) + \frac{1}{\varepsilon^2}\nabla p = 0 \,, \tag{5.130a}$$

$$\frac{1}{\gamma p}(\partial_t + \boldsymbol{u} \cdot \nabla)p + \nabla \cdot \boldsymbol{u} = 0, \quad \boldsymbol{m} = \rho \boldsymbol{u} \,, \tag{5.130b}$$

with the initial conditions

$$\rho(0,\boldsymbol{x}) = \rho_0(\boldsymbol{x}), \quad \boldsymbol{u}(0,\boldsymbol{x}) = \boldsymbol{u}_0(\boldsymbol{x}), \quad p(0,\boldsymbol{x}) = p_0(\boldsymbol{x}) \quad \text{in } \Omega_0 \,, \tag{5.130c}$$

and the boundary condition

$$\boldsymbol{u} \cdot \boldsymbol{n}_t = \boldsymbol{u}_\Omega \cdot \boldsymbol{n}_t \quad \text{on } \partial\Omega_t \,, \tag{5.130d}$$

where $\boldsymbol{n}_t$ is the outward normal to the boundary $\partial\Omega_t$. The limit of (5.130a,b) as ε (proportional to the square of the Mach number) tends to zero is a singular limit, and Ali (2003, Sect. 3), following Meister (1999), performed first a single-scale asymptotic analysis (where, in the asymptotic expansions, the terms proportional to $\varepsilon^n, n = 0,1,2\ldots$, are functions only of t and x) in order to study the basic mechanism of the singular limit in (5.130a,b). In this simple case, two results were derived. (i) At the first two orders in ε, the pressure p is a function of time only: $p = p^{(0)}(t) + \varepsilon p^{(1)}(t) + O(\varepsilon^2)$, where $p^{(0)}(t) = C_0|\Omega_t|^{-\gamma}$ and $p^{(1)}(t) = C_1 p^0(t)$; $|\Omega_t|$ is the measure of the domain Ω_t, and C_0 and C_1 are constants. (ii) The leading-order density and velocity $(\rho^{(0)}, \boldsymbol{u}^{(0)})$ satisfy the system

$$\frac{1}{\rho^{(0)}}\frac{\mathrm{D}\rho^{(0)}}{\mathrm{D}t} = -\frac{1}{|\Omega_t|}\frac{\mathrm{d}|\Omega_t|}{\mathrm{d}t},$$

$$\rho^{(0)}\frac{\mathrm{D}\boldsymbol{u}^{(0)}}{Dt} + \nabla p^{(2)} = 0 \,,$$

$$\nabla \cdot \boldsymbol{u}^{(0)} = \frac{1}{|\Omega_t|}\frac{d|\Omega_t|}{\mathrm{d}t} \,, \tag{5.131a}$$

where $\mathrm{D}/\mathrm{D}t = \partial/\partial t + \boldsymbol{u}^{(0)} \cdot \nabla$. Some obvious restrictions on the initial data in Ω_0 are

$$\boldsymbol{u}_0(\boldsymbol{x}) = \boldsymbol{u}_0^{(0)}(\boldsymbol{x}) + O(\varepsilon), \quad p_0(\boldsymbol{x}) = \text{const} + O(\varepsilon^2), \quad \nabla \cdot \boldsymbol{u}_0^{(0)} = \text{const} \,. \tag{5.131b}$$

According to Ali (2003), 'we expect new effects, which are not described by the single scale theory, to arise when the restriction $\nabla \cdot \boldsymbol{u}_0^{(0)}$ = constant does not hold'. Ali, in the next sections of his paper (Ali 2003, pp. 2027–2037), considers a two-timescale (t and $\tau = t/\varepsilon$) analysis, with *sublinearity* conditions for the terms $U^{(i)}(t,\tau,\boldsymbol{x}) = (\rho^{(i)}, \boldsymbol{u}^{(i)}, p^{(i)})(t,\tau,\boldsymbol{x})$ in asymptotic expansions in powers of ε such that

$$\lim_{\tau\to\infty} \frac{1}{\tau} U^{(i)}(t,\tau,\boldsymbol{x}) = 0, \quad i = 0,1,2,\dots\ , \tag{5.132a}$$

and for the fast-time average of $U^{(i)}(t,\tau,\boldsymbol{x})$,

$$\langle U^{(i)}\rangle(t,\boldsymbol{x}) = \lim_{\tau\to\infty} \frac{1}{\tau} \int_0^t U^{(i)}(t,t',\boldsymbol{x})\mathrm{d}t'\ . \tag{5.132b}$$

Ali writes (if the limit (5.131b) exists and is finite) the usual decomposition,

$$\langle U^{(i)}\rangle + \delta U^{(i)}, \quad \text{with } \langle \delta U^{(i)}\rangle = 0\ . \tag{5.132c}$$

Ali (2003) gives a pertinent discussion concerning the initial and boundary conditions for the derived approximate model equations inherited from the starting initial and boundary data (5.130c,d) for the Euler inviscid, compressible equations (5.130a,b), and observes that in general, it is not possible to uniquely assign initial data for the derived acoustic equations with respect to the fast time τ, but for a simple case of motion this problem can be solved. The paper of Ali (2003) is supplemented by an appendix on weakly non-isentropic flow, that is, flows with constant entropy at order zero, which are produced by constant initial-density distributions. In the case of asymptotics with single time and length scales, first considered in Ali (2003), some obvious restrictions on the initial data in Ω_0 (see (5.130c)) are $\boldsymbol{u}_0(\boldsymbol{x}) = \boldsymbol{u}_0^{(0)}(\boldsymbol{x}) + O(\varepsilon)$ and $p_0(\boldsymbol{x}) = \text{const} + O(\varepsilon^2)$, with the constraint $\nabla\cdot\boldsymbol{u}_0^{(0)} = \text{const}$, when the expansions for $(\boldsymbol{u}(t,\boldsymbol{x}), p(t,\boldsymbol{x}))$ are $\boldsymbol{u}^{(0)}(t,\boldsymbol{x}) + \varepsilon\boldsymbol{u}^{(1)}(t,\boldsymbol{x}) + O(\varepsilon^2)$ and $p(t,\boldsymbol{x}) = p^{(0)}(t) + \varepsilon p^{(1)}(t) + O(\varepsilon^2)$. Therefore, we expect new effects, which are not described by the single-scale theory, to arise when the restriction $\nabla\cdot\boldsymbol{u}_0^{(0)}$ $=$ const does not hold. Namely, in the two-timescale analysis, the average-free components $(\delta\rho^{(1)},\ \delta\boldsymbol{u}^{(1)}, \delta p^{(2)})$ are *coupled* to $(\langle\rho^{(1)}\rangle, \langle\boldsymbol{u}^{(1)}\rangle)$ through the averaged first-order density $\langle\rho^{(1)}\rangle$ appearing on the right-hand side of the average equation for $\langle\boldsymbol{u}^{(1)}\rangle$. A significant case (in fact, considered in Sects. 5.3.1 and 5.3.2 above) that leads to *decoupling* is induced by a constant initial-density distribution; when the leading-order density $\rho^{(0)}$ is inversely proportional to the volume $|\Omega_t|$, the averaged first-order density $\langle\rho^{(1)}\rangle$ vanishes. One the other hand, in general we must expect some sort of instantaneous jump from the initial data (5.130c) to possibly different initial data, adjusted to the asymptotic equations resulting from a two-timescale analysis, because the low-Mach-number limit for generic data is not uniform in a time interval containing zero, unless the initial data are compatible with the zero-Mach-number equations. In particular, if the initial pressure is not constant in Ω_0, then the convergence as ε tends to zero cannot be uniform in the neighbourhood of

$t = 0$. Ali (2003) showed how to assign initial and boundary conditions for all the averaged quantities appearing in the two-timescale expansion, written with an error of $O(\varepsilon^2)$.

Concerning the appropriate initial conditions for the corresponding average-free functions, Ali (2003, pp. 2035–2036) observed that the slow time t and the fast time $\tau = t/M$ are linearly related ($t = \varepsilon\tau$) by the small parameter ε. Since the parameter ε is arbitrary, it might be meaningful to regard the fast time τ as an independent variable and to assign, for $\tau = 0$, 'initial' data that depend on t. As a most unfortunate consequence of this approach, arbitrary 'initial' data would be compatible with the initial data (5.130c). Ali (2003, p. 2029), in the case of a two-timescale analysis, derived, for the average-free parts of the first-order pressure and of the leading-order velocity $\delta p^{(1)}$ and $\delta \boldsymbol{u}^{(0)}$, a linear acoustic system which can be replaced by the following wave equation (when $r = 1$, the flow is weakly non-isentropic):

$$\frac{\partial^2 \delta p^{(1)}}{\partial \tau^2} - \nabla \cdot (r \, \nabla \delta p^{(1)}) = 0 \,, \tag{5.133a}$$

where $r(t, \boldsymbol{x}) = \gamma p^{(0)}(t)/\rho^{(0)}(t, \boldsymbol{x})$. As the boundary condition, we write

$$\frac{\partial \delta p^{(1)}}{\partial n_t} = 0 \quad \text{on } \partial \Omega_t \,, \tag{5.133b}$$

and as the initial data,

$$\delta p^{(1)}(t, 0, \boldsymbol{x}) = \pi_\tau(\boldsymbol{x}) \quad \text{and} \quad \left(\frac{\partial \delta p^{(1)}}{\partial \tau}\right)_{\tau=0} = -\gamma p^{(0)}(t) \, \Delta\omega_\tau(\boldsymbol{x}) \quad \text{in } \Omega_t \,. \tag{5.133c}$$

This shows that the fast acoustics can be completely resolved for at least one class of motions (weakly isentropic, so that $\rho^{(0)}(t, \boldsymbol{x}) = \rho^{(0)}(t)$) of the boundary $\partial\Omega_t$. Concerning $\pi_\tau(\boldsymbol{x})$ and $\Delta\omega_\tau(\boldsymbol{x})$, these two functions of $\boldsymbol{x}$ necessarily satisfy the condition (Ali 2003, p. 2036)

$$\int \pi_\tau(\boldsymbol{x}) \, \mathrm{d}v = 0, \quad \int \Delta\omega_\tau(\boldsymbol{x}) \, \mathrm{d}v = 0, \quad \text{in } \Omega_t.$$

Finally, following the conclusions of Ali (2003, pp. 2037–2038), we mention that the analysis performed by Ali is not conclusive, since the theory presented is not capable of providing a full resolution of high-frequency acoustics. Nevertheless, the representation derived in Sect. 6 of his paper gives a hint of a partial theoretical understanding of the acoustic modes generated by the motion of a boundary. Obviously, the main key point is that one fast time variable is not sufficient to describe the sequence of modes produced by a generic motion of the boundary. Thus we need to extend the above theory of Ali to include a family of fast time variables nonlinearly related to the slow time and (eventually?) to the space variables. Whether the number of independent fast variables for each term of the asymptotic expansion should increase with the order of the term is an open question. This extension, mentioned by Ali (2003), which has theoretical interest in itself and is a necessary step for the development of efficient numerical schemes for low-Mach-number flows in time-dependent bounded domains (as is the case in combustion problems), was formally realized in Sects. 5.3.1 and 5.3.2 of this book, and was discovered (Guiraud and Zeytounian 1980) 25 years ago!

5.4 Low-Mach-Number Flows in a Time-Dependent Cavity: the Dissipative Case

In the dissipative case, in place of the Euler equations, we consider the NSF equations (3.48) in the cavity Ω for the velocity vector $\boldsymbol{u}$ and the thermodynamic functions p, ρ and T:

$$\frac{\mathrm{D}\rho}{\mathrm{D}t} + \rho(\nabla \cdot \boldsymbol{u}) = 0 ,$$

$$\rho\frac{D\boldsymbol{u}}{\mathrm{D}t} + \frac{1}{\gamma M^2}\nabla p = \frac{1}{Re}\left[\Delta\boldsymbol{u} + \frac{1}{3}\nabla(\nabla \cdot \boldsymbol{u})\right] ,$$

$$\rho\frac{\mathrm{D}T}{\mathrm{D}t} + (\gamma - 1)p\,\nabla \cdot \boldsymbol{u} = \frac{\gamma}{Pr\,Re}\Delta T + \gamma(\gamma-1)\frac{M^2}{Re}\Phi(\boldsymbol{u}) ,$$

$$p = \rho T . \tag{5.134a}$$

The initial conditions (as in the inviscid case) correspond to the state of rest,

$$t = 0 : \quad \boldsymbol{u} = 0 \quad \text{and}\ p = \rho = T = 1 . \tag{5.134b}$$

It is necessary to specify two conditions on the time-deformable wall of the cavity $\partial\Omega(t) = \Sigma(t)$, namely, first, the no-slip condition for the velocity $\boldsymbol{u}$,

$$\boldsymbol{u} = \boldsymbol{u}_\Sigma(t, \boldsymbol{P}) \quad \text{on}\ \Sigma(t) , \tag{5.134c}$$

where we assume that the boundary $\Sigma(t)$ is deformed over time via a given 'boundary velocity' $\boldsymbol{u}_\Sigma(t,\boldsymbol{P}) = \boldsymbol{v}_\Sigma(t,\boldsymbol{P}) + w_\Sigma(t,\boldsymbol{P})\boldsymbol{n}$, and, second, the thermal boundary condition

$$T = T_\mathrm{w}(t) + \Lambda^* M^\alpha \Theta(t, P), \quad \alpha > 0, \quad \Lambda^* = O(1), \quad \boldsymbol{P} \in \Sigma(t) , \tag{5.134d}$$

where Λ^* is a measure of the given thermal field $\Theta(t,\boldsymbol{P})$ on $\Sigma(t)$, and $\alpha > 0$ is a positive scalar. We observe that the thermal condition (5.134d) is compatible with the initial condition (5.134b) for the temperature T ($T_\mathrm{w}(0) \equiv 1$, when we work with dimensionless quantities).

5.4.1 The Case $Re = O(1)$ and $Pr = O(1)$ Fixed and $M \downarrow 0$. The Navier–Fourier Average Model Problem

In this case, when we use a multiple-timescale technique, we obtain from (5.134a), in place of the system of equations (5.90),

$$D\rho^* + M\left[\frac{\partial \rho^*}{\partial t} + \nabla \cdot (\rho^* \boldsymbol{u}^*)\right] = 0\,, \tag{5.135a}$$

$$\begin{aligned}\frac{1}{\gamma}\nabla p^* + M\rho^* D\boldsymbol{u}^* + M^2\Big[\rho^*\left(\frac{\partial \boldsymbol{u}^*}{\partial t} + (\boldsymbol{u}^* \cdot \nabla)\boldsymbol{u}^*\right)\\ - \frac{1}{Re}(\Delta \boldsymbol{u}^* + \frac{1}{3}\nabla(\nabla \cdot \boldsymbol{u}^*)\Big] = 0\,,\end{aligned} \tag{5.135b}$$

$$\begin{aligned}\rho^* DT^* + M\Big[\rho^*\left(\frac{\partial T^*}{\partial t} + \boldsymbol{u}^* \cdot \nabla T^*\right) + (\gamma - 1)p\nabla \cdot \boldsymbol{u}^*\\ - \frac{\gamma}{Pr\; Re}\Delta T^* + \gamma(\gamma-1)\frac{M^2}{Re}\Phi(\boldsymbol{u}^*)\Big]\,,\end{aligned} \tag{5.135c}$$

$$p^* = \rho^* \mathrm{T}^*\,. \tag{5.135d}$$

We then write

$$U = (\boldsymbol{u}, p, \rho, T) \to U^* = U_0^* + MU_1^* + M^2 U_2^* + \dots\,, \tag{5.136a}$$

where

$$U^* = \langle U^* \rangle + U^{*\prime}\,. \tag{5.136b}$$

A Simple Leading-Order Solution for the Thermodynamic Functions

We consider the expansion (5.136a) and derive from (5.135a–c), at the leading order,

$$D\rho_0^* = 0, \quad \nabla p_0^* = 0, \quad DT_0^* = 0\,. \tag{5.137a}$$

From (5.135d) we have: $p_0^* = \rho_0^* T_0^*$, and we obtain also

$$Dp_0^* = 0\,. \tag{5.137b}$$

As a consequence,

$$\rho_0^* = \rho_0^*(t, \boldsymbol{x}), \quad T_0^* = T_0^*(t, \boldsymbol{x}), \quad p_0^* = p_0^*(t)\,. \tag{5.137c}$$

Then, from (5.135a–c), for the order M, we obtain the following equations:

$$D\rho_1^* + \frac{\partial \rho_0^*}{\partial t} + \nabla \cdot (\rho_0^* \boldsymbol{u}_0^*) = 0\,,$$

$$\frac{1}{\gamma}\nabla p_1^* + \rho_0^* D\boldsymbol{u}_0^* = 0\,,$$

$$\rho_0^* DT_1^* + \rho_0^*\frac{\partial T_0^*}{\partial t} + \boldsymbol{u}_0^* \,.\, \nabla T_0^* + (\gamma-1)p_0^*(t)\,\nabla\,.\,\boldsymbol{u}_0^* = \frac{\gamma}{Pr\;Re}\Delta T_0^*\,. \quad (5.138)$$

Now, if we consider (5.136b), we can derive, thanks to (5.137c) and the second equation of (5.138),

$$D\langle\rho_1^*\rangle = 0, \quad D\langle p_1^*\rangle = 0, \quad D\langle T_1^*\rangle = 0, \quad D\langle \boldsymbol{u}_0^*\rangle = 0\,, \quad (5.139a)$$

and as a consequence, from (5.138), we obtain the following three average equations:

$$\frac{\partial \rho_0^*}{\partial t} + \nabla\,.\,(\rho_0^*\langle\boldsymbol{u}_0^*\rangle) = 0\,,$$

$$\rho_0^*\frac{\partial T_0^*}{\partial t} + \langle\boldsymbol{u}_0^*\rangle \,.\, \nabla T_0^* + (\gamma-1)p_0^*(t)\,\nabla\,.\,\langle\boldsymbol{u}_0^*\rangle = \frac{\gamma}{Pr\;Re}\Delta T_0^*\,,$$

$$\nabla\langle p_1^*\rangle = 0\,. \quad (5.139b)$$

For $T_0^*(t,\boldsymbol{x})$, the solution of the second equation in the above system (5.139b), we have as conditions, according to (5.134b,d),

$$T_0^*(t=0,\boldsymbol{x}) = 1 \quad \text{and } T_0^*(t,\boldsymbol{x}) = T_\mathrm{w}(t) \quad \text{on } \Sigma(t)\,, \quad (5.140)$$

since $\Lambda^* = O(1)$ and $\alpha > 0$ when $M \to 0$. However, if we take into account that $T_0^* = p_0^*(t)/\rho_0^*$, then, in place of the second equation (for T_0^*) in the system (5.139b), thanks to the first equation of (5.139b), we can write the equation for $T_0^*(t,\boldsymbol{x})$ in the following form:

$$\left\{\frac{\partial}{\partial t} + \langle\boldsymbol{u}_0^*\rangle\,.\,\nabla\right\}\,\log\left[\frac{T_0^*}{(p_0^*(t))^{(\gamma-1)/\gamma}}\right] = \frac{1}{Pr\;Re}\Delta\left[\frac{T_0^*}{p_0^*(t)}\right]\,. \quad (5.141a)$$

From (5.141a), if $T_0^*(t,\boldsymbol{x}) \equiv T_\mathrm{w}(t)$ then the right-hand side of the equation, proportional to $(1/Pr\;Re)$, is zero, and

$$\left\{\frac{\partial}{\partial t} + \langle\boldsymbol{u}_0^*\rangle\,.\,\nabla\right\}\log\left[\frac{T_0^*}{(p_0^*(t))^{(\gamma-1)/\gamma}}\right] = 0\,;$$

otherwise,

$$\left\{\frac{\partial}{\partial t} + \langle\boldsymbol{u}_0^*\rangle\,.\,\nabla\right\}\frac{p_0^*(t)}{(\rho_0^*)^\gamma} = 0\,. \quad (5.141b)$$

Finally, since $p_0^*(t)/(\rho_0^*)^\gamma \equiv 1$ at $t=0$, we have also, by continuity (when $M \downarrow 0$, shocks are transferred to a large distance $O(1/M)$ in space and in time),

$$p_0^*(t) = (\rho_0^*(t))^\gamma, \quad \text{with } T_0^*(t) = (p_0^*(t))^{(\gamma-1)/\gamma}\,, \quad (5.142a)$$

$$p_0^*(t) = T_\mathrm{w}(t)\rho_0^*(t)\,. \quad (5.142b)$$

The Solution for the Fluctuations $\rho_1^{*\prime}$ and $u_0^{*\prime}$

From the system of equations (5.138), with the above resuts (5.142) we can derive the following 'acoustic-type' equations for $\rho_1^{*\prime}$, $T_1^{*\prime}$ and $\boldsymbol{u}_0^{*\prime}$:

$$D\left[\frac{\rho_1^{*\prime}}{\rho_0^*(t)}\right] + \nabla \cdot \boldsymbol{u}_0^{*\prime} = 0 ,$$

$$\frac{1}{\gamma}\nabla\left[\frac{p_1^{*\prime}}{\rho_0^*(t)}\right] + D\boldsymbol{u}_0^{*\prime} = 0 , \tag{5.143}$$

$$D\left[\frac{T_1^{*\prime}}{T_{\mathrm{w}}(t)}\right] + (\gamma - 1)\, \nabla \cdot \boldsymbol{u}_0^{*\prime} = 0 .$$

However, from the dimensionless equation of state (5.135d) for our thermally perfect gas, together with (5.136b) and (5.142b), we have the following for the thermodynamic fluctuations:

$$\frac{p_1^{*\prime}}{p_0^*(t)} = \frac{\rho_1^{*\prime}}{\rho_0^*(t)} + \frac{T_1^{*\prime}}{T_{\mathrm{w}}(t)} . \tag{5.144a}$$

From (5.143), we derive also that

$$D\left[\frac{T_1^{*\prime}}{T_{\mathrm{w}}(t)}\right] = (\gamma - 1)D\left[\frac{\rho_1^{*\prime}}{\rho_0^*(t)}\right] , \tag{5.144b}$$

$$D\left[\frac{p_1^{*\prime}}{p_0^*(t)}\right] = \gamma D\left[\frac{\rho_1^{*\prime}}{\rho_0^*(t)}\right] , \tag{5.144c}$$

and, as a consequence,

$$\frac{p_1^{*\prime}}{p_0^*(t)} = \gamma\frac{\rho_1^{*\prime}}{\rho_0^*(t)} , \tag{5.145a}$$

$$\frac{T_1^{*\prime}}{T_{\mathrm{w}}(t)} = (\gamma - 1)\frac{\rho_1^{*\prime}}{\rho_0^*(t)} . \tag{5.145b}$$

Finally, for $\rho_1^{*\prime}/\rho_0^*(t)$ and $\boldsymbol{u}_0^{*\prime}$, we obtain the following linear (inviscid) equations of acoustics, where the usual time derivative is replaced by the operator D:

$$D\left[\frac{\rho_1^{*\prime}}{\rho_0^*(t)}\right] + \nabla \cdot \boldsymbol{u}_0^{*\prime} = 0 , \tag{5.146a}$$

$$T_{\mathrm{w}}(t)\, \nabla\left[\frac{\rho_1^{*\prime}}{\rho_0^*(t)}\right] + D\boldsymbol{u}_0^{*\prime} = 0 . \tag{5.146b}$$

For the system (5.146a,b), we write the following slip condition:

$$\boldsymbol{u}_0^{*\prime} \cdot \boldsymbol{n} = 0 \quad \text{on } \Sigma(t) . \tag{5.146c}$$

Rigorously speaking, the above slip condition (5.146c) is not obvious, and must be obtained from matching with a model dissipative system for the fluctuations valid near the wall $\Sigma(t)$, where a no-slip condition (a consequence of (5.134c)) is assumed. Here, this perplexing question is not considered.

The solution of the system of equations (5.146a,b), according to (5.87) and (5.102), is

$$\boldsymbol{u}_0^{*\prime} = \sum_{n\geq 1} [A_n(t)\; Cn - B_n(t)\; Sn]\; \boldsymbol{U}_n \;, \tag{5.147a}$$

$$\rho_1^{*\prime} = \rho_0^*(t) \left[\frac{\rho_0^*(t)}{p_0^*(t)}\right]^{1/2} \sum_{n\geq 1} [A_n(t)\; Sn + B_n(t)\; Cn] R_n \;, \tag{5.147b}$$

and for $p_1^{*\prime}$ and $T_1^{*\prime}$ we have the two relations (5.145a,b).

The Analysis of the Equations for $\boldsymbol{u}_1^*$, p_2^*, ρ_2^* and T_2^*

Now, from (5.135a–d), at the order M^2, we derive the following second-order approximate equations:

$$D\left[\frac{\rho_2^*}{\rho_0^*(t)}\right] + \nabla \cdot \boldsymbol{u}_1^* + G = 0 \;, \tag{5.148a}$$

$$\nabla\left[\frac{p_2^*}{\gamma \rho_0^*(t)}\right] + D\boldsymbol{u}_1^* + \boldsymbol{F} = 0 \;, \tag{5.148b}$$

$$DT_2^* + (\gamma - 1)\frac{p_0^*}{\rho_0^*(t)} \nabla \cdot \boldsymbol{u}_1^* + H = 0 \;, \tag{5.148c}$$

$$p_2^* = T_2^* + T_{\mathrm{w}}(t)\frac{\rho_2^*}{\rho_0^*(t)} + \frac{\rho_1^*}{\rho_0^*(t)} T_1^* \;, \tag{5.148d}$$

where

$$G = \frac{\partial[\rho_1^*/\rho_0^*(t)]}{\partial t} + \nabla \cdot \left\{\frac{\rho_1^*}{\rho_0^*(t)}\boldsymbol{u}_0^*\right\} \;, \tag{5.149a}$$

$$\begin{aligned}\boldsymbol{F} = {} & \frac{\partial \boldsymbol{u}_0^*}{\partial t} + (\boldsymbol{u}_0^* \cdot \nabla)\boldsymbol{u}_0^* + \frac{\rho_1^*}{\rho_0^*(t)} D\boldsymbol{u}_0^* \\ & - \frac{1}{Re}\frac{1}{\rho_0^*(t)}[\Delta \boldsymbol{u}_0^* + \frac{1}{3}\nabla(\nabla \cdot \boldsymbol{u}_0^*)] \;,\end{aligned} \tag{5.149b}$$

$$\begin{aligned}H = {} & \frac{\partial T_1^*}{\partial t} + \boldsymbol{u}_0^* \cdot \nabla T_1^* + \frac{\rho_1^*}{\rho_0^*(t)} \frac{\mathrm{d}T_{\mathrm{w}}(t)}{\mathrm{d}t} \\ & + (\gamma - 1)\frac{p_1^*}{\rho_0^*(t)} \nabla \cdot \boldsymbol{u}_0^* + \frac{\rho_1^*}{\rho_0^*(t)} DT_1^* \\ & - \frac{\gamma}{Pr\; Re}\frac{1}{\rho_0^*(t)} \Delta T_1^* \,.\end{aligned} \tag{5.149c}$$

Obviously, if we write:

$$G = \langle G \rangle + G', \quad \boldsymbol{F} = \langle \boldsymbol{F} \rangle + \boldsymbol{F}' \quad \text{and } H = \langle H \rangle + H' ,$$

then we obtain from (5.148a–c) with (5.142b) the following averaged reduced system:

$$\begin{aligned} \nabla \left[\frac{\langle p_2^* \rangle}{\gamma \rho_0^*(t)} \right] + \langle \boldsymbol{F} \rangle = 0 , \\ \nabla . \langle \boldsymbol{u}_1^* \rangle + \langle G \rangle = 0 , \\ \langle H \rangle - (\gamma - 1) T_{\mathrm{w}}(t) \langle G \rangle = 0 . \end{aligned} \tag{5.150}$$

The above average system (5.150) deserves a careful analysis, but here we are interested mainly in obtaining a Navier–Fourier nearly-incompressible-type average model system for $\langle \boldsymbol{u}_0^* \rangle$, $\langle p_2^* \rangle$ and $\langle T_1^* \rangle$.

Navier–Fourier-Type, Nearly Incompressible Average Equations

From the first average equation of the system (5.139b), with (5.142), we obtain a 'continuity equation'

$$\frac{\mathrm{d} \log \rho_0^*}{\mathrm{d}t} + \nabla \,.\, \langle \boldsymbol{u}_0^* \rangle = 0. \tag{5.151}$$

Then, we use the first average equation in (5.150), where $\langle \boldsymbol{F} \rangle$ is given by

$$\begin{aligned} \langle \boldsymbol{F} \rangle = \frac{\partial \langle \boldsymbol{u}_0^* \rangle}{\partial t} + (\langle \boldsymbol{u}_0^* \rangle \,.\, \nabla) \langle \boldsymbol{u}_0^* \rangle + \left\langle \frac{\rho_1^{*\prime}}{\rho_0^*(t)} D \boldsymbol{u}_0^{*\prime} \right\rangle \\ + \langle (\boldsymbol{u}_0^{*\prime} \,.\, \boldsymbol{\nabla}) \boldsymbol{u}_0^{*\prime} \rangle - \frac{1}{Re} \frac{1}{\rho_0^*(t)} \Delta \langle \boldsymbol{u}_0^* \rangle , \end{aligned} \tag{5.152}$$

because $\nabla \,.\, \langle \boldsymbol{u}_0^* \rangle$ is a function of the slow time t only, according to (5.151), and we derive a Navier-type equation for $\langle \boldsymbol{u}_0^* \rangle$,

$$\frac{\partial \langle \boldsymbol{u}_0^* \rangle}{\partial t} + (\langle \boldsymbol{u}_0^* \rangle . \nabla) \langle \boldsymbol{u}_0^* \rangle + \nabla \Pi = \frac{1}{Re} \frac{1}{\rho_0^*(t)} \Delta \langle \boldsymbol{u}_0^* \rangle , \tag{5.153}$$

Here, we have taken into account (5.119), which gives the terms $\langle (\boldsymbol{u}_0^{*\prime} \,.\, \nabla) \boldsymbol{u}_0^{*\prime} \rangle$ and $\langle [\rho_1^{*\prime} / \rho_0^*(t)] D \boldsymbol{u}_0^{*\prime} \rangle$ as a function of the acoustic amplitudes (A_n, B_n). The function Π in (5.153) is the same that in (5.120) and is given by (5.121). Finally, a Fourier-type equation for $\langle \mathrm{T}_1^* \rangle$ is derived directly from the last equation of the averaged system (5.150) when we take into account that:

$$\begin{aligned} \langle G \rangle = \nabla \,.\, \left\langle \frac{\rho_1^{*\prime}}{\rho_0^*(t)} \boldsymbol{u}_0^{*\prime} \right\rangle + \frac{\partial}{\partial t} \left[\frac{\langle \rho_1^* \rangle}{\rho_0^*(t)} \right] \\ + \nabla \left\{ \left[\frac{\langle \rho_1^* \rangle}{\rho_0^*(t)} \right] \langle \boldsymbol{u}_0^* \rangle \right\} , \end{aligned} \tag{5.154a}$$

$$\langle H\rangle = \frac{\partial\langle T_1^*\rangle}{\partial t} + \langle \boldsymbol{u}_0^*\rangle \,.\, \nabla\langle T_1^*\rangle + \frac{\mathrm{d}T_\mathrm{w}(t)}{\mathrm{d}t}\frac{\langle \rho_1^*\rangle}{\rho_0^*(t)} + \langle \boldsymbol{u}_0^{*\prime} . \boldsymbol{\nabla}\, T_1^{*\prime}\rangle$$
$$- \frac{\gamma}{Pr\ Re}\frac{1}{\rho_0^*(t)}\Delta\langle T_1^*\rangle + (\gamma-1)\left\langle \frac{p_1^{*\prime}}{\rho_0^*(t)}\nabla \,.\, \boldsymbol{u}_0^{*\prime}\right\rangle$$
$$+ (\gamma-1)\frac{\langle p_1^*\rangle}{\rho_0^*(t)}\boldsymbol{\nabla} \,.\, \langle \boldsymbol{u}_0^*\rangle + \left\langle \frac{\rho_1^{*\prime}}{\rho_0^*(t)} DT_1^{*\prime}\right\rangle \,. \tag{5.154b}$$

Namely, as a consequence of the relations:

$$\langle \boldsymbol{u}_0^{*\prime} \,.\, \nabla T_1^{*\prime}\rangle = 0, \quad \left\langle \frac{p_1^{*\prime}}{\rho_0^*(t)}\nabla \,.\, \boldsymbol{u}_0^{*\prime}\right\rangle = 0\,, \tag{5.155a}$$

$$\left\langle \frac{\rho_1^{*\prime}}{\rho_0^*(t)} DT_1^{*\prime}\right\rangle = 0, \quad \left\langle \frac{\rho_1^{*\prime}}{\rho_0^*(t)} \boldsymbol{u}_0^{*\prime}\right\rangle = 0\,, \tag{5.155b}$$

the following average equation for $\langle T_1^*\rangle$ can be derived:

$$\left[\frac{\partial}{\partial t} + \langle \boldsymbol{u}_0^*\rangle \,.\, \nabla\right]\left\{\gamma\langle T_1^*\rangle - (\gamma-1)\frac{\langle p_1^*\rangle}{\rho_0^*(t)}\right\} = \frac{\gamma}{Pr\ Re}\frac{1}{\rho_0^*(t)}\Delta\langle T_1^*\rangle$$
$$+ (\gamma-1)\frac{\mathrm{d}\rho_0^*(t)}{\mathrm{d}t}\frac{\langle T_1^*\rangle}{\rho_0^*(t)} - \gamma\frac{\mathrm{d}T_\mathrm{w}(t)}{\mathrm{d}t}\frac{\langle \rho_1^*\rangle}{\rho_0^*(t)} \,. \tag{5.156}$$

We observe also that

$$\nabla\langle p_1^*\rangle = 0, \quad \int_{\Omega(t)}\langle \rho_1^*\rangle \mathrm{d}v = 0\,, \tag{5.157a}$$

$$\frac{\langle \rho_1^*\rangle}{\rho_0^*(t)} = \frac{\langle p_1^*\rangle}{p_0^*(t)} - \frac{\langle T_1^*\rangle}{T_\mathrm{w}(t)} \,. \tag{5.157b}$$

For the average system of three equations (5.151), (5.153) and (5.156), together with (5.157), we write the boundary conditions on the wall (see (5.134c,d)) as

$$\langle \boldsymbol{u}_0^*\rangle = \boldsymbol{u}_\Sigma \quad \text{and } \langle T_1^*\rangle = \Lambda^*\Theta \quad \text{on } \Sigma(t)\,, \tag{5.158}$$

when we choose the value $\alpha = 1$ in (5.134d). The initial conditions are simply

$$t = 0: \quad \langle \boldsymbol{u}_0^*\rangle = 0 \quad \text{and } \langle T_1^*\rangle = 0\,. \tag{5.159}$$

The model problem derived here, of the equations (5.151), (5.153) and (5.156) for $\langle \boldsymbol{u}_0^*\rangle, \langle T_1^*\rangle$ and Π, with the two conditions (5.157) for $\langle p_1^*\rangle$ and $\langle \rho_1^*\rangle$ and subject to the conditions (5.158) on the wall of the cavity $\Sigma(t)$, and (5.159) at $t = 0$, deserves further investigation for application to combustion phenomena.

5.4.2 The Case $Re \gg 1$ and $M \ll 1$. Viscous Damping of Acoustic Oscillations

The study of the viscous damping of fast acoustic oscillations is a difficult problem. Obviously, the inviscid theory developed in Sect. 5.3 does not give us the possibility to investigate this damping process, and when $Re = O(1)$ is fixed, as in Sect. 5.4.1 above, damping also does not occur. On the other hand, if we are dealing with a slightly viscous flow (large Reynolds number, $Re \gg 1$), we must start, from the NSF equations (5.134a) in place of the Euler equations (5.80) analysed in Sect. 5.3, and bring into the analysis a second small parameter $\varepsilon^2 = 1/Re \ll 1$, the inverse of the Reynolds number. We must then expect that the fast acoustic oscillations will be damped out. Unfortunately, a precise analysis of this damping phenomenon, for instance when a general similarity rule

$$\varepsilon^2 = M^\beta, \quad \beta > 0\,, \tag{5.160}$$

is assumed, appears not to be an easy task and raises many questions.

The Acoustic-Type Inhomogeneous Equations

Below, we introduce a new very-slow-time derivative $\delta\mathcal{D}(\delta \ll 1)$, where $\mathcal{D}$ is a second differential operator operating through a family of very slow times, and modify (5.84) accordingly:

$$\frac{\partial U}{\partial t} = \frac{1}{M} DU^* + \frac{\partial U^*}{\partial t} + \delta\mathcal{D}U^*\,, \tag{5.161}$$

where we use U^* to represent the solution U expressed through the variety of timescales (fast, slow and very slow). With (5.161) in place of (5.135), we can derive a new starting system of equations for the functions $\boldsymbol{u}^*$ and (p^*, ρ^*, T^*), dependent on the family of *fast* times (via the operator $M^{-1}DU^*$), on the *slow* time t and on the family of *very slow* times (via the operator $\delta\mathcal{D}U^*$). Namely, we have the following NSF system:

$$D\rho^* + M\left[\frac{\partial \rho^*}{\partial t} + \boldsymbol{u}^* . \nabla\rho^* + \rho^*\nabla . \boldsymbol{u}^*\right] + \delta M\mathcal{D}\rho^* = 0\,, \tag{5.162a}$$

$$\frac{1}{\gamma}\nabla p^* + M\rho^* D\boldsymbol{u}^* + M^2\left[\rho^*\left(\frac{\partial \boldsymbol{u}^*}{\partial t} + (\boldsymbol{u}^* . \nabla)\boldsymbol{u}^*\right) - \varepsilon^2(\Delta\boldsymbol{u}^* + \frac{1}{3}\nabla(\nabla . \boldsymbol{u}^*)\right]$$
$$+ \delta M^2\rho^*\mathcal{D}\boldsymbol{u}^* = 0\,, \tag{5.162b}$$

$$\rho^* DT^* + M\left[\rho^*\left(\frac{\partial T^*}{\partial t} + \boldsymbol{u}^* . \nabla T^*\right) + (\gamma - 1)p\nabla . \boldsymbol{u}^* - \varepsilon^2\frac{\gamma}{Pr})\Delta T^*\right]$$
$$+ \delta M\rho^*\mathcal{D}T^* - \gamma(\gamma - 1)\varepsilon^2 M^3\Phi(\boldsymbol{u}^*) = 0\,, \tag{5.162c}$$

$$p^* = \rho^* T^* \,. \tag{5.162d}$$

Our purpose is the derivation of an acoustic-type inhomogeneous system (with the very-slow-time derivative operator $\mathcal{D}$) of equations for the terms $\boldsymbol{u}^*_{\mathrm{A}}$ and $(p^*_{\mathrm{A}}, \rho^*_{\mathrm{A}}, T^*_{\mathrm{A}})$ in the expansions for $\boldsymbol{u}^*$ and (p^*, ρ^*, T^*)

$$\boldsymbol{u}^* = \boldsymbol{u}^*_0 + M\boldsymbol{u}^*_1 + \ldots + \delta M \boldsymbol{u}^*_{\mathrm{A}} + \ldots \,, \tag{5.163a}$$

$$(p^*, \rho^*, T^*) = (p^*_0, \rho^*_0, T^*_0) + M(p^*_1, \rho^*_1, T^*_1) + M^2(p^*_2, \rho^*_2, T^*_2) + \ldots + \delta M^2(p^*_{\mathrm{A}}, \rho^*_{\mathrm{A}}, T^*_{\mathrm{A}}) + \ldots \,, \tag{5.163b}$$

where we assume that

$$\text{order } \delta > M \,. \tag{5.163c}$$

From the above NSF system (5.162), together with (5.163a,b), we obtain the required equations, at the order δM^2, for $\boldsymbol{u}^*_{\mathrm{A}}$ and $(p^*_{\mathrm{A}}, \rho^*_{\mathrm{A}}, T^*_{\mathrm{A}})$:

$$D\rho^*_{\mathrm{A}} + \rho^*_0 \nabla \cdot \boldsymbol{u}^*_{\mathrm{A}} + \mathcal{D}\rho^*_1 = 0 \,,$$

$$\frac{1}{\gamma}\nabla p^*_{\mathrm{A}} + \rho^*_0 D\boldsymbol{u}^*_{\mathrm{A}} + \rho^*_0 \mathcal{D}\boldsymbol{u}^*_0 = 0 \,,$$

$$\rho^*_0 DT^*_{\mathrm{A}} + (\gamma - 1)p_0 \nabla \boldsymbol{u}^*_{\mathrm{A}} + \rho^*_0 \mathcal{D}T^*_1 = 0 \,, \tag{5.164a}$$

where

$$p^*_{\mathrm{A}} = \rho^*_0 \rho^*_{\mathrm{A}} + T^*_0 T^*_{\mathrm{A}} \,. \tag{5.164b}$$

In the equations (5.164), the secular terms related to $\mathcal{D}\boldsymbol{u}^*_0$, $\mathcal{D}\rho^*_1$ and $\mathcal{D}T^*_1$ cannot be eliminated unless the (slip) boundary condition on the wall for the component of the velocity $\delta M w^*_{\mathrm{A}} \equiv \delta M(\boldsymbol{u}^*_{\mathrm{A}} \cdot \boldsymbol{n})$ in (5.163a) in the direction of the normal $\boldsymbol{n}$ towards the inside of the cavity Ω (with a deformable wall $\partial\Omega = \Sigma(t)$ depending on the slow time t) is *inhomogeneous*. Such an inhomogeneity may be provided by a boundary-layer analysis, the inhomogeneity appears to be a Stokes layer of thickness $\chi^2 = \varepsilon^2 M$ (according to (5.162b,c)). As a consequence, evaluating the flux outward from this Stokes layer, we find (see (5.175) below) that the value of $\delta M w^*_{\mathrm{A}}$ on Σ is given by $\Delta w^{*\infty}_{\mathrm{S0}}$. Finally, from the matching between the acoustic and Stokes-layer components of the normal velocity components, we find

$$\delta M = \chi \quad \Rightarrow \delta = \frac{\chi}{M} = (Re\ M)^{-1/2} \,, \tag{5.165}$$

and in this case, for the system of acoustic-type inhomogeneous equations (5.164a), we write the following inhomogeneous condition:

$$w^*_{\mathrm{A}} = \boldsymbol{u}^*_{\mathrm{A}} \cdot \boldsymbol{n} = {w^*}^{\infty}_{\mathrm{S0}} \quad \text{on } \Sigma \,. \tag{5.166}$$

The Associated Stokes-Layer Equations

For the derivation of the Stokes-layer equations and the determination of $w^{*\infty}_{\mathrm{S}0}$, we write, for a space point $\boldsymbol{M}_{\mathrm{S}}$ inside the cavity near the wall $\Sigma(t)$,

$$\boldsymbol{M}_{\mathrm{S}} = \boldsymbol{P} + z\boldsymbol{n}, \quad \text{where } \boldsymbol{P}(s_1, s_2) \in \Sigma(t) \quad \text{and } z = \chi\zeta\,, \tag{5.167a}$$

and where (s_1, s_2, ζ) are a local curvilinear system of orthogonal coordinates for the Stokes layer. In such a case, we obtain the following relations:

$$\frac{\partial}{\partial t} + \boldsymbol{u}^* \,.\, \nabla = \frac{\partial}{\partial t} + \boldsymbol{v}^*_{\mathrm{S}} \,.\, \boldsymbol{D} + w^*_{\mathrm{S}}\frac{\partial}{\partial \zeta} + O(\chi)\,,$$

$$\nabla \,.\, \boldsymbol{u}^* = \boldsymbol{D} \,.\, \boldsymbol{v}^*_{\mathrm{S}} + \frac{\partial w^*_{\mathrm{S}}}{\partial \zeta} + O(\chi), \quad \Delta = \frac{1}{\chi^2}\frac{\partial^2}{\partial \zeta^2} + O\left(\frac{1}{\chi}\right). \tag{5.167b}$$

With (5.167), the following (dominant dissipative) equations for the Stokes-layer functions $\boldsymbol{v}^*_S, w^*_S, \rho^*_S, p^*_S$ and T^*_S, can be derived from the NSF system (5.162):

$$D\rho^*_{\mathrm{S}} + M\left[\frac{\partial \rho^*_{\mathrm{S}}}{\partial t} + v^*_{\mathrm{S}} \,.\, \boldsymbol{D}\rho^*_{\mathrm{S}} + w^*_{\mathrm{S}}\frac{\partial \rho^*_{\mathrm{S}}}{\partial \zeta} + \rho^*_{\mathrm{S}}\boldsymbol{D} \,.\, \boldsymbol{v}^*_{\mathrm{S}} + \rho^*_{\mathrm{S}}\frac{\partial w^*_{\mathrm{S}}}{\partial \zeta}\right]$$
$$+ \delta M\mathcal{D}\rho^*_{\mathrm{S}} = O(\chi)\,, \tag{5.168a}$$

$$\frac{1}{\gamma}\boldsymbol{D}p^*_{\mathrm{S}} + M\rho^*_{\mathrm{S}}Dv^*_{\mathrm{S}} - M^2\frac{\varepsilon^2}{\chi^2}\frac{\partial^2 v^*_{\mathrm{S}}}{\partial \zeta^2} = O(M^2)$$
$$+ O(\delta M^2) + O(\chi M^2) + O\left(\frac{\varepsilon^2 M^2}{\chi}\right), \tag{5.168b}$$

$$\frac{\partial p^*_{\mathrm{S}}}{\partial \zeta} = O(\chi^2) + O(\varepsilon^2)\,, \tag{5.168c}$$

$$\rho^*_{\mathrm{S}}DT^*_{\mathrm{S}} + M\Bigg\{\rho^*_{\mathrm{S}}\left(\frac{\partial T^*_{\mathrm{S}}}{\partial t} + v^*_{\mathrm{S}} \,.\, \boldsymbol{D}\mathbf{T}^*_{\mathrm{S}} + w^*_{\mathrm{S}}\frac{\partial T^*_{\mathrm{S}}}{\partial \zeta}\right)$$
$$+ (\gamma - 1)p^*_{\mathrm{S}}\left(\boldsymbol{D} \,.\, \boldsymbol{v}^*_{\mathrm{S}} + \frac{\partial w^*_{\mathrm{S}}}{\partial \zeta}\right) - \frac{\varepsilon^2}{\chi^2}\frac{\gamma}{Pr}\frac{\partial^2 T^*_{\mathrm{S}}}{\partial \zeta^2}\Bigg\}$$
$$+ O(\delta M) + O\left(\frac{M\varepsilon^2}{\chi}\right) + O\left(\frac{\varepsilon^2 M^3}{\chi^2}\right), \tag{5.168d}$$

$$p^*_{\mathrm{S}} = \rho^*_{\mathrm{S}}T^*_{\mathrm{S}}\,. \tag{5.168e}$$

With the expansions

$$\boldsymbol{v}^*_{\mathrm{S}} = \boldsymbol{v}^*_{\mathrm{S}0} + \ldots, \quad w^*_{\mathrm{S}} = w^*_{\mathrm{S}0} + \ldots\,, \tag{5.169a}$$

$$(\rho^*_{\mathrm{S}}, p^*_{\mathrm{S}}, T^*_{\mathrm{S}}) = (\rho^*_{\mathrm{S}0}, p^*_{\mathrm{S}0}, T^*_{\mathrm{S}0}) + M(\rho^*_{\mathrm{S}1}, p^*_{\mathrm{S}1}, T^*_{\mathrm{S}1}) + \ldots\,, \tag{5.169b}$$

where $p^*_{\mathrm{s}0} = p^*_{0|\Sigma}$, we derive first a problem for $T^*_{\mathrm{S}0}$,

$$\rho^*_{S0} DT^*_{S0} - \frac{\gamma}{Pr}\frac{\partial_2 T^*_{S0}}{\partial\zeta^2} = 0 ,$$

$$\zeta = 0 : T^*_{S0} = T_w(t) , \tag{5.170}$$

$$\zeta \to \infty : T^*_{S0} \to T^*_{0|\Sigma} ,$$

where we consider the thermal condition (5.134d) with $\alpha = 1$. In this case, we find that, to leading order in the Stokes layer, $(\rho^*_{S0}, p^*_{S0}, T^*_{S0})$ have the same values as in the interior (corresponding to $\zeta \to \infty$) and, in particular, $T^*_{S0} \equiv T^*_{0|\Sigma}$. The Stokes layer structure is felt to $O(M)$ for $(\rho^*_{S1}, p^*_{S1}, T^*_{S1})$, while it is felt to leading order for the velocity vector $(\boldsymbol{v}^*_{S0}, w^*_{S0})$. Since, at order M, according to (5.168b), $(1/\gamma)\boldsymbol{D}p^*_{S1} + \rho^*_0 D\boldsymbol{v}^{*\infty}_{S0} = 0$ when $\zeta \to \infty$, then, using $\boldsymbol{v}^*_{S0} = \boldsymbol{v}^{*\infty}_{S0} + \boldsymbol{v}^{*\prime}_{S0}$, we derive the following problem for $\boldsymbol{v}^{*\prime}_{S0}$:

$$\rho^*_0 \boldsymbol{D}\boldsymbol{v}^{*\prime}_{S0} - \frac{\partial^2 \boldsymbol{v}^{*\prime}_{S0}}{\partial\zeta^2} = 0 , \tag{5.171a}$$

$$\zeta = 0 : \quad \boldsymbol{v}^{*\prime}_{S0} = -\boldsymbol{v}^{*\infty}_{S0}, \quad \text{and } \zeta \to \infty : \quad \boldsymbol{v}^{*\prime}_{S0} \to 0 , \tag{5.171b}$$

where (see, (5.134c) and (5.147a)) $\boldsymbol{v}^{*\infty}_{S0} = \boldsymbol{u}_{|\Sigma} + \langle \boldsymbol{u}^*_0 \rangle_\Sigma + \sum_{n\geq 1}[A_n\, Cn - B_n\, Sn]\boldsymbol{U}_{n|\Sigma}$.

On the other hand, again from (5.168) at order M, we have:

$$D\rho^*_{S1} + \frac{\partial \rho^*_0}{\partial t} + \rho^*_0 \left[\boldsymbol{D} \boldsymbol{.} \boldsymbol{v}^*_{S0} + \frac{\partial w^*_{S0}}{\partial\zeta}\right] = 0, \quad \frac{\partial p^*_{S1}}{\partial\zeta} = 0 ,$$

$$\rho^*_0 DT^*_{S1} + \rho^*_0 \frac{\partial T^*_0}{\partial t} + \left(\gamma - 1)p^*_0 (D \boldsymbol{.} \boldsymbol{v}^*_{S0} + \frac{\partial w^*_{S0}}{\partial\zeta}\right)$$
$$- \frac{\gamma}{Pr}\frac{\partial^2 T^*_{S1}}{\partial\zeta^2} = 0 , \tag{5.172}$$

when we take into account the relation $\chi^2 = \varepsilon^2 M$. From the last equation of (5.172), we derive the following relation when $\zeta \to \infty$:

$$\rho^*_0 DT^{*\infty}_{S1} + \rho^*_0 \frac{\partial T^*_0}{\partial t} + (\gamma - 1)p^*_0 \left(\boldsymbol{D} \boldsymbol{.} \boldsymbol{v}^{*\infty}_{S0} + \frac{\partial w^*_{S0}}{\partial\zeta}\right) = 0 ,$$

For the perturbation of the temperature $T^{*'}_{S1} = T^*_{S1} - T^{*\infty}_{S1}$, we obtain

$$\rho^*_0 DT^{*'}_{S1} + (\gamma - 1)p^*_0 \left(\boldsymbol{D} \boldsymbol{.} \boldsymbol{v}^*_{S0} + \frac{\partial w^*_{S0}}{\partial\zeta}\right) - \frac{\gamma}{Pr}\frac{\partial^2 T^{*\prime}_{S1}}{\partial\zeta^2} = 0.$$

However, using, with: $\rho^*_{S1} = \rho^*_{S1} - \rho^{*\infty}_{S1}$ and $T^*_0 \rho^{*\prime}_{S1} + \rho^*_0 T^{*\prime}_{S1} = 0$, we can derive also

$$(\gamma - 1)\frac{p^*_0}{\rho^*_0}\left(\boldsymbol{D} \boldsymbol{.} \boldsymbol{v}^{*\infty}_{S0} + \frac{\partial w^*_{S0}}{\partial\zeta}\right) = -(\gamma - 1)\frac{p^*_0}{\rho^*_0} D\rho^{*\prime}_{S1} = (\gamma - 1)\left(\frac{p^*_0}{\rho^*_0}\right) DT^{*\prime}_{S1}$$

and we can write the following problem for $T^{*\prime}_{\mathrm{S}1}$:

$$\rho^*_0 DT^{*\prime}_{\mathrm{S}1} - \frac{1}{Pr}\frac{\partial^2 T^{*\prime}_{\mathrm{S}1}}{\partial\zeta^2} = 0\,, \tag{5.173a}$$

$$\zeta = 0: \quad T^{*'}_{\mathrm{S}1} = \Lambda^*\Theta - T^{*\infty}_{\mathrm{S}1}, \quad \text{and } \zeta \to \infty: \quad T^{*\prime}_{\mathrm{S}1} \to 0\,, \tag{5.173b}$$

where $T^{*\infty}_{\mathrm{S}1} = (\gamma-1)\sum_{n\geq 1}[A_n Cn - B_n Sn]R_{n|\Sigma}$. Finally, from the first equation of (5.172), we derive the following formula:

$$w^{*\infty}_{\mathrm{S}0} = -\boldsymbol{D}\,.\int_0^\infty \boldsymbol{v}^{*'}_{\mathrm{S}0}\,\mathrm{d}\zeta + D\int_0^\infty \frac{T^{*\prime}_{\mathrm{S}1}}{T^*_0}\mathrm{d}\zeta\,. \tag{5.174}$$

From a straightforward, but long and tedious, calculation (given by J.-P. Guiraud) we find, via the solution of the two problems (5.171) for $\boldsymbol{v}^{*\prime}_{\mathrm{S}0}$ and (5.173) for $T^{*\prime}_{\mathrm{S}1}$, a formula for the determination of $w^{*\infty}_{\mathrm{S}0}$. Namely, in (5.166) we have

$$w^{*\infty}_{\mathrm{S}0} = \Sigma_{n\geq 1}[(A_n + B_n)\,Cn + (A_n - B_n)Sn]\Phi_{n|\Sigma}\,, \tag{5.175}$$

where (see (5.89))

$$\Phi_{n|\Sigma} = \left[2\omega_n\sqrt{p^*_0/\rho^*_0}\right]^{-1/2}\left[\boldsymbol{D}\,.\,\boldsymbol{U}_{n|\Sigma}\right] - (\gamma-1)\left[\omega_n\frac{\sqrt{p^*_0/\rho^*_0}}{2\,Pr}\right]^{1/2} R_{n|\Sigma}\,; \tag{5.176}$$

here we take into account only the oscillatory part in the solutions of the problems (5.171), and (5.173), which eventually admits secular terms. We observe that in (5.175), each pair of amplitudes (A_n, B_n) is a function of a very slow time τ_n such that

$$\tau_n = \delta\theta_n(t) \quad \text{and} \quad \frac{\mathrm{d}\tau_n}{\mathrm{d}t} = \delta\frac{\mathrm{d}\theta_n}{\mathrm{d}t}\,, \tag{5.177a}$$

and we write

$$\mathcal{D}A_n = \frac{\mathrm{d}\theta_n}{\mathrm{d}t}\frac{\mathrm{d}A_n}{\mathrm{d}\tau_n}, \quad \mathcal{D}B_n = \frac{\mathrm{d}\theta_n}{\mathrm{d}t}\frac{\mathrm{d}B_n}{\mathrm{d}\tau_n}\,. \tag{5.177b}$$

We stress again that the above inhomogeneous condition (5.175) is associated with the system of equations (5.164a), together with (5.164b), for the functions $\boldsymbol{u}^*_{\mathrm{A}}$ and (p^*_{A}, ρ^*_{A}, T^*_{A}). Obviously, the homogeneous counterpart of the system (5.164) with (5.175) is identical to that which governs $\boldsymbol{u}^{*\prime}_0$, $p^{*\prime}_1$ and $\rho^{*\prime}_1$ (see, for instance, (5.101a,b,c,e)).

Damping Process

First, from the system of equations (5.164) we can derive two equations for $\boldsymbol{u}^*_{\mathrm{A}}$ and $\rho^*_{\mathrm{A}}/\rho^*_0$, with the terms $\mathcal{D}\rho^*_1$ and $\mathcal{D}\boldsymbol{u}^*_0$. Then, we take into account the solution (5.87) for the fluctuating parts (which generates secular terms) and use also the relations (5.177b). With the condition (5.175), where $\Phi_{n|\Sigma}$ is given by (5.176), we derive an acoustic-type inhomogeneous problem for $\boldsymbol{u}^*_{\mathrm{A}}$ and $\rho^*_{\mathrm{A}}/\rho^*_0$:

$$D\left(\frac{\rho^*_{\mathrm{A}}}{\rho^*_0}\right) + \nabla \cdot \boldsymbol{u}^*_{\mathrm{A}} + \sqrt{\frac{\rho^*_0}{p^*_0}} \sum_{n\geq 1} \left[\theta_n' \frac{\mathrm{d}A_n}{\mathrm{d}\tau_n} Sn + \theta_n' \frac{\mathrm{d}B_n}{\mathrm{d}\tau_n} Cn\right] R_n = 0 ,$$

$$D\boldsymbol{u}^*_{\mathrm{A}} + \frac{p^*_0}{\rho^*_0} \nabla \left(\frac{\rho^*_{\mathrm{A}}}{\rho^*_0}\right) + \sum_{n\geq 1} \left[\theta_n' \frac{\mathrm{d}A_n}{\mathrm{d}\tau_n} Cn - \theta_n' \frac{\mathrm{d}B_n}{\mathrm{d}\tau_n} Sn\right] \boldsymbol{U}_n = 0 ,$$

$$\boldsymbol{u}^*_{\mathrm{A}|\Sigma} \cdot \boldsymbol{n} = \sum_{n\geq 1} [(A_n + B_n)\, Cn + (A_n - B_n)\, Sn] \Phi_{n|\Sigma} . \tag{5.178a}$$

We look for a solution of the above problem (5.178a) in the following form:

$$\left(\boldsymbol{u}^*_{\mathrm{A}}, \frac{\rho^*_{\mathrm{A}}}{\rho^*_0}\right) = \sum_{n\geq 1} \left\{ \left[\boldsymbol{u}^{*(1)}_{\mathrm{A}n}, \sqrt{\frac{\rho^*_0}{p^*_0}} \rho^{*(1)}_{\mathrm{A}n}\right] Cn + \left[\boldsymbol{u}^{*(2)}_{\mathrm{A}n}, \sqrt{\frac{\rho^*_0}{p^*_0}} \rho^{*(2)}_{\mathrm{A}n}\right] Sn \right\} , \tag{5.178b}$$

We find the following system for $\boldsymbol{u}^{*(1)}_{An}$ and $\rho^{*(1)}_{An}$:

$$\nabla \cdot \boldsymbol{u}^{*(1)}_{\mathrm{A}n} + \omega_n \rho^{*(1)}_{\mathrm{A}n} + \sqrt{\frac{\rho^*_0}{p^*_0}} \theta_n' \frac{\mathrm{d}B_n}{\mathrm{d}\tau_n} R_n = 0 ,$$

$$-\omega_n \boldsymbol{u}^{*(1)}_{\mathrm{A}n} + \nabla \rho^{*(1)}_{\mathrm{A}n} - \sqrt{\frac{\rho^*_0}{p^*_0}} \theta_n' \frac{\mathrm{d}B_n}{\mathrm{d}\tau_n} \boldsymbol{U}_n = 0 ,$$

$$\boldsymbol{u}^{*(1)}_{\mathrm{A}n|\Sigma} \cdot \boldsymbol{n} = (A_n + B_n) \Phi_{n|\Sigma} . \tag{5.179}$$

A quite analogous system is obtained for $\boldsymbol{u}^{*(2)}_{\mathrm{A}n}$ and $\rho^{*(2)}_{\mathrm{A}n}$, with A_n in place of B_n in two equations of (5.179) and $(A_n - B_n)$ in place of $(A_n + B_n)$ in condition of (5.179). Now, these two systems are soluble only if a compatibility condition for each system is satisfied. For instance, the system (5.179) is soluble only if

$$\sqrt{\frac{\rho^*_0}{p^*_0}} \theta_n' \frac{\mathrm{d}B_n}{\mathrm{d}\tau_n} = (A_n + B_n) \int_{\Sigma} R_n \Phi_{n|\Sigma}\, \mathrm{d}S . \tag{5.180}$$

Finally, we have, as a compatibility condition, the following system:

$$\theta_n' \frac{\mathrm{d}A_n}{\mathrm{d}\tau_n} + \phi_n (A_n - B_n) = 0, \quad \theta_n' \frac{\mathrm{d}B_n}{\mathrm{d}\tau_n} + \phi_n (A_n + B_n) = 0 . \tag{5.181a}$$

With the choice

$$\theta_n' = \frac{d\theta_n}{\mathrm{d}t} \equiv \phi_n(t) \quad \Rightarrow \theta_n(t) = \int_0^t \phi_n(t')\mathrm{d}t' , \tag{5.181b}$$

where $\phi_n(t) > 0$ because $\phi_n(t) = -\sqrt{(p_0^*/\rho_0^*)} \int_\Sigma R_n \Phi_{n|\Sigma} \, \mathrm{d}S$, and with the term

$$\int_\Sigma R_n(\boldsymbol{D} \cdot \boldsymbol{U}_{n|\Sigma}) \, \mathrm{d}S = -\int_\Sigma \omega_n |\boldsymbol{U}_{n|\Sigma}|^2 \mathrm{d}S \,,$$

the systems (5.181a,b), for $\theta_n(t)$ and (5.177a) for τ_n lead to law for the damping of the oscillations, namely

$$(A_n \,; B_n) = \alpha_n \exp\left[-\delta \int_0^t \phi_n(t')\mathrm{d}t'\right] \{\cos[\theta_n(t) + \varphi_n] \,; -\sin[\theta_n(t) + \varphi_n])\} \,. \tag{5.182}$$

5.5 Comments

Further investigation is necessary to understand how viscous damping operates when the constraint (5.165) is not satisfied. Other points should be investigated as well – one of these is the behaviour of the Rayleigh layer. For a deeper investigation of dissipative effects in the case of a time-dependent cavity, it is necessary to consider the case where a similarity rule (5.160) between two small parameter ε and M applies. We observe that a Rayleigh layer emerges in the solution of the problems (5.171) and (5.173) because the terms $-[\boldsymbol{u}_\Sigma + \langle \boldsymbol{u}_0^* \rangle_{|\Sigma}]$ and $\Lambda^* \Theta$, respectively, occur in the condition for $\zeta = 0$. The investigation of the evolution of the Rayleigh layers in these two problems with time is difficult. If the Stokes layer corresponds to acoustic (oscillating) eigenfunctions of the cavity, the Rayleigh layer corresponds to the above-mentioned terms in the condition for $\zeta = 0$. The thickness of the Stokes layer is given by $\chi = \sqrt{M/Re}$ and is independent of the time (the behaviour for a large time does not have any influence on the Stokes layer). Concerning the Rayleigh layer, however, its thickness grows as the square root of the time (see, for instance, the solutions (5.46a,b), together with (5.46c)). Obviously, a deeper analysis of the interaction of these two boundary layers is required. Concerning the theory of combustion, we mention the book by Liñan and Williams (1993) devoted to fundamental aspects of combustion, and also the book by Warnatz, Maas and Dibble (1996), where the physical and chemical aspects are considered, with reference to modelling, simulation and experiments. A classic book on acoustics is the book by Blokhintsev (1956), which considers the acoustics of non-homogeneous moving media. In Crighton et al. (1996), the reader can find some modern methods in analytical acoustics. Pertinent theoretical and applied papers concerning various aspects of combustion theory can also be found in the journals *Combustion Science and Technology*, *Combustion Theory Modelling* and *Combustion and Flame*.

References

G. Ali, Low Mach number flows in time-dependent domain. SIAM J. Appl. Math., **63**(6) (2003), 2020–2041.

D. I. Blokhintsev, *Acoustics of Nonhomogeneous Moving Media*. NACA TM 1399, 1956 (Translation from Russian).

D. G. Crighton, A. P. Dowling, J. E. Ffowcs Williams, M. Hecki and F. G. Leppington, *Modern Methods in Analytical Acoustics*. Springer, 1996.

J.-P. Guiraud and R. Kh. Zeytounian, Long time evolution of acoustic waves and the concept of incompressibility. In: *Proceedings of the BAIL I Conference*, Dublin, June 1980, edited by. Boole Press, Dublin, 1980, pp. 297–300.

A. Liñan and E. A. Williams, *Fundamental Aspects of Combustion*. Oxford, 1993.

A. Meister, Asymptotic single and multiple scale expansion in the low mach number limit. SIAM J. Appl. Math., **60**(6) (1999), 256–271.

B. Müller, Low-Mach-number asymptotics of the Navier–Stokes equations. J. Engrg. Math., **34** (1998), 97–109.

A. B. Vasilieva and B. F. Boutousov, *Asymptotic Methods in Singular Perturbation Theory*. Vischaya Chkola, Moscow, 1990 (in Russian).

J. Warnatz, U. Maas and R.W. Dibble, *Combustion: Physical and Chemical Fundamentals, Modelling and Simulation, Experiments, Pollutant Formation*. Springer, 1996.

R. Kh. Zeytounian and J.-P. Guiraud, *Evolution d'ondes acoustiques dans une enceinte et concept d'incompressibilité*. IRMA, Lille, 1980, Vol. 2, Fasc. 4(4), pp. IV-1–IV-16. See also R. Kh. Zeytounian and J.-P. Guiraud. C. R., Acad. Sci. Paris, **290B** (1980), 75–78.

6

Slow Atmospheric Motion as a Low-Mach-Number Flow

First, in Sect. 6.1, we discuss in detail various facets of the applicability of the Boussinesq approximation to slow atmospheric motion. In Sect. 6.2, we give an asymptotic derivation of the Boussinesq lee-wave-model inviscid equations, from the system of Euler equations (1.40) derived in Sect. 1.7.1. In Sect. 6.3, the corresponding *dissipative* Boussinesq equations are derived for the free-circulation (breeze) problem, from the hydrostatic dissipative system of equations (1.57) derived in Sect. 1.7.4. Section 6.4 is devoted to various complementary remarks related to derived Boussinesq model equations. Quasi-non-divergent flows involve some intriguing features tied to the phenomenon of blocking, well known (see Drazin 1961) in stratified flows at low Froude number, and this might explain why such flows have not been used in weather prediction (see, for instance, Phillips 1970 and Monin 1972). However, fortunately, blocking is avoided whenever the main low-Mach-number approximation is coupled (matched in the framework of the MAE) with another one – for instance, with an inner-in-time approximation close to the initial time and an outer approximation valid in the far field. This has been shown by Zeytounian and Guiraud (1984) (see also Zeytounian 1990, Chap. 12), and in Sect. 6.5 we perform a systematic asymptotic investigation of this quasi-non-divergent model. On the other hand, the case of a low Rossby/Kibel number in the framework of the Kibel primitive equations (1.54), derived in Sect. 1.7.3, when the sphericity parameter δ and the hydrostatic parameter ε are both small and the f_0-plane approximation is used, is also taken into account. In Sect. 6.6, from (1.54), a low-Mach-number asymptotic investigation with the similarity rule (2.15) gives us the possibility to justify the quasi- geostrophic model.

If various asymptotic investigations have given (one hundred years after the well-known statement of the approximation in Boussinesq 1903) some satisfactory answers to various questions about the justification of the Boussinesq approximation for modelling atmospheric mesoscale motions driven by the dynamic action of relief, thermal non-homogeneity of the ground, buoyancy and Coriolis forces, unfortunately this is not the case for the

R. Kh. Zeytounian: *Topics in Hyposonic Flow Theory*, Lect. Notes Phys. **672**, 163–228 (2006)
www.springerlink.com

quasi-non-divergent model. Concerning the case of small Kibel and Mach numbers (quasi-geostrophic and ageostrophic models derived in Sect. 1.7.4), a complete asymptotic theory in the framework of the hydrostatic dissipative system of equations (1.57) has been developed by Guiraud and Zeytounian (1980).

6.1 Some Comments on the Boussinesq Approximation and the Derivation of the Boussinesq Equations

Concerning the justification of the Boussinesq approximation, we observe that Eckart and Ferris (1956) pointed out that the Boussinesq approximate original equations (Boussinesq 1903, p. 174) are quite similar to a set of equations introduced by Oberbeck (1879). In spite of this, Lord Rayleigh (1916) referred to Boussinesq (1903) for the equations that he used in his theoretical approach to the classical Bénard (1900) thermal problem (considered in the Sect. 7.2 of this book); see, for instance, the books by Chandrasekhar (1961) and Drazin and Reid (1981). In the book by Joseph (1976, Chap. 8, pp. 4–5), the approximate equations resulting from the Boussinesq approximation are called the 'Oberbeck–Boussinesq' (OB) equations, and in Landau and Lifshitz (1988, Sect. 56), the reader can find an 'ad hoc' derivation of these OB equations (known as the 'free-convection equations' in Russia). The reader is likewise directed to the books of Phillips (1977, Sect. 2.4) and Kotchin, Kibel and Roze (1963, Part 1, Sects. 36 and 38) for an 'ad hoc' way of obtaining approximate equations, via the Boussinesq approximation, for oceanic and atmospheric motions. On the other hand, the purpose of the paper of Batchelor (1953) is to expound certain (pertinent) generalities about the motion of a fluid of variable density subject to gravity, for which the Mach number of the fluid is everywhere small.

From a physical point of view, according to Joseph (1976, Chap. 8, pp. 4–5), the crucial features of the Boussinesq approximation, which leads at the leading order to the OB approximate equations in the case of a non-homogeneous, heat-conducting viscous, weakly expansible liquid in a gravity field, are the following. (i) The variation of the density perturbation is neglected in the mass continuity equation and in the equation for the horizontal motion, but (ii) this density perturbation is taken into account in the equation for the vertical motion through its influence as a buoyancy term. (iii) The influence of the pressure perturbation on the buoyancy and in the equation of energy (written for the temperature perturbation) can be neglected. (iv) The influence of the perturbation of pressure in the equation of state can be also neglected, and the rate of viscous dissipation can be neglected in the equation for the temperature perturbation. In fact, this Boussinesq approximation works well only for thermal convective viscous motion with a temperature difference of a few degrees or less between the lower and upper surfaces when a horizontal liquid layer is considered, as in the Rayleigh–Bénard (RB) problem. This can be formally justified by an (ad hoc) dimensional scale analysis; see, for instance, Spiegel and Veronis (1960), Mihaljan (1962), Malkus (1964), Dutton and Fichtl (1969), Perez Gordon and Velarde (1975, 1976), and Mahrt (1986).

6.1.1 Asymptotic Approaches

For a rigorous, asymptotic derivation of the approximate OB equations for the RB problem in the framework of the full thermal Bénard problem (for a weakly dilatable/expansible liquid) with a deformable free surface subject to a temperature-dependent surface tension, see, for instance, Zeytounian (1998) and Velarde and Zeytounian (2002). In this book, we consider only the case of a perfect gas. In this case $p = R\rho T$, where $R = C_p - C_v$ is the gas constant and $\gamma = C_p/C_v$, both constant, and T is the absolute temperature; this is the case for atmospheric motion if we assume that the air is dry.

The Case of the 'Lee Waves' Inviscid Problem

In the case of lee wave motion, a significant similarity rule can be written between the Mach number M and the Boussinesq number Bo, which is the ratio of the characteristic vertical length scale H_{c} of the lee wave motion to the height $H_{\mathrm{s}} = RT_{\mathrm{c}}/g$ of a homogeneous hydrostatic atmosphere. Namely (see (1.13g)), we write

$$\frac{Bo}{M} = B^* = 0(1) . \tag{6.1}$$

The question of the possibility of an asymptotically consistent derivation of approximate (Boussinesq) equations according to the Boussinesq approximation (as proposed in 1903) was posed by Paul Germain at the end of 1967, when the present author (being, at that time, a research engineer in the Aerodynamics Department of ONERA, Chatillon, France) was writing his doctoral thesis, based on his theoretical and numerical results on the lee waves, atmospheric phenomena, which were obtained in the Hydrometeorological Center, Moscow, during the period 1961–1966 under the supervision of Il'ya Afanasievitch Kibel. However, in 1967, unfortunately, the author did not have the possibility to work seriously on this question, and only beginning in October 1972 (being then Titular Fluid Mechanics Professor at the University of Lille I, France) did he work purposefully on this problem.

As a consequence, in the author's DSc thesis, defended in March 1969 (Zeytounian 1969), he considered as a starting (exact!) model, in place of the Euler compressible, non-viscous, adiabatic equations, the *isochoric* equations (very intensively used for the study of stratified flows; see, for instance, the monograph by Yih 1980), corresponding to conservation of the density (in place of the specific entropy (see Sect. 6.4) along trajectories. In such a case, the isochoric flow has a divergenceless velocity (the incompressible equation of continuity applies), but the density is not constant in a stratified, non-homogeneous flow. In this particular case, the non-Boussinesq nonlinear effects are proportional to the parameter $\lambda = (1/2)\beta H_s$, assumed small, when in a homogeneous atmosphere (in hydrostatic equilibrium; see (1.38)) the dimensionless density,

$$\rho_s(z_s) \approx \exp(-\beta z_s) , \tag{6.2}$$

is a decreasing exponential, where z_s is the 'standard' altitude. Since the Froude number $Fr = U_c/(gH_s)^{1/2}$ based on H_s, where U_c is a reference velocity, is also always a small parameter for atmospheric motion, it is necessary that

$$D = \frac{\lambda}{(Fr)^2} = O(1) \tag{6.3}$$

(which is a similarity rule), where D is the Scorer–Dorodnitsyn parameter in the theory of lee waves. We observe here that in the case of an isochoric flow we have $H_c \equiv H_s$ and $Bo \equiv 1$, and obviously the isochoric equations are more complicated than the corresponding limiting Boussinesq model equations (see, for instance, Sect. 6.2). The reader can find in a recent book (Zeytounian 2002a, Sects. 4.6, 5.3 and 5.4)) various pieces of information concerning isochoric fluid flows and isochoric 2D steady lee waves. Finally, after some partial results (see Zeytounian 1972), the author derived in spring 1973 the equations for a Boussinesq viscous, heat-conducting model, thanks to the introduction of the above Boussinesq number Bo, the similarity rule (6.1)) and the relation (identity)

$$B_o = \gamma\left(\frac{M}{Fr}\right)^2 \tag{6.4}$$

between Bo, M and the Froude number Fr defined above, from the full (exact) NSF equations (for a viscous, compressible, heat-conducting, baroclinic, stratified perfect gas), via an asymptotic, consistent, logical approach (see, for instance, Zeytounian 1990, Chap. 8). This result was presented at the 11th Symposium on Advanced Problems and Methods in Fluid Mechanics (Sopot-Kamienny Potok, Poland, 3–8 September 1973). In Bois (1976, 1979, 1984), the reader can find an asymptotic theory of Boussinesq waves in a stratified atmosphere, and it is noted there that the Boussinesq limit,

$$Bo \downarrow 0 \text{ and } M \downarrow 0 \quad \text{such that } \frac{Bo}{M} = B^* = O(1) , \tag{6.5}$$

is, in fact, the most significant limit (in the non-viscous case) in comparison of two other particular limits (more degenerate and therefore less significant) of flows at a small Mach number and at a small Boussinesq number, in the presence of gravity, namely, the quasi-non-divergent (Monin–Charney) limit and the classical incompressible limit (see, for instance, Zeytounian 2002a, pp. 150–151 and also (6.9) below).

The Case of 'Free' Atmospheric Circulation

Concerning the justification of the Boussinesq approximation in the case of free atmospheric circulation (for instance in the case of breezes), it is necessary to take into account the influence of the Coriolis force and also to write a boundary condition for the temperature, expressing the influence of a localized thermal inhomogenity on a flat ground surface. For example, such a condition might have the form

$$T = T_s(0) + \Delta T^0\, \Theta\,, \tag{6.6}$$

where $T_s(0)$ is the value of the temperature of the standard atmosphere (a function only of the altitude z_s) on the flat ground surface, and Θ is a given (known) function of time and of the position in a bounded region D_c (of diameter L_c) on the ground surface (in Sect. 6.3, we state this thermal condition (6.6) more precisely). As a consequence, we have the possibility to introduce the following new (local) characteristic vertical scale:

$$h_c = \frac{R\,\Delta T^0}{g} \ll L_c, \quad \text{such that } \tau_0 = \frac{\Delta T^0}{T_s(0)} \ll 1\,. \tag{6.7}$$

On the one hand, a height $h_c \approx 10^3\,\text{m}$ is significant for a breeze phenomenon, but, on the other hand, when we take the Coriolis force into account in the dynamic dissipative, unsteady equations for the atmosphere, the characteristic horizontal length scale for the breeze phenomenon is $L_c \approx 10^5$ m, as a consequence of $Ro = O(1)$. Obviously, in such a case we must consider $\lambda = h_c/L_c$ as a long-wave, hydrostatic, small parameter (analogous to our ε defined by the second relation of (1.13c), and (see Sect. 6.3) the approximate model limit equations for the free atmospheric circulation are *hydrostatic*, boundary-layer-type Boussinesq equations. Namely, it is necessary to perform first a long-wave, hydrostatic limiting process

$$\lambda \downarrow 0 \quad \text{and } Re \uparrow \infty, \quad \text{such that } \lambda^2\, Re = Re_\perp = O(1)\,, \tag{6.8a}$$

and then to take the Boussinesq limit,

$$\tau_0 \downarrow 0 \quad \text{and } M \downarrow 0, \quad \text{such that } \frac{\tau_0}{M} = 1\,, \tag{6.8b}$$

which gives us the possibility to evaluate the characteristic velocity U_c, after a judicious choice (in accordance with the breeze problem considered) of $L_c, T_s(0)$ and the constant kinematic viscosity ν_c in the reference Reynolds number $Re = U_c L_c/\nu_c$.

6.1.2 Some Comments

Curiously, the rigorous dimensionless-analysis approach, with the introduction of consistent similarity rules (see, for instance, (6.1) and (6.8a,b)) in the framework of an asymptotic modelling process related to a small main Mach number M in the case of the atmosphere, gives us the possibility to derive very easily (and accurately) the associated leading-order approximate limit equations via the Boussinesq approximation. This is performed in Sects. 6.2 and 6.3, without any complicated transformations or 'ad hoc' estimates, as was usually done in the early papers cited above. Thanks to this asymptotic, consistent, logical approach, we have the possibility to derive also, rationally, higher-order *well-balanced* approximate equations containing various non-Boussinesq effects. However, a fundamental (rather open) problem, in the framework of the justification of the Boussinesq approximation, is the deep discussion of the various limitations of this Boussinesq approximation, which are not always easy to identify. Section 6.4 is an attempt to provide the elements of such a discussion, with, hopefully, enough novelty in the arguments or at least their presentation and interpretation, from the point of view of asymptotic modelling. We observe that only during the last 30 years has the systematic use of consistent dimensional analysis and asymptotic methods in the modelling of fluid flow phenomena (see, for instance, Zeytounian 2002b) given us the possibility to reveal the asymptotic character of the atmospheric Boussinesq approximation for a perfect gas in the framework of low-Mach-number asymptotics. We hope that the sceptical reader will now finally be convinced of the asymptotic status of two Boussinesq systems of model equations, derived via the Boussinesq approximation in the framework of the exact atmospheric equations.

6.2 An Asymptotic, Consistent Justification of the Boussinesq Approximation for Lee Wave Phenomena in the Atmosphere

Our starting point is the Euler equations written in the form (1.40) for the velocity vector $\boldsymbol{u} = (u, v\, w)$ and the thermodynamic perturbations π, ω and θ defined in (1.39). We observe that in these equations, the Strouhal number is unity ($St = L_c/t_c U_c \equiv 1$) and, obviously, $1/\gamma M^2 \gg 1$ in the usual atmospheric ('low-speed hyposonic') motions under consideration. We note also that Batchelor (1953, p. 228), considered, in fact, the *reciprocal* ($Bo^{-1} = p_s(0)/ gH_c\rho_s(0)$) of the Boussinesq number *Bo; Bo* itself was introduced first in Zeytounian (1974). Naturally, when we choose H_s for H_c, then $Bo \equiv 1$, but in the case of the Boussinesq approximation the asymptotic analysis performed below shows that, if $M \ll 1$, then *necessarily* $Bo \ll 1$! This is a fundamental observation which gives us the possibility to derive, rigorously by an asymptotic method, the corresponding Boussinesq-limit model equations. The emergence of the above Boussinesq number was the first key step (in the early investigations at the end of the 1960s) in the framework of an asymptotic, consistent derivation of leading-order model equations 'à la Boussinesq'. If we assume that Bo is, like the Mach number M, a small parameter, it is necessary in general to consider the following three limiting cases:

$$\text{(i)} \quad Bo \text{ fixed}, \quad M \downarrow 0 \quad \text{and then } Bo \downarrow 0 ,$$

$$\text{(ii)} \quad M \text{ fixed}, \quad Bo \downarrow 0 \quad \text{and then } M \downarrow 0 ,$$

$$\text{(iii)} \quad M \downarrow 0 \text{ and } Bo \downarrow 0, \quad \text{such that } Bo = M^{\alpha} B^* = 0(1) , \tag{6.9}$$

where the scalar $a > 0$. The Boussinesq model equations associated with the assertion/approximation of Boussinesq (1903) can be derived, with the particular, significant choice $\alpha = 1$, under the Boussinesq limiting process corresponding to case (iii), with (6.4), and for γ and the time–space variables $(t, \boldsymbol{x})$ fixed.

From the Boussinesq similarity rule (6.1), we have the following constraint on the characteristic vertical length scale H_{B}, the thickness of the Boussinesq layer in which the lee waves atmospheric motion occurs:

$$B^* = O(1) \quad \Rightarrow H_{\mathrm{B}} \approx \frac{U_c}{g} \left[\frac{RT_s(0)}{\gamma} \right]^{1/2} . \tag{6.10}$$

As a consequence, for the usual meteorological values of U_{c} and $T_{\mathrm{s}}(0)$, we obtain for H_{B} *only* the value $10^3\,\mathrm{m} \ll H_{\mathrm{s}} \approx 10^4\,\mathrm{m}$ (!), and this is a strong restriction on the application of the Boussinesq equations to atmospheric motions – in particular, in the case the prediction of 'lee waves' in the whole troposphere around and downstream of a mountain.

In Sect. 6.4, we discuss the implications of this restriction. Now, we rewrite the dimensionless (exact) Euler equations (1.40) derived in Sect. 1.7.1 for $\boldsymbol{u} = (u, v, w), \pi, \omega$ and θ, in the following form:

$$\frac{\mathrm{D}\omega}{\mathrm{D}t} + (1+\omega)\, \nabla \cdot \boldsymbol{u} = (1+\omega) \frac{Bo}{T_{\mathrm{s}}(z_{\mathrm{s}})} \left[1 + \frac{\mathrm{d}T_{\mathrm{s}}(z_{\mathrm{s}})}{\mathrm{d}z_{\mathrm{s}}} \right] (\boldsymbol{u} \cdot \boldsymbol{k}) ; \tag{6.11a}$$

$$(1+\omega) \frac{\mathrm{D}\boldsymbol{u}}{\mathrm{D}t} + \frac{T_{\mathrm{s}}(z_{\mathrm{s}})}{\gamma M^2} \nabla \pi - (1+\omega) \frac{Bo}{\gamma M^2} \theta \boldsymbol{k} = 0 ; \tag{6.11b}$$

$$(1+\omega) \frac{\mathrm{D}\theta}{\mathrm{D}t} - \frac{\gamma - 1}{\gamma} \frac{\mathrm{D}\pi}{\mathrm{D}t} + (1+\pi)\, Bo\, N^2(z_{\mathrm{s}}) (\boldsymbol{u} \cdot \boldsymbol{k}) = 0 ; \tag{6.11c}$$

$$\pi = \omega + (1+\omega)\theta , \tag{6.11d}$$

where the parameter

$$N^2(z_{\mathrm{s}}) = \frac{1}{T_{\mathrm{s}}(z_{\mathrm{s}})} \left(\frac{\gamma - 1}{\gamma} + \frac{\mathrm{d}T_{\mathrm{s}}(z_{\mathrm{s}})}{\mathrm{d}z_{\mathrm{s}}} \right) \tag{6.12}$$

take into account the stratification of the hydrostatic reference-state atmosphere and is related to the dimensionless Väisälä internal frequency.

The consideration of the above full, dimensionless Euler exact equations for $\boldsymbol{u}, \pi, \theta$ and ω is the second key step in the asymptotic, rational derivation of the Boussinesq model equations.

6.2.1 Asymptotic Derivation of Boussinesq Lee-Waves-Model Inviscid Equations When $M \to 0$

For atmospheric motions, the Mach number M is always a small parameter (we have a low-speed/hyposonic flow), and it is very natural to consider an asymptotic solution of the above dimensionless Euler equations (6.11), together with (6.12), for $\boldsymbol{u}, \pi, \omega$ and θ, when $M \ll 1$. Namely, we write the following asymptotic expansions:

$$\boldsymbol{u} = \boldsymbol{u}_{\mathrm{B}} + M^c \boldsymbol{u}^* + \cdots , \tag{6.13a}$$

$$(\omega, \theta) = M^e(\omega_{\mathrm{B}}, \theta_{\mathrm{B}}) + M^{e+1}(\omega^*, \theta^*) + \cdots , \tag{6.13b}$$

$$\pi = M^b \pi_{\mathrm{B}} + M^{b+1}\pi^* + \cdots , \tag{6.13c}$$

where c, e and b are positve real scalars. In this case, we obtain the following four dominant equations, for $\boldsymbol{u}_{\mathrm{B}}, \omega_{\mathrm{B}}, \theta_{\mathrm{B}}$ and π_{B}, from the above (exact Euler) equations (6.11):

$$\nabla \cdot \boldsymbol{u}_{\mathrm{B}} = -M^e \left[\frac{\mathrm{D}\omega_{\mathrm{B}}}{\mathrm{D}t} + \omega_{\mathrm{B}} \, \nabla \cdot \boldsymbol{u}_{\mathrm{B}}\right]$$

$$+ \frac{Bo}{T_{\mathrm{s}}(Bo\, z)} \left[1 + \frac{\mathrm{d}T_{\mathrm{s}}(z_{\mathrm{s}})}{\mathrm{d}z_{\mathrm{s}}}\right] (\boldsymbol{u}_{\mathrm{B}} \cdot \boldsymbol{k}) + O(Bo\, M^e) \, , \tag{6.14a}$$

$$\frac{D\theta_{\mathrm{B}}}{\mathrm{D}t} + \frac{Bo}{M^e} N^2(Bo\, z)(\boldsymbol{u}_{\mathrm{B}} \cdot \boldsymbol{k})$$

$$= M^{b-e}\left\{\frac{\gamma - 1}{\gamma}\frac{\mathrm{D}\pi_{\mathrm{B}}}{\mathrm{D}t} - Bo\, \pi_{\mathrm{B}} N^2(Bo\, z)(\boldsymbol{u}_{\mathrm{B}} \cdot \boldsymbol{k})\right\} + O(Bo^2) \, , \tag{6.14b}$$

$$\omega_{\mathrm{B}} + \theta_{\mathrm{B}} = M^e \omega_{\mathrm{B}} \theta_{\mathrm{B}} + M^{b-e} \pi_{\mathrm{B}} \, , \tag{6.14c}$$

$$\frac{\mathrm{D}\boldsymbol{u}_{\mathrm{B}}}{\mathrm{D}t} + M^{b-2}\frac{T_{\mathrm{s}}(Boz)}{\gamma}\nabla \pi_{\mathrm{B}} - \frac{1}{\gamma} Bo\, M^{e-2} \theta_{\mathrm{B}} \boldsymbol{k}$$

$$= -M^e \omega_{\mathrm{B}} \left[\frac{\mathrm{D}\boldsymbol{u}_{\mathrm{B}}}{\mathrm{D}t} + \frac{1}{\gamma} Bo\, M^{e-2} \theta_{\mathrm{B}} \boldsymbol{k}\right] , \tag{6.14d}$$

and we observe that $z_{\mathrm{s}} = Bo\, z$, such that $-\mathrm{d}T_{\mathrm{s}}(z_{\mathrm{s}})/\mathrm{d}z_{\mathrm{s}} = \Gamma_{\mathrm{s}}(Bo\, z)$.

In the above equations, M is the main small parameter, but we have the possibility also to consider Bo as a small parameter. In fact, the limiting case (with t and $\boldsymbol{x}$ fixed)

$$M \to 0 \quad \text{with } Bo \to 0, \quad \text{and } N^2(0) = O(1), \quad \gamma = 0(1) \tag{6.15}$$

corresponds to the adiabatic, stratified, atmospheric inviscid Boussinesq case for the lee waves problem. First, from (6.14d), it seems reasonable to assume that

$$b = 2 \quad \text{and also } Bo\, M^{e-2} = O(1)\ , \tag{6.16a}$$

when we consider the above limiting process (6.15), because, obviously, the terms linked to π_{B} and θ_{B} must be present in the approximate limiting equation (the significant Boussinesq equation) for $\boldsymbol{u}_{\mathrm{B}}$ when M and Bo both tend to zero, γ being $O(1)$. Now, from (6.14b), when Bo tends to zero and $N^2(0) = O(1)$, it is necessary to take into account (at least) the term proportional to $(Bo/M^e)N^2(0)$, which is responsible for the stratification. As a consequence, obviously,

$$\frac{Bo}{M^e} = O(1)\ . \tag{6.16b}$$

Then, from (6.16b) and the second relation of (6.16a), we have necessarily

$$2 - e = e \quad \Rightarrow e = 1\ , \tag{6.16c}$$

and we obtain from (6.16b) the Boussinesq similarity rule (6.1). As a consequence of (6.16c) and the first relation of (6.16a), if we consider the limiting process (6.15), together with (6.1) for t and $\boldsymbol{x}$ fixed, and the expansions

$$\boldsymbol{u} = \boldsymbol{u}_{\mathrm{B}} + M u^* + \dots, \quad (\omega, \theta) = M(\omega_{\mathrm{B}}, \theta_{\mathrm{B}}) + M^2(\omega^*, \theta^*) + \dots\ , \tag{6.17a}$$

$$\pi = M^2\left[\pi_{\mathrm{B}} + M^2\pi^* + \dots\right]\ , \tag{6.17b}$$

then we derive, finally, the following equations for the inviscid, adiabatic Boussinesq lee-wave approximate dimensionless model for $\boldsymbol{u}_{\mathrm{B}}, \omega_{\mathrm{B}}, \theta_{\mathrm{B}}$ and π_{B}:

$$\begin{gathered}
\nabla \cdot \boldsymbol{u}_{\mathrm{B}} = 0, \\
\frac{\mathrm{D_B}\boldsymbol{u}_{\mathrm{B}}}{\mathrm{D}t} + \nabla\left(\frac{\pi_{\mathrm{B}}}{\gamma}\right) - \frac{B^*}{\gamma}\theta_{\mathrm{B}}\boldsymbol{k} = 0\ , \\
\frac{\mathrm{D_B}\theta_{\mathrm{B}}}{\mathrm{D}t} + B^* N^2(0)(\boldsymbol{u}_{\mathrm{B}} \cdot k) = 0 \\
\omega_{\mathrm{B}} = -\theta_{\mathrm{B}}\ ,
\end{gathered} \tag{6.18}$$

where $\mathrm{D_B}/\mathrm{D}t = \partial/\partial t + \boldsymbol{u}_{\mathrm{B}} \cdot \nabla$. We observe that, in dimensionless form, we have obviously $T_{\mathrm{s}}(0) \equiv 1$, but, in general, $N^2(0)$ is different from zero. The choice $c = 1$ in (6.13a) follows when we want to derive the second-order 'à la Boussinesq' limit model equations for $\boldsymbol{u}^*, \omega^*, \theta^*$ and π^*. Namely, it is necessary to observe that the above way to derive the inviscid Boussinesq model equations (6.18) is the only rational way to obtain a consistent derivation of second-order, 'à la Boussinesq', linear model equations (but with non- Boussinesq effects) for the terms with asterisks in the asymptotic expansions (6.17). The above Boussinesq equations were first derived in Zeytounian (1974), and the reader can find a tentative 'theory' of the Boussinesq approximation for atmospheric motion in Zeytounian (1990, Chap. 8). In fact, in Germain (1986, pp. 225–226), the reader can find the above Boussinesq equations (6.18), but with a viscous (incompressible, 'à la Navier', since we have $\nabla \cdot \boldsymbol{u}_{\mathrm{B}} = 0$) term of the form $(1/Re)\,\nabla^2 \boldsymbol{u}_{\mathrm{B}}$ on the right-hand side of the second equation (for $\boldsymbol{u}_{\mathrm{B}}$), and a convective term of the form $(1/Pe)\,\nabla^2\theta_{\mathrm{B}}$, where Pe is the Péclet number (the product of the Reynolds number Re with the Prandtl number Pr) on the right-hand side of the third equation (for θ_{B}). See also (for the inviscid case) Kotchin, Kibel and Roze (1963, Sect. 38, pp. 561–565).

Two Degenerate Cases: $B^* \to \infty$ and $B^* \to 0$

It is interesting to note, on the one hand, that if

$$B^* \to \infty \Leftrightarrow M \ll Bo, \quad Bo \text{ fixed}, M \to 0 \quad \text{and then } Bo \to 0\,, \tag{6.19a}$$

then in this case we derive, in place of the Boussinesq equations (6.18), the following very degenerate 'quasi-non-divergent' system of dimensionless equations for the horizontal velocity vector $V_{\mathrm{qnd}} = (u_{\mathrm{qnd}}, v_{\mathrm{qnd}})$ and for $\pi_{\mathrm{qnd}}/\gamma$:

$$\begin{gathered}\frac{\partial \pi_{\mathrm{qnd}}}{\partial z} = 0, \quad w_{\mathrm{qnd}} = 0, \quad \theta_{\mathrm{qnd}} = 0\,,\\ \boldsymbol{D} \cdot \boldsymbol{V}_{\mathrm{qnd}} = 0\,,\\ \frac{\partial \boldsymbol{V}_{\mathrm{qnd}}}{\partial t} + (\boldsymbol{V}_{\mathrm{qnd}} \cdot \boldsymbol{D})\,\boldsymbol{V}_{\mathrm{qnd}} + \boldsymbol{D}\left(\frac{\pi_{\mathrm{qnd}}}{\gamma}\right) = 0\,.\end{gathered} \tag{6.19b}$$

If, on the other hand,

$$B^* \to 0 \Leftrightarrow M \gg Bo, \quad M \text{ fixed}, Bo \to 0 \text{ and then } M \to 0\,, \tag{6.20a}$$

then in this case we derive, in place of the Boussinesq equations (6.18), the classical 'incompressible' Navier dimensionless model equations for u_i, v_i, w_i and π_i:

$$\begin{gathered}\frac{\partial u_i}{\partial x} + \frac{\partial v_i}{\partial y} + \frac{\partial w_i}{\partial z} = 0\,,\\ \frac{\mathrm{D}_i \boldsymbol{u}_i}{\mathrm{D}t} + \nabla\left(\frac{\pi_i}{\gamma}\right) = 0\,,\end{gathered} \tag{6.20b}$$

$$\text{with} \quad \frac{\mathrm{D}_i \theta_i}{\mathrm{D}t} = 0\,, \tag{6.20c}$$

where $\omega_i = -\theta_i$, and $\mathrm{D}_i/\mathrm{D}t = \partial/\partial t + u_i \cdot \nabla$. So, in both cases, we derive from the dimensionless dominant Euler equations (6.14) a less significant limit system than the inviscid, adiabatic Boussinesq approximate equations (6.18).

6.2.2 The Case of Steady Two-Dimensional Flow and the Long Problem

When we consider a steady adiabatic two-dimensional flow in the plane (x, z) in which gravity acts, such that $\mathrm{D_B}/\mathrm{D}t \equiv u_\mathrm{B}\,\partial/\partial x + w_\mathrm{B}\,\partial/\partial z$ and $\partial u_\mathrm{B}/\partial x + \partial w_\mathrm{B}/\partial z = 0$, we have the possibility to introduce a stream function $\psi_\mathrm{B}(x, z)$ such that

$$u_\mathrm{B} = \frac{\partial \psi_\mathrm{B}}{\partial z}, \quad w_\mathrm{B} = -\frac{\partial \psi_\mathrm{B}}{\partial x} . \tag{6.21}$$

In this case (see, for instance, Kotchin, Kibel and Roze 1963, pp. 563–564), the following single equation for $\psi_\mathrm{B}(x, z)$ can be derived: from the steady 2D Boussinesq system of equations ((6.18), for $u_\mathrm{B}, w_\mathrm{B}, \pi_\mathrm{B}$ and θ_B, with $\partial/\partial t = 0$):

$$\frac{\partial^2 \psi_\mathrm{B}}{\partial x^2} + \frac{\partial^2 \psi_\mathrm{B}}{\partial z^2} - \frac{B^*}{\gamma} z \frac{\mathrm{d}H(\psi_\mathrm{B})}{\mathrm{d}\psi_\mathrm{B}} = F(\psi_\mathrm{B}) , \tag{6.22}$$

where $H(\psi_\mathrm{B})$ and $F(\psi_\mathrm{B})$ are two arbitrary functions of ψ_B only. In the particular case of the airflow over a mountain, if we have a 2D steady, uniform, constant flow in the direction of $x > 0$ with $\pi_\mathrm{B} = \theta_\mathrm{B} \equiv 0$ at infinity upstream of the mountain, where $x \to -\infty$, we can derive the following linear Helmholtz equation for the function $\Delta_\mathrm{B}(x, z) \equiv z - \psi_\mathrm{B}$:

$$\frac{\partial^2 \Delta_\mathrm{B}}{\partial x^2} + \frac{\partial^2 \Delta_\mathrm{B}}{\partial z^2} + \frac{B^{*2}}{\gamma} N^2(0) \Delta_\mathrm{B} = 0 . \tag{6.23}$$

The dominant feature of (6.23), from a mathematical point of view, is that its linearity is not related to any particular hypothesis about the small perturbations in the steady 2D Boussinesq equation (6.22). However, from the exact slip condition on the mountain, simulated by the dimensionless equation

$$z = \delta \eta(\alpha x) , \tag{6.24}$$

where δ is an amplitude parameter and $\alpha > 0$, we must write the following boundary (slip, nonlinear) condition for Δ_B on the surface of the mountain:

$$\Delta_\mathrm{B}(x, \delta\eta(\alpha x)) = \delta \eta(\alpha x) . \tag{6.25a}$$

At infinity upstream, we have the behaviour condition

$$\Delta_\mathrm{B}(-\infty, z) = 0 . \tag{6.25b}$$

Because of the emergence of steady lee waves at infinity downstream, there is only the possibility to assume one physically realistic boundary condition, namely

$$|\Delta_\mathrm{B}(+\infty, z)| < \infty . \tag{6.25c}$$

The problem (6.23)–(6.25c) is the classical problem of Long (1953), if also

$$\Delta_\mathrm{B}(x, z = 1) = 0 . \tag{6.25d}$$

However, in reality (see Sect. 6.4), this condition (6.25d) does not emerge from the exact dimensionless formulation of the lee waves problem, considered in the framework of the 2D steady Euler equations for the whole troposphere and where the upper boundary condition is assumed to be a slip condition on the tropopause (on $z = 1/Bo$).

6.3 The Free-Circulation (Breeze) Problem

For the modelling of atmospheric free-circulation phenomena, and to justify the Boussinesq approximation in this case, it is necessary to start from the full non-adiabatic, viscous, unsteady, compressible NSF equations for the atmosphere, written in a rotating system of spherical coordinates (as in Zeytounian 1991, Chap. 2). More precisely (see Sect. 1.7.2), in a coordinate frame rotating with the Earth, we consider the full NSF equations from the beginning and take into account the Coriolis force ($2\boldsymbol{\Omega} \wedge \boldsymbol{u}$), the gravitational acceleration (modified by the centrifugal force) $\boldsymbol{g}$ and the effect of thermal radiation Q, where $\boldsymbol{u} = (\boldsymbol{v}, w)$ is the (relative) velocity vector as observed in the Earth's frame, rotating with an angular velocity $\boldsymbol{\Omega}$. The thermodynamic functions are, again, ρ (the atmospheric air density), p (the atmospheric air pressure) and T (the absolute temperature of the dry air). First, we assume that

$$\frac{\mathrm{d}\mathsf{R}_s}{\mathrm{d}z_s} = \rho_s Q, \quad \text{where } \mathsf{R}_s = \mathsf{R}_s(T_s(z_s)) \,, \tag{6.26}$$

and we note (again) that $\rho_s(z_s)$ and $T_s(z_s)$ in (6.26) are (as in Sect. 1.7.1) the density and temperature in the hydrostatic reference state (functions only of the altitude z_s; see (1.38)), and that Q, which represents a heat source (thermal radiation), is assumed to be a function only of this hydrostatic reference state via the 'standard' temperature $T_s(z_s)$. In doing this, we consider only a mean, standard, distribution for Q and ignore variations from it in the perturbed atmosphere in motion (namely, in 'free circulation'). However, in this case the temperature $T_s(z_s)$ for the hydrostatic reference state (the 'standard atmosphere') satisfies (in place of $\mathrm{d}T_s/\mathrm{d}z_s$ being a known function of z_s) the following equation, written here in dimensional form:

$$k_c \frac{\mathrm{d}T_s}{\mathrm{d}z_s} + \mathsf{R}_s(T_s(z_s)) = 0 \,, \tag{6.27}$$

where $k_c = \text{const}$ (in a simplified case).

6.3.1 More Concerning the Hydrostatic Dissipative Equations

Here, we give some complementary details concerning the derivation of the hydrostatic dissipative equations. First, as the vector of rotation of the Earth Ω is directed from south to north parallel to the axis joining the poles, it can be expressed as follows:

$$\boldsymbol{\Omega} = \Omega_0 \boldsymbol{e}, \quad \text{where } \boldsymbol{e} = (\boldsymbol{k} \sin\phi + \boldsymbol{j} \cos\phi) \,, \tag{6.28}$$

and where ϕ is the algebraic latitude of the point P^0 of observation on the Earth's surface around which the atmospheric free circulation is analysed ($\phi > 0$ in the northen hemisphere). The unit vectors directed to the east, to the north and to the zenith, the latter in the direction opposite to the vector $\boldsymbol{g} = -g\boldsymbol{k}$ (the force of gravity), are denoted by $\boldsymbol{i}, \boldsymbol{j}$ and $\boldsymbol{k}$, respectively. It is helpful to employ spherical coordinates λ, ϕ, r, and in this case u, v, w denote the corresponding relative velocity components parallel to these directions, i.e. of increasing azimuth (λ), latitude (ϕ) and radius (r), respectively. However, it is very convenient here to introduce the transformation (1.43) where ϕ^0 is a reference latitude (for $\phi^0 \approx 45°$, we have $a_0 \approx 6367\,\text{km}$, which is the mean radius of the Earth), and the origin of this right-handed curvilinear coordinate system lies on the Earth's surface (for flat ground, where $r = a_0$) at latitude ϕ^0 and longitude $\lambda = 0$. We assume, therefore, that the atmospheric free-circulation phenomenon occurs in a mid-latitude mesoscale region (with a horizontal scale $L_c \approx 10^5\,\text{m}$), distant from the equator, around some central latitude ϕ^0, and therefore that $\sin\phi^0, \cos\phi^0$ and $\tan\phi^0$ are all of order unity. For a *small* parameter $\delta = L_c/a_0$, in the leading order, x and y are the Cartesian coordinates in the f^0-plane (tangent) approximation. To use this approximation, it is necessary to introduce non-dimensional variables and functions, and perform a careful dimensional analysis (see, for instance, Zeytounian 1991, Chap. 2). Then, when $\delta = 0$ (ϕ being in this case ϕ^0) and as a consequence of the long-wave, hydrostatic limiting process (6.8a), with

$$t' = \frac{t}{1/\Omega_0}, \quad x' = \frac{x}{L_c}, \quad y' = \frac{y}{L_c} \quad \text{and } z' = \frac{z}{h_c}, \quad \text{fixed}, \tag{6.29}$$

a set of dimensionless hydrostatic dissipative equations for the dimensionless horizontal velocity $\boldsymbol{v}' = (u', v')$, the dimensionless vertical velocity w' and the dimensionless thermodynamic functions p', ρ', T' (as a function of the dimensionless time and space variables t', x', y', z') are derived. Namely (when the dynamic viscosity $\mu = \mu_c = \text{const}$ in the Stokes relation, and the conductivity coefficient $k = k_c = \text{const}$), we derive a set of dimensionless quasi-hydrostatic dissipative equations (see (1.57), for instance). Here, more precisely, we have (with the primes omitted and with a Strouhal number $St = \Omega_0 L_c/U_c$)

$$\rho\left\{St\frac{\mathrm{D}\boldsymbol{v}}{\mathrm{D}t} + \frac{1}{Ro}(\boldsymbol{k}\wedge\boldsymbol{v}) + \frac{1}{\gamma M^2}\boldsymbol{D}p = \frac{1}{Re_\perp}\frac{\partial^2\boldsymbol{v}}{\partial z^2},\right.$$

$$\frac{\partial p}{\partial z} + \tau_0\rho = 0\,,$$

$$St\,\frac{\mathrm{D}\rho}{\mathrm{D}t} + \rho\left\{\frac{\partial w}{\partial z} + (\boldsymbol{D}\,.\,\boldsymbol{v})\right\} = 0,$$

$$\rho\,St\,\frac{\mathrm{D}T}{\mathrm{D}t} - \frac{\gamma-1}{\gamma}St\,\frac{\mathrm{D}p}{\mathrm{D}t} = \frac{1}{Pr\,Re_\perp}\left\{\frac{\partial^2 T}{\partial z^2}\right.$$

$$\left. + (\gamma-1)\,Pr\,M^2\left|\frac{\partial\boldsymbol{v}}{\partial z}\right|^2 + \tau_0^2\sigma\frac{\mathrm{dR_s}(T_s)}{\mathrm{d}z_s}\right\},$$

$$p = \rho T\,, \tag{6.30}$$

where, in accordance with (6.8a), $Re_\perp = \lambda^2\, Re$ with $\lambda = \frac{h_c}{L_c}$, and h_c is defined by the first relation in (6.7). We observe that we have taken into account here the fact that our Boussinesq number is (see (6.7)) given by

$$\tau_0 = \frac{\Delta T^\circ}{T_s(0)} \equiv \frac{h_c}{RT_s(0)/g} \ . \tag{6.31}$$

In the above hydrostatic dissipative equations (6.30),

$$St\frac{\mathrm{D}}{\mathrm{D}t} = St\frac{\partial}{\partial t} + \boldsymbol{v}\,.\,\boldsymbol{D} + w\frac{\partial}{\partial z}, \quad \boldsymbol{D} = \frac{\partial}{\partial x}\boldsymbol{i} + \frac{\partial}{\partial y}\boldsymbol{j}, \quad \boldsymbol{k}\,.\,\boldsymbol{D} = 0\,. \tag{6.32}$$

In the first equation of (6.30), for the horizontal velocity vector, the Rossby number Ro is $U_c/f^0 L_c$, where $f^0 = 2\Omega_0 \sin\phi^0$, and the Mach number M is $U_c/(\gamma R T_s(0))^{1/2}$. As the Prandtl number, we have $Pr = C_p \mu_c/k_c$, where k_c and μ_c are the values of k and μ on the flat ground surface, and $Re = U_c L_c/\nu_c$ is the Reynolds number, where the kinematic viscosity $\nu_c = \mu_c/\rho_s(0)$ is assumed constant. In the fourth equation of (6.30), for T the dimensionless parameter σ is a measure of the (standard) heat source term, and from this equation in the case of a standard atmosphere (in hydrostatic equilibrium), we obtain the dimensionless form of (6.27) for the dimensionless standard temperature $T_s(z_s)$ (written without the primes). Namely,

$$\frac{\mathrm{d}T_s}{\mathrm{d}z_s} + \sigma \mathsf{R}_s = 0, \quad \text{since } z_s = \tau_0 z\ , \tag{6.33}$$

which is an equation defining $T_s(z_s)$ via $\mathsf{R}_s = \mathsf{R}_s(T_s(z_s))$, when suitable boundary conditions are assumed for z_s. The determination of $T_s(z_s)$ is, in reality a complicated problem (see Kibel 1957, Sect. 1.4, in the original Russian edition), but for our purpose here this problem does not have any influence, because (with dimensionless quantities) z_s tends to zero as τ_0 tends to zero, when z is fixed and $T_s(0) = 1$. Now, if we assume that the Strouhal number is of order unity, then the characteristic time is such that

$$Ro \approx \frac{1}{2\ \sin\phi^0} \approx 1\ , \tag{6.34a}$$

and in this case unsteadiness and the Coriolis force are both operative in the free-circulation problem. However, it is necessary to evaluate ΔT^0 (for instance in (6.6)) and U_c via the physical data of the free-circulation problem. On the one hand, the similarity rule $\lambda^2\ Re = Re_\perp \approx 1$, used with the limiting process (6.8a), when we take into account the definition of the characteristic vertical scale $h_c = R\ \Delta T^0/g$, gives us the possibility to evaluate ΔT^0 in the boundary condition (6.6); on the other hand, the second similarity rule in the Boussinesq limit (6.8b), $\tau_0/M = 1$, when we take into account the definition of $\tau_0 = \Delta T^0/T_s(0)$, gives us the possibility to evaluate the characteristic velocity

U_c, after making a judicious choice of L_c, ν_c and the standard temperature on the ground surface $z_s = 0, T_s(0)$. Namely,

$$\Delta T^0 = \frac{g}{R}\sqrt{\frac{\nu_c L_c}{U_c}}, \quad U_c = \Delta T^0 \sqrt{\frac{\gamma R}{T_s(0)}} . \tag{6.34b}$$

For the equations (6.30), we consider as the thermal condition on a flat, thermally non-homogeneous region D_c of the ground surface the following dimensionless thermal boundary condition (see (6.6)):

$$\begin{aligned} T &= 1 + \tau_0 \Theta(t, x, y), \quad \text{for } (x, y) \in D_c, \\ T &\equiv 1, \quad \text{if } (x, y) \notin D_c , \end{aligned} \tag{6.35}$$

again written without primes.

6.3.2 The Boussinesq 'Free'-Circulation Problem

Finally, from the Boussinesq free-circulation limiting process (6.8b), with $\tau_0 = M$ when both tend to zero, and the following asymptotic expansions with respect to the low Mach number M,

$$\boldsymbol{v} = \boldsymbol{v}_{\text{FC}} + O(M), \quad w = w_{\text{FC}} + O(M), \quad p = p_s(Mz)[1 + M^2 \pi_{\text{FC}} + \cdots] , \tag{6.36a}$$

$$T = T_s(Mz)\,[1 + M\theta_{\text{FC}} + \cdots], \quad \rho = \rho_s(Mz)\,[1 + M\omega_{\text{FC}} + \cdots] , \tag{6.36b}$$

we derive a set of hydrostatic, viscous, non-adiabatic Boussinesq equations for the limiting functions $\boldsymbol{v}_{\text{FC}}, w_{\text{FC}}, \pi_{\text{FC}}$ and θ_{FC} governing free-circulation phenomena:

$$\begin{aligned}
&\boldsymbol{D} \cdot \boldsymbol{v}_{\text{FC}} + \frac{\partial w_{\text{FC}}}{\partial z} = 0 , \\
&St \frac{\partial \boldsymbol{v}_{\text{FC}}}{\partial t} + (\boldsymbol{v}_{\text{FC}} \cdot \boldsymbol{D})\boldsymbol{v}_{\text{FC}} + w_{\text{FC}} \frac{\partial \boldsymbol{v}_{\text{FC}}}{\partial z} + \frac{1}{Ro}(\boldsymbol{k} \wedge \boldsymbol{v}_{\text{FC}}) \\
&\qquad + \boldsymbol{D}\left(\frac{\pi_{\text{FC}}}{\gamma}\right) = Gr_{\perp}^{-1/2} \frac{\partial^2 \boldsymbol{v}_{\text{FC}}}{\partial z^2} , \\
&\frac{\partial \pi_{\text{FC}}}{\partial z} = \theta_{\text{FC}}, \quad \omega_{\text{FC}} = -\theta_{\text{FC}} , \\
&St \frac{\partial \theta_{\text{FC}}}{\partial t} + (\boldsymbol{v}_{\text{FC}} \cdot \boldsymbol{D})\theta_{\text{FC}} + w_{\text{FC}} \frac{\partial \theta_{\text{FC}}}{\partial z} + \Lambda(0) w_{\text{FC}} \\
&\qquad = Gr_{\perp}^{-1/2} \frac{\partial^2 \theta_{\text{FC}}}{\partial z^2} ,
\end{aligned} \tag{6.37}$$

where

$$\Lambda(0) = \frac{\gamma - 1}{\gamma} + \left(\frac{\mathrm{d}T_s}{\mathrm{d}z_s}\right)_0 \quad \text{and } Gr_{\perp} \equiv Re_{\perp}^2 . \tag{6.38}$$

The above model equations (6.37) and (6.38) for free-circulation phenomena can be considered as an inner (boundary-layer) degenerate case of the full atmospheric NSF equations, where $\delta = 0$ under the limiting process (6.8) together with (6.36). However, because the corresponding outer degenerate case gives the trivial zero solution (as a consequence of the local character of the free circulation/convection), we obtain in fact the behaviour of the free circulation far from the assumed flat ground surface $z = 0$ (when $z \uparrow \infty$, there is no outer solution). Therefore, we must consider the following boundary conditions for the model equations (6.37), where we take into account also (6.35):

$$\begin{aligned} z \uparrow \infty: &\quad (\boldsymbol{v}_{\mathrm{FC}}, w_{\mathrm{FC}}, \pi_{\mathrm{FC}}, \theta_{\mathrm{FC}}) = 0 \,, \\ z = 0: &\quad \boldsymbol{v}_{\mathrm{FC}} = w_{\mathrm{FC}} = 0, \quad \theta_{\mathrm{FC}} = \Theta(t, x, y) \quad \text{when } (x, y) \in D \,, \\ &\quad (\boldsymbol{v}_{\mathrm{FC}}, w_{\mathrm{FC}}, \pi_{\mathrm{FC}}, \theta_{\mathrm{FC}}) = 0, \quad \text{when} \quad (x, y) \notin D \,. \end{aligned} \tag{6.39}$$

Concerning the initial conditions at $t = 0$ for $\boldsymbol{v}_{\mathrm{FC}}$ and θ_{FC}, it is necessary, in fact, to consider an unsteady adjustment problem (see Sect. 6.4) if we want to study the initial strongly unsteady transition process of the free circulation.

6.4 Complementary Remarks Relating to Derived Boussinesq Model Equations

6.4.1 The Problem of Initial Conditions

If we consider the two main model systems of equations (6.18) and (6.37), derived above in Sects. 6.2 and 6.3 via the Boussinesq approximation, then we observe that in these approximate systems (derived when the time is fixed and $O(1)$) we have an equation with a divergenceless velocity in place of the full unsteady, compressible continuity equation, and only two derivatives with respect to time: for $\boldsymbol{u}_{\mathrm{B}}$ and θ_{B} in (6.18), and for $\boldsymbol{v}_{\mathrm{FC}}$ and θ_{FC} in (6.37). On the other hand, in the full unsteady NSF equations, for a compressible, viscous, heat-conductive system, we have *three* derivatives with respect to time: for the velocity vector $\boldsymbol{u}$, the density ρ and the temperature T. As a consequence, if we want to solve a pure initial-value, or Cauchy (prediction), problem (in the L^2 norm, for example), it is necessary (for well-posedness) to have a complete set of initial conditions (data)

$$t = 0: \quad \boldsymbol{u} = \boldsymbol{u}^0(\boldsymbol{x}), \quad \rho = \rho^0(\boldsymbol{x}), \quad T = T^0(\boldsymbol{x}) \,, \tag{6.40}$$

where $\rho^0(\boldsymbol{x}) > 0$ and $T^0(x) > 0$. Now, if we consider, for instance, the system of Boussinesq equations (6.18), then we have the possibility to assume only *two* pieces of initial data, namely

$$t = 0: \quad \boldsymbol{u}_{\mathrm{B}} = \boldsymbol{u}^0_{\mathrm{B}}(\boldsymbol{x}) \quad \text{and } \theta_{\mathrm{B}} = \theta^0_{\mathrm{B}}(\boldsymbol{x}) \,, \tag{6.41a}$$

and we observe that for divergenceless Boussinesq flows, it is also necessary that

$$\text{the boundary integral } \int \boldsymbol{u} \cdot \boldsymbol{n} \, \mathrm{d}\sigma \text{ vanishes and } \nabla \cdot \boldsymbol{u}_{\mathrm{B}}^{0} = 0 \,. \tag{6.41b}$$

Naturally, this last constraint has no analogue for compressible flows, because of the occurence of the unsteady term $\partial\rho/\partial t$ in the compressible continuity equation. Indeed, the consideration of an unsteady adjustment problem is also necessary for obtaining initial conditions for the hydrostatic dissipative system (6.30), which is an intermediate approximate system for the derivation (via the Boussinesq limit (6.8b)) of the free-circulation model equations (6.37). More precisely, for the hydrostatic dissipative system (6.30), the singularity (close to the initial time) is strongly related to the hydrostatic balance (the second equation in (6.30)), which replaces the full unsteady equation for the vertical velocity w. As a consequence, the initial conditions for the free-circulation equations (6.37) for $\boldsymbol{v}_{\mathrm{FC}}$ and θ_{FC} are derived via two unsteady adjustment problems! In Guiraud and Zeytounian (1982), the problem of the adjustment to the hydrostatic balance was investigated, and the major conclusion was that the initial conditions for $\boldsymbol{v}$ and T in the system of hydrostatic equations (6.30), with (in the non-dissipative case) $Re_{\perp} = \infty$, may be derived from a full set of initial conditions for the non-hydrostatic Euler equations, under the f^0-plane approximation, by solving a one-dimensional unsteady problem of vertical motion and matching. After that, it is necessary to consider a second unsteady adjustment problem, to incompressibility, to obtain consistent initial conditions for $\boldsymbol{v}_{\mathrm{FC}}$ and θ_{FC} in the system of equations (6.37) for free circulation. Unfortunately, until now this second adjustment problem has not been considered, and this is also the case for the problem of adjustment to hydrostatic balance (the second equation in the hydrostatic dissipative system (6.30)) in the dissipative case when $Re_{\perp} = O(1)$. Below, we consider mainly the derivation of initial conditions for the Boussinesq equations (6.18).

Adjustment to Boussinesq State

If we consider our approximate simplified Boussinesq equations (6.18) derived above in Sect. 6.2, then it is necessary first to elucidate the problem of the adjustment to a Boussinesq flow – this unsteady adjustment to incompressibility being a result of the generation, dispersion and damping of the fast internal acoustic waves generated close to $t = 0$. More precisely, the obtaining of consistent initial conditions (at $t = 0$) for the limiting model Boussinesq equations is, in fact, a consequence of a matching between two asymptotic representations: the main one (Boussinesq, with t fixed) at $t = 0$ and a local one (acoustic, near $t = 0$, with $\tau = t/M$ fixed, when $\tau \to +\infty$):

$$\text{Lim [Boussinesq, at } t = 0] = \text{Lim [Acoustic, when } \tau \to +\infty] \,. \tag{6.42}$$

In order to solve such a problem, it is necessary to introduce an initial layer in the vicinity of $t = 0$ by distorting the timescale and the unknowns, which are initially undefined. In Zeytounian (1991, Chap. 5), the reader can find the solution to this problem for the Boussinesq equations, but the result is valid only when we assume the following initial conditions for the exact Euler equations (6.11):

$$t = 0 : \boldsymbol{u} = \boldsymbol{u}^0 = (\boldsymbol{u}^0, \boldsymbol{v}^0, \boldsymbol{w}^0), \quad \pi = M\pi^0, \quad \omega = M\omega^0 \quad \text{and } \theta = M\theta^0 \,, \tag{6.43a}$$

where the initial data are given functions of x, y and z, and the initial velocity vector is assumed to be of the following form:

$$\boldsymbol{u}^0 = \nabla\phi^0 + \nabla \wedge \boldsymbol{\psi}^0 \,. \tag{6.43b}$$

In this case, we introduce first a dimensionless short (acoustic) time $\tau = t/M$, using $\boldsymbol{u}_{\mathrm{A}}, \pi_{\mathrm{A}}, \omega_{\mathrm{A}}, \theta_{\mathrm{A}}$ to represent $\boldsymbol{u}, \pi, \omega, \theta$ considered as functions of the short time τ and of x, y, z. Then we set up a local (close to initial time) asymptotic expansion in place of the Boussinesq expansion (6.17), namely

$$(\boldsymbol{u}_{\mathrm{A}}, \pi_{\mathrm{A}}, \omega_{\mathrm{A}}, \theta_{\mathrm{A}}) = (\boldsymbol{u}_{\mathrm{A0}}, 0, 0, 0) + M(\pi_{\mathrm{A1}}, \omega_{\mathrm{A1}}, \theta_{\mathrm{A1}}) + \dots \,, \tag{6.44a}$$

and we observe that the (acoustic) expansion (6.44a) is the only consistent local asymptotic expansion matched to the Boussinesq expansion (6.17) when we assume the data (6.43a) as the initial condition for the Euler exact equations (6.11). In this case, from Euler exact equations together with

$$\boldsymbol{u}_{\mathrm{A0}} = \nabla\phi_{\mathrm{A0}} + \nabla \wedge \boldsymbol{\psi}_{\mathrm{A0}} \,, \tag{6.44b}$$

We derive the following initial-value acoustic adjustment problem for the velocity potential function $\phi_{A0}(\tau, x, y, z)$:

$$\frac{\partial^2 \phi_{\mathrm{A0}}}{\partial \tau^2} - \Delta\phi_{\mathrm{A0}} = 0, \quad \tau = 0 : \quad \phi_{\mathrm{A0}} = \phi^0, \quad \frac{\partial \phi_{\mathrm{A0}}}{\partial \tau} = -\frac{\pi^0}{\gamma} \,. \tag{6.45a}$$

We also obtain the two relations

$$\frac{\partial \phi_{\mathrm{A0}}}{\partial \tau} + \omega_{\mathrm{A1}} = \omega^0 - \frac{\pi^0}{\gamma} \qquad \text{and } \boldsymbol{\psi}_{\mathrm{A0}} \equiv \boldsymbol{\psi}^0 \,. \tag{6.45b}$$

The solution of (6.45a) is straightforward, and from it we obtain

$$\tau \uparrow \infty : \quad |\phi_{\mathrm{A0}}| \downarrow 0 \quad \text{and} \quad \left|\frac{\partial \phi_{\mathrm{A0}}}{\partial \tau}\right| \downarrow 0 \,, \tag{6.46}$$

which is a classical result for the d'Alembert acoustic-wave problem in an exterior domain (see Wilcox 1975).

Finally, as a consequence of (6.45b) and (6.46) and the matching condition

$$\lim{}_{\mathrm{A}}(\tau \uparrow \infty) = \lim{}_{\mathrm{B}}(t \downarrow 0) \,, \tag{6.47}$$

we obtain the following initial conditions for the Boussinesq equations (6.18):

$$t = 0: \quad \boldsymbol{u}_{\mathrm{B}} = \nabla \wedge \boldsymbol{\psi}^0, \quad \theta_{\mathrm{B}} = \frac{1}{\gamma}\pi^0 - \omega^0 \,. \tag{6.48}$$

The reader should keep in mind that no initial conditions are required on ω_{B} and π_{B}; more precisely, from the second of the Boussinesq equations (6.18), the initial value of π_{B} may be computed once the initial value of $\boldsymbol{u}_{\mathrm{B}}$ is known, according to the first condition in (6.48) and

$$\omega_{\mathrm{B}} = \omega^0 - \frac{1}{\gamma}\pi^0 \quad \text{for} \quad t = 0 \,. \tag{6.49}$$

What happens if the initial values of π, ω and θ, the solutions of the exact starting Euler equations (6.11), are different from the data $M\pi^0, M\omega^0$, and $M\theta^0$, as this is assumed in (6.43a), is not known, and this open problem deserves further investigation.

6.4.2 The Problem of the Upper Condition in the Boussinesq Case

The problem of an upper boundary condition at the top of the troposphere for the 2D steady Boussinesq equation (6.23) is an intriguing problem. If we assume that an upper flat plane (at altitude H_{s}), considered as the tropopause bounds the lee wave phenomenon (which occurs in the whole troposphere), then we must write the following dimensionless upper boundary conditions for the exact Euler equations (6.11):

$$\theta = 0, \quad w = 0 \quad \text{on } z = \frac{1}{Bo} \,, \tag{6.50}$$

where $w = \boldsymbol{u} \,.\, \boldsymbol{k}$. These equations are very singular when $Bo \to 0$. We observe that the dimensionless equation for the top of the troposphere is assumed as: $z = H_{\mathrm{c}}/H_{\mathrm{s}} = 1/Bo$, and as a consequence the outer region is bounded by the plane $\zeta = 1/B^*$, where $B^* = Bo/M = O(1)$ and $\zeta = Mz$. As a consequence, in reality, for the 2D steady Boussinesq equation (6.23), we obtain as the upper condition the 'paradoxical' behaviour condition

$$\Delta_{\mathrm{B}}(x, z \uparrow +\infty) = +\infty \,. \tag{6.51}$$

Obviously, the 'infinity in altitude' relative to z for the Boussinesq/Long model problem (6.23) with the conditions (6.25 a,b,c) must be understood as a behaviour condition relative to the 'inner' vertical (Boussinesq) coordinate z (using dimensionless quantities), which is matched to an outer coordinate $\zeta = Mz$.

This outer vertical dimensionless coordinate ζ, taking into account the upper condition at the top of the troposphere $H_{\mathrm{s}} = RT_{\mathrm{s}}(0)/g \gg H_{\mathrm{B}}$ (given by (6.10)), is lost in the framework of the Boussinesq 'inner' problem.

The Radiation (Sommerfeld) Condition

In an unbounded atmosphere (which is, in fact here, a boundary-layer-type inner region), it is necessary to impose a radiation Sommerfeld condition on Δ_{B}, the solution of the Helmholtz equation (6.23); namely:

$$\Delta_{\mathrm{B}} \approx \left[\frac{2K_0}{\pi r}\right]^{1/2} \sin\theta \,\mathrm{Real}\left\{G(\cos\theta)\exp\left[\mathrm{i}\left(K_0 r - \frac{\pi}{4}\right)\right]\right\} , \tag{6.52}$$

when $r = (x^2 + z^2)^{1/2} \to \infty$, where $K_0 \equiv (B^{*2}/\gamma)N^2(0)$, and the function $G(\cos\theta)$ is arbitrary and depends on the form of the relief simulated by $z = \delta\eta(\alpha x)$. So, to satisfy the condition on the behaviour at infinity upstream (for $x \to -\infty$), the following condition must also be imposed:

$$G(\cos\theta) = 0 \quad \text{for } \cos\theta < 0 . \tag{6.53}$$

The polar coordinates r, θ, in the upper half-plane $z > 0$ are defined such that $x = r\cos\theta$ and $z = r\sin\theta$. The inner Boussinesq model problem is, in fact, the problem considered by Miles (1968) and also by Kozhevnikov (1963, 1968), with the conditions (6.52) and (6.53), which express the condition that 'no waves are radiated inwards'. In Bois (1984), which gives an asymptotic theory of lee waves in an unbounded atmosphere, it is shown that, within a generalized Boussinesq approximation, the linearized atmospheric flow over relief can, in realistic cases, be approximated by a confined flow.

The Result of Guiraud and Zeytounian (1979)

In a paper by Guiraud and Zeytounian (1979), the associated outer problem is also considered, with an upper slip condition on the rigid plane $\zeta = 1/B^*$, and it is shown that the upper and lower boundaries of the troposphere alternately reflect internal short-wavelength gravity waves excited by the lee waves of the inner (Boussinesq) approximation, with a wavelength of the order of the Mach number, on the scale of the outer region. As a consequence, there is a double scale built into the solution and we must take care of it. The important point of the analysis of Guiraud and Zeytounian (1979) is that: *these short-wavelength gravity waves propagate downstream and that no feedback occurs on the inner Boussinesq flow close to the mountain* (to the lowest order at least). As a consequence, we should understand the imposed upper boundary at the top of the troposphere as an *artificial one*, having asymptotically *no effect* on the *inner Boussinesq flow, which is the only really interesting flow.*

6.4.3 Non-Boussinesq Effects: Isochoric and Deep-Convection Approximate Equations

Obviously, our approach to the derivation of the model equations 'à Boussinesq' (6.18) and (6.37) gives us the possibility to derive the associated *well-balanced* 'second-order' linear approximate equations, which include various non-Boussinesq (compressible) effects, but here, in the present book, we do not consider such a laborious (but straightforward!) derivation of second-order models. Naturally, this derivation certainly deserves a further accurate analysis to take account of weakly compressible (non-Boussinesq) effects. Below, we discuss briefly two 2D steady model equations which generalize the Boussinesq 2D steady equation (6.23) and take into account some non-Boussinesq effects.

Isochoric Model Equations

If we consider an Eulerian (non-viscous) motion with a constant internal energy per unit mass – an *isochoric* flow, where $e = e_{\mathrm{c}} \equiv \mathrm{const}$ – then we obtain, from the thermodynamic adiabatic equation for the energy, the result, that this flow is necessarily *divergenceless*:

$$\nabla \cdot \boldsymbol{u}_{\mathrm{Is}} = 0 \,. \tag{6.54a}$$

As a consequence, from the continuity equation, in place of the conservation of specific entropy, we obtain the conservation of the isochoric density,

$$\frac{\mathrm{D}\rho_{\mathrm{Is}}}{\mathrm{D}t} = 0 \,. \tag{6.54b}$$

With the compressible momentum equation,

$$\rho_{\mathrm{Is}}\frac{\mathrm{D}\boldsymbol{u}_{\mathrm{Is}}}{\mathrm{D}t} + \nabla p_{\mathrm{Is}} = \rho_{\mathrm{Is}}\boldsymbol{g} \,, \tag{6.54c}$$

we have an isochoric system of three equations (6.54), for $\boldsymbol{u}_{\mathrm{Is}}, \rho_{\mathrm{Is}}$ and p_{Is}.

We observe that, in the framework of the Euler compressible, non-viscous dimensionless equations for a perfect gas (see, for instance, (1.30a)), the isochoric equations can also be derived under the limiting process

$$M \to 0 \quad \text{and}\ \gamma \to \infty, \quad \text{such that}\ \gamma M^2 = M^* = O(1) \,. \tag{6.55}$$

We have $\gamma \to \infty$ because, for a perfect 'isochoric' gas, $C_{\mathrm{p}} = O(1)$ but $C_v \to 0$, and in this case we have the relation (in dimensional notation)

$$T_{\mathrm{Is}} = \frac{p_{\mathrm{Is}}}{C_p \rho_{\mathrm{Is}}} \tag{6.54d}$$

for the isochoric temperature. In Yih (1980), monograph the reader can find a theory of stratified flows based mainly on the above three isochoric equations (6.54a,b,c); sometimes (in the work of many authors) the distinction between the isochoric and Boussinesq equations is not clearly made. In fact, the difference is twofold. For instance, on the one hand, in the steady 2D inviscid case, we can derive, in place of the dimensionless linear Helmholtz–Boussinesq equations (6.23), the following dimensionless quasi-nonlinear equation for the (isochoric) function $\Delta_{\mathrm{Is}}(x,z)$ (see, for instance, Zeytounian (2002a), pp. 104–108, and also Sect. 7.3.2 in the present book):

$$\frac{\partial^2 \Delta_{\mathrm{Is}}}{\partial x^2}+\frac{\partial^2 \Delta_{\mathrm{Is}}}{\partial z^2}+\frac{Bo^2}{M^*}N_{\mathrm{Is}}(Bo\,z)\Delta_{\mathrm{Is}} = \frac{Bo}{2}\left[2\frac{\partial \Delta_{\mathrm{Is}}}{\partial z}-\left(\frac{\partial \Delta_{\mathrm{Is}}}{\partial x}\right)^2-\left(\frac{\partial \Delta_{\mathrm{Is}}}{\partial z}\right)^2\right], \tag{6.56}$$

where

$$N_{\mathrm{Is}}(z_{\mathrm{s}})=\frac{1}{T_{\mathrm{s}}(z_{\mathrm{s}})}\left[1+\frac{\mathrm{d}T_{\mathrm{s}}(z_{\mathrm{s}})}{\mathrm{d}z_{\mathrm{s}}}\right], \quad z_{\mathrm{s}}=Bo\,z\,.$$

On the other hand, this isochoric equation (6.56) is valid throughout the whole thickness of the troposphere (Bo being unity). Moreover, when $Bo \to 0$ we again recover a Helmholtz equation, as in the case of the 2D steady Boussinesq case (see (6.23)), but for this it is necessary to assume that

$$Bo = B^* M \to 0 \quad \text{and } \gamma = O(1)\,. \tag{6.57}$$

With (6.57) we derive again, from (6.56), the Helmholtz equation (6.23) for $\lim \Delta_{\mathrm{Is}}$ when $Bo \to 0$, but with $K_{\mathrm{Is}} = (B^{*2}/\gamma)N^2_{\mathrm{Is}}(0)$ in place of the coefficient $K_0 = B^{*2}/\gamma)N^2(0)$.

Deep-Convection Model Equations

When $Bo = O(1)$, we have also the possibility to assume a constraint on the parameter $N^2(z_{\mathrm{s}})$ given by (6.12). Indeed in the whole thickness of the troposphere, the temperature gradient in dimensionless form for $T_{\mathrm{s}}(z_{\mathrm{s}})$, $-\mathrm{d}T_{\mathrm{s}}(z_{\mathrm{s}})/\mathrm{d}z_{\mathrm{s}}$, is very close to $(\gamma-1)/\gamma$, and as a consequence we can assume that

$$-\frac{\mathrm{d}T_{\mathrm{s}}(z_{\mathrm{s}})}{\mathrm{d}z_{\mathrm{s}}}=\frac{\gamma-1}{\gamma}+M^{\beta}\chi(z_{\mathrm{s}})\,, \tag{6.58}$$

where $\beta > 0$ is arbitrary and $|\chi(z_{\mathrm{s}})| = O(1)$.

In such a case, (6.14b) becomes

$$\frac{\mathrm{D}\theta_{\mathrm{D}}}{\mathrm{D}t}-\frac{Bo}{M^e}M^{\beta}\frac{\chi(Bo\,z)}{T_{\mathrm{s}}(Bo\,z)}(\boldsymbol{u}_{\mathrm{D}}\,.\,k) -M^{b-e}\frac{\gamma-1}{\gamma}\frac{\mathrm{D}\pi_{\mathrm{D}}}{\mathrm{D}t}=O(Bo\,M^{\beta})+O(Bo^2)\,, \tag{6.59a}$$

when we make the replacements

$$\boldsymbol{u}_{\mathrm{B}} \to \boldsymbol{u}_{\mathrm{D}}, \theta_{\mathrm{B}} \to \theta_{\mathrm{D}}, \quad \pi_{\mathrm{B}} \to \pi_{\mathrm{D}} \quad \text{and } \omega_{\mathrm{B}} \to \omega_{\mathrm{D}}$$

in the asymptotic expansions (6.13). On the other hand, in place of (6.14a), using (6.58), we write

$$\nabla \cdot \boldsymbol{u}_{\mathrm{D}} = -M^e \left[\frac{\mathrm{D}\omega_{\mathrm{D}}}{\mathrm{D}t} + \omega_{\mathrm{D}} \nabla \cdot \boldsymbol{u}_{\mathrm{D}} \right] + Bo \frac{1}{\gamma T_{\mathrm{s}}(Bo\, z)} (\boldsymbol{u}_{\mathrm{D}} \cdot \boldsymbol{k}) + O(Bo\, M^e) + O(Bo\, M^\beta) \,. \tag{6.59b}$$

The equations (6.14c) and (6.14d) are unchanged:

$$\omega_{\mathrm{D}} + \theta_{\mathrm{D}} = M^{b-e} \pi_{\mathrm{D}} + O(M^e) \,, \tag{6.59c}$$

$$\frac{\mathrm{D}\boldsymbol{u}_{\mathrm{D}}}{\mathrm{D}t} + M^{b-2} \frac{T_{\mathrm{s}}(Bo\, z)}{\gamma} \nabla \pi_{\mathrm{D}} - \frac{1}{\gamma} Bo\, M^{e-2} \theta_{\mathrm{D}} \boldsymbol{k} = O(M^e) \,. \tag{6.59d}$$

Now, from (6.59a), when $Bo = O(1)$, if we want to take into account the weak stratification term proportional to $\chi(Bo\, z)$, then it is necessary to choose $\beta = 2$, which is consistent with regard to the buoyancy term and the pressure term in (6.59d) only if $e = b = 2$. In this case, in place of the first and fourth Boussinesq equations (6.18), we derive the following two limiting deep-convection equations; namely

$$\nabla \cdot \boldsymbol{u}_{\mathrm{D}} = Bo \frac{1}{\gamma T_{\mathrm{s}}(Bo\, z)} (\boldsymbol{u}_{\mathrm{D}} \cdot \boldsymbol{k}) \tag{6.60a}$$

and

$$\omega_{\mathrm{D}} + \theta_{\mathrm{D}} = \pi_{\mathrm{D}} \,. \tag{6.60b}$$

Then, when $Bo = O(1)$ and $M \to 0$, with

$$\frac{N^2(Bo\, z)}{M^2} = -\frac{\chi(Bo\, z)}{T_{\mathrm{s}}(Bo\, z)} = O(1) \,, \tag{6.61}$$

we derive, in place of the second and third Boussinesq equations (6.18), the following two (non-Boussinesq) equations for $\boldsymbol{u}_{\mathrm{D}}$ and θ_{D}:

$$\frac{\mathrm{D}_{\mathrm{D}}\boldsymbol{u}_{\mathrm{D}}}{\mathrm{D}t} + \frac{T_{\mathrm{s}}(Bo\, z)}{\gamma} \nabla \pi_{\mathrm{D}} - \frac{1}{\gamma} Bo\, \theta \boldsymbol{k} = 0 \,, \tag{6.60c}$$

$$\frac{\mathrm{D}_{\mathrm{D}}\theta_{\mathrm{D}}}{\mathrm{D}t} - \frac{\gamma - 1}{\gamma} \frac{\mathrm{D}_{\mathrm{D}}\pi_{\mathrm{D}}}{\mathrm{D}t} - Bo \frac{\chi(Bo\, z)}{T_{\mathrm{s}}(Bo\, z)} (\boldsymbol{u}_{\mathrm{D}} \cdot \boldsymbol{k}) = 0 \,, \tag{6.60d}$$

where $Bo = O(1)$ and $\mathrm{D}_{\mathrm{D}}/\mathrm{D}t = \partial/\partial t + \boldsymbol{u}_{\mathrm{D}} \cdot \nabla$. For the functions $\boldsymbol{u}_{\mathrm{D}}, \omega_{\mathrm{D}}, \theta_{\mathrm{D}}$ and π_{D}, we have the above four deep-convection equations (6.60) in place of the Boussinesq equations (6.18), which are valid throughout the whole troposphere. The term $Bo[\chi(Bo\, z)/T_{\mathrm{s}}(Bo\, z)](\boldsymbol{u}_{\mathrm{D}} \cdot k)$ in (6.60d) takes into account a weak stratification relative to the basic standard temperature,

$$T_{\mathrm{s}}(Bo\, z) = 1 - \frac{Bo}{\gamma}(\gamma - 1) z, \quad \text{when } M \to 0 \,, \tag{6.62}$$

and in the equation (6.60d) for the perturbation of the temperature θ_{D} we have also a term dependent on the pressure perturbation π_{D}. This perturbation of the pressure π_D is also present in the determination of the perturbation of the density $\omega_D = \pi_D - \theta_D$, from (6.60b), and from (6.60a) we say that deep tropospheric convection is no longer associated with a divergenceless velocity vector $\boldsymbol{u}_D$. The deep-convection equations (6.60) can be derived from the exact Euler equations (6.11) via the following asymptotic expansion:

$$\boldsymbol{u} = \boldsymbol{u}_D + \cdots , \quad (\omega, \theta, \pi) = M^2(\omega_D, \theta_D, \pi_D) + \dots , \tag{6.63}$$

together with (6.61), when $Bo = O(1)$ and $M \to 0$. In particular, for a steady, 2D (so that $\boldsymbol{u}_D = (u_D, w_D)$), lee waves problem in the whole troposphere, we can introduce a stream function $\Psi_D(x, z)$ such that

$$(u_D; w_D) = \left[1 - \frac{(\gamma-1)Bo}{\gamma} z\right]^{-1/(\gamma-1)} \left(-\frac{\partial \Psi_D}{\partial z}; +\frac{\partial \Psi_D}{\partial x} \right) , \tag{6.64}$$

and we derive the following integral from (6.60d):

$$\theta_D - \frac{\gamma-1}{\gamma}\pi_D - Bo \int \frac{\chi(Bo\, z)}{T_s(Bo\, z)} \mathrm{d}z = \Theta(\Psi_D) . \tag{6.65a}$$

Then, from (6.60c), using also (6.60a), we derive a second, vorticity, integral,

$$\left[1 - \frac{(\gamma-1)Bo}{\gamma} z\right]^{-1/(\gamma-1)} \Omega_D - \frac{Bo}{\gamma} z \frac{\mathrm{d}\Theta(\Psi_D)}{\mathrm{d}\Psi_D} = \omega(\Psi_D) . \tag{6.65b}$$

Finally, via a straightforward calculation, the following single equation for Ψ_D is derived:

$$\frac{\partial^2 \Psi_D}{\partial x^2} + \frac{\partial^2 \Psi_D}{\partial z^2} + \frac{Bo}{\gamma}\left[1 - \frac{(\gamma-1)Bo}{\gamma} z\right]^{-1} \frac{\partial \Psi_D}{\partial z}$$
$$= \left[1 - \frac{(\gamma-1)Bo}{\gamma} z\right]^{-2/(\gamma-1)} \left\{ \omega(\Psi_D) + \frac{Bo}{\gamma} z \frac{\mathrm{d}\Theta(\Psi_D)}{\mathrm{d}\Psi_D} \right\} . \tag{6.66}$$

which is an extended form of the Boussinesq equation (6.22). In Pekelis (1976), the reader can find some numerical results for deep lee waves in the troposphere, which are computed via an equation very similar to (6.66), but derived from the 'anelastic' equations of Ogura and Phillips (1962). We observe also that, in the 2D steady case, the Boussinesq (6.23), isochoric (6.56) and deep-convection (6.60) model equations have been derived in Zeytounian (1979) directly from a single exact vorticity equation, deduced from an exact Euler steady, compressible, 2D, non-viscous, adiabatic dimensionless system of equations for a thermally perfect gas (see, for instance, Sect. 7.3.2 in this book). Concerning the Boussinesq case and the well-known Long (1955) problem, the reader can finds in Zeytounian (1969) various results of numerical computations of a flow downstream of an obstacle, and also a discussion concerning parasitic solutions and their filtering via an algorithm, based on the proposition that: *if the flow downstream of the obstacle is established then the flow should hardly be perturbed, globally, far enough upstream.*

6.5 Low-Mach-Number ($M \ll 1$) Asymptotics of the Hydrostatic Model Equations, With $Ki = O(1)$

The atmospheric motion within a thin sheet spread over a plane tangential to the Earth's surface can be investigated very well in the framework of the classical hydrostatic approximation when the horizontal length scale of the motion L_c is much more than the thickness of the troposphere, which can be characterized by the height H_s of a homogeneous hydrostatic atmosphere. In such a case

$$Bo \equiv 1, \quad \varepsilon = \frac{H_c}{L_c} \ll 1 , \tag{6.67a}$$

where $H_c \leq H_s$ is the characteristic vertical length of the 'tangential' motion. Now, if

$$H_s \ll L_c \ll a_0, \quad \text{or } \delta \ll 1 , \tag{6.67b}$$

where $a_0 \approx 6367\,\text{km}$ is the Earth's radius, then, for the horizontal velocity $\boldsymbol{v} = (u, v)$ (we write simply $\boldsymbol{v}$ in place of $\boldsymbol{v}_\perp$), for the local height H of an isobaric surface above the flat ground surface ($z = H(t, x, y, p)$), for the total derivative of the pressure $\varpi = \mathrm{D_p}/\mathrm{D_t}$ (which plays the role of the vertical velocity component) and for the temperature T (all functions of the 'time space–pressure coordinates' t, x, y, p), a system of dissipative hydrostatic equations can derived when we take into account the following similarity rule:

$$\varepsilon^2 \, Re = Re_\perp . \tag{6.67c}$$

We observe that when the Kibel number Ki is $O(1)$, we do not have the possibility to take into account the 'β-effect' (which is a trace of the sphericity) because the ratio δ/Ki tends to zero as $\delta \to 0$. With the above assumptions, we write the following dissipative hydrostatic equations (when $St \equiv 1$, we have $Ro \equiv Ki$):

$$\boldsymbol{D} \,.\, \boldsymbol{v} + \frac{\partial \varpi}{\partial p} = 0 , \tag{6.68a}$$

$$\frac{\partial \boldsymbol{v}}{\partial t} + (\boldsymbol{v} \,.\, \boldsymbol{D})\boldsymbol{v} + \varpi \frac{\partial \boldsymbol{v}}{\partial p} + \frac{1}{Ki}(\boldsymbol{k} \wedge \boldsymbol{v}) + \frac{1}{\gamma M^2} \boldsymbol{D} H$$
$$= \frac{1}{Re_\perp} \frac{\partial}{\partial p} \left[\mu \rho \frac{\partial \boldsymbol{v}}{\partial p} \right] , \tag{6.68b}$$

$$\frac{\partial T}{\partial t} + \boldsymbol{v} \,.\, \boldsymbol{D} T + \varpi K(T) = \frac{1}{\mathrm{Re}_\perp} \left\{ \frac{\partial}{\partial p} \left[\frac{\mu}{Pr} \rho \frac{\partial T}{\partial p} \right] \right.$$
$$\left. - \sigma \frac{\mathrm{dR_s}}{\mathrm{d}p} \right\} + O\left(\frac{M^2}{Re_\perp} \right) , \tag{6.68c}$$

$$T = -p \frac{\partial H}{\partial p} , \tag{6.68d}$$

where

$$K(T) \equiv \frac{\partial T}{\partial p} - \frac{\gamma - 1}{\gamma}\frac{T}{p}\,, \tag{6.68e}$$

and $\boldsymbol{D}$ is the horizontal gradient with respect to (x, y). The Kibel primitive (inviscid, adiabatic) equations are obtained when $1/Re_{\perp} \equiv 0$, namely

$$\boldsymbol{D}\,.\,\boldsymbol{v} + \frac{\partial \varpi}{\partial \mathrm{p}} = 0\,, \tag{6.69a}$$

$$\frac{\partial \boldsymbol{v}}{\partial t} + (\boldsymbol{v}\,.\,\boldsymbol{D})\boldsymbol{v} + \varpi\frac{\partial \boldsymbol{v}}{\partial p} + \frac{1}{Ki}(\boldsymbol{k} \wedge \boldsymbol{v}) + \frac{1}{\gamma M^2}\boldsymbol{D}H = 0\,, \tag{6.69b}$$

$$\frac{\partial T}{\partial t} + \boldsymbol{v}\,.\,\boldsymbol{D}T + \varpi K(T) = 0\,, \tag{6.69c}$$

$$T = -p\frac{\partial H}{\partial p}\,. \tag{6.69d}$$

It is useful (in the field of dynamic meteorology) to extract two scalar equations from the above Kibel primitive system of equations (6.69): one for

$$\Omega = \boldsymbol{k}\,.\,(\boldsymbol{D} \wedge \boldsymbol{v}) = \frac{\partial v}{\partial x} - \frac{\partial u}{\partial y}, \quad \boldsymbol{v} = (u, v)\,, \tag{6.70a}$$

and one for the horizontal divergence,

$$D = \boldsymbol{D}\,.\,\boldsymbol{v} = \frac{\partial u}{\partial x} + \frac{\partial v}{\partial y}\,, \tag{6.70b}$$

and we find, in place of the two equations (6.69a,b), the following set of three scalar equations for D and Ω:

$$D + \frac{\partial \varpi}{\partial p} = 0\,, \tag{6.70c}$$

$$\left(\frac{\partial}{\partial t} + \boldsymbol{v}\,.\,\boldsymbol{D} + \varpi\frac{\partial}{\partial p}\right)D + D^2 + 2\left(\frac{\partial v}{\partial x}\frac{\partial u}{\partial y} - \frac{\partial u}{\partial x}\frac{\partial v}{\partial y}\right)$$
$$+ \frac{\partial u}{\partial p}\frac{\partial \varpi}{\partial x} + \frac{\partial v}{\partial p}\frac{\partial \varpi}{\partial y} - Ki^{-1}\Omega + \frac{1}{\gamma M^2}\boldsymbol{D}^2 H = 0\,, \tag{6.70d}$$

$$\left(\frac{\partial}{\partial t} + \boldsymbol{v}\,.\,\boldsymbol{D} + \varpi\frac{\partial}{\partial p}\right)\Omega + D\Omega + 2\left(\frac{\partial \varpi}{\partial x}\frac{\partial v}{\partial p} - \frac{\partial \varpi}{\partial y}\frac{\partial u}{\partial p}\right) + Ki^{-1}\,D = 0\,. \tag{6.70e}$$

We observe that the Kibel primitive equations (6.69) are, in fact, a system of three equations for $\boldsymbol{v}$, H and ϖ, if we use the relation (6.69d) for T in (6.69c); namely, in place of (6.69c), we derive the following equation (6.69c′) for H:

$$\left(\frac{\partial}{\partial t} + \boldsymbol{v}\,.\,\boldsymbol{D}\right)\left(\frac{\partial H}{\partial p}\right) + \varpi K(H) = 0, \quad K(H) \equiv \frac{\partial^2 H_0}{\partial p^2} + \frac{1}{\gamma p}\frac{\partial H_0}{\partial p}\,. \tag{6.69c′}$$

For this system, we must write a *slip condition* on the ground. For the case of a flat ground surface, we have the condition

$$\left[\frac{\partial}{\partial t} + \boldsymbol{v} \;.\, \boldsymbol{D} + \varpi \frac{\partial}{\partial p}\right] H = 0 \quad \text{on}\, H = 0\,. \tag{6.71a}$$

We must consider the pressure on the flat ground, $p_{\text{ground}} = P_{\text{g}}(t, x, y)$, as one unknown. From $\varpi = \mathrm{D}p/\mathrm{D}t$, we derive the following equation for the determination of $P_{\text{g}}(t, x, y)$:

$$\frac{\partial P_{\text{g}}}{\partial t} + \boldsymbol{v} \,.\, \boldsymbol{D} P_{\text{g}} = \varpi(t, x, y, P_{\text{g}})\,. \tag{6.71b}$$

and the slip condition (6.71a) can be written for

$$p = P_{\text{g}}(t, x, y), \quad \text{provided that} \;\; H(t, x, y, P_{\text{g}}(t, x, y)) = 0\,. \tag{6.71c}$$

Below, our main small parameter is the Mach number $M \ll 1$. However, in the dissipative case we have also the parameter $1/Re_{\perp}$ on the right-hand side of (6.68b,c). If we assume that $1/Re_{\perp} \ll 1$ (for the case where $U_{\text{c}} \gg L_{\text{c}} v_{\text{c}}/H_{\text{c}}^2$) then it is necessary to write a similarity rule between the two small parameters M and $1/Re_{\perp}$,

$$Re_{\perp} = Re_{\perp}^{*}\, M^{-\alpha}, \quad Re_{\perp}^{*} = O(1), \quad \alpha > 0\,. \tag{6.72}$$

In the dissipative case, it is necessary to add to the slip condition (6.71a) two new boundary conditions. Namely, for a flat ground surface, together with (6.71a), we have also

$$\boldsymbol{v} = 0, \quad \frac{\mu}{Pr} p \frac{\partial \log T}{\partial p} = \sigma \mathsf{R}_{\text{s}}(p)\,. \tag{6.73}$$

In order to complete the definition of our two (Eulerian and dissipative) problems to be solved, we must add initial conditions, namely

$$t = 0: \quad \boldsymbol{v} = \boldsymbol{v}^0(x, y, p), \quad H = H^0(x, y, p), \quad P_{\text{g}} = P_{\text{g}}^0(x, y)\,, \tag{6.74}$$

but we observe that the initial data $\boldsymbol{v}^0, H^0$ and P_{g}^0 are, in general, different for the Eulerian (6.69) and dissipative (6.68) cases. Finally, some behaviour conditions are also necessary at infinity, first, when $p \uparrow 0$ at high altitude, and second, with respect to the horizontal coordinates x and y, along the isobaric surfaces. Concerning the conditions with respect to x and y, this is a very difficult question and could be answered convincingly only by considering a matching with a solution on an extended horizontal scale. The consistent (asymptotically significant) formulation of the condition at $p = 0$ is an another delicate question. However, obviously, the derivation of consistent low-Mach-number model equations from the systems (6.68) and (6.69) is, in fact, independent of these behaviour conditions although these conditions are especially important in the process of the (numerical) solution of the derived model equations. Before we consider the derivation and discussion (in Sect. 6.5.1) of the low-Mach-number, Monin–Charney model, according to Monin (1961) and Charney (1962). We observe that from the dissipative equation (6.68c) for T, thanks to the thermal condition in (6.73), we have the possibility to assume that, in general, for the leading term in the asymptotic expansion of

$$U = (\boldsymbol{v}, T, H, \varpi) = U_0 + MU_1 + M^2U_2 + \ldots + M^\alpha U_\alpha + \ldots\,, \tag{6.75a}$$

we have

$$U_0 = (\boldsymbol{v}_0(t,x,y,p), T_0(p), H_0(p), \varpi_0 = 0)\,, \tag{6.75b}$$

where

$$H_0(p) = -\int_{P_{\mathrm{g}}}^{p} \frac{T_0(p')}{p'}\mathrm{d}p'\,, \tag{6.75c}$$

since, according to (6.71c), $H_0\ (P_{\mathrm{g}}) = 0$. We assume here that

$$K(T_0(p)) = \frac{\mathrm{d}T_0}{\mathrm{d}p} - \frac{\gamma-1}{\gamma}\frac{T_0}{p} = -\frac{K_0(p)}{p}\,, \tag{6.76a}$$

where

$$K_0(p) = \frac{\gamma-1}{\gamma}T_0(p) - p\frac{\mathrm{d}T_0(p)}{\mathrm{d}p} \tag{6.76b}$$

is different from zero and is assumed to be positive. Rigorously, the above relation (6.75b) with (6.75c), for U_0 in the low-Mach-number expansion (6.75a), is a direct consequence of the requirement that secularities do not appear during a sufficiently long period of time (since the perturbation of the temperature T_α must be bounded). Indeed, we can derive for T_α, at the order M^α, the equation

$$\frac{\partial T_\alpha}{\partial t} + \boldsymbol{v}_0 \,.\, \boldsymbol{D}T_\alpha + \varpi_0 K(T_\alpha) + \varpi_\alpha K(T_0)$$
$$= \frac{1}{Re^*_\perp}\left\{\frac{\partial}{\partial p}\left[\frac{\mu_0}{Pr}\rho_0\frac{\partial T_0}{\partial p}\right] - \sigma\frac{\mathrm{dR_s}}{\mathrm{d}p}\right\} \tag{6.68c'}$$

from the equation (6.68c), when we use the similarity rule (6.72). As a consequence, it follows that the right-hand side of this approximate non-adiabatic equation (6.68c′) for T_α must be zero.

6.5.1 The Monin–Charney Quasi-Non-Divergent Model

We assume now that $M \ll 1$ but $Ki = O(1)$, and consider in the framework of the Kibel primitive (inviscid, adiabatic) equations the limiting process

$$\mathrm{M} \downarrow 0 \quad \text{with}\, t, x, y \text{ and } p \text{ fixed}\,, \tag{6.77a}$$

together with the expansion

$$
\begin{aligned}
(\boldsymbol{v}, \varpi, H, T, P_{\mathrm{g}}) = (\boldsymbol{v}_0, T_0(p), H_0(p), \varpi_0 = 0, P_{\mathrm{g}0} = 1) \\
+ M^2(\boldsymbol{v}_2, \varpi_2, H_2, T_2, P_{\mathrm{g}2}) \\
+ M^4(\boldsymbol{v}_4, \varpi_4, H_4, T_4, P_{\mathrm{g}4}) + \dots \,,
\end{aligned}
\tag{6.77b}
$$

where (since we are working with dimensionless quantities)

$$
H_0(P_{\mathrm{g}0}) = 0 \quad \Rightarrow P_{\mathrm{g}0} = 1 \,. \tag{6.78}
$$

Taking it for granted that (6.77b) holds, we find from the two equations (6.69a,b) that v_0 and H_2 must satisfy the following system of two model equations:

$$
\boldsymbol{D} \,.\, \boldsymbol{v}_0 = 0 \,, \tag{6.79a}
$$

$$
\frac{\partial \boldsymbol{v}_0}{\partial t} + (\boldsymbol{v}_0 \,.\, \boldsymbol{D})\mathbf{v}_0 + \frac{1}{Ki}(\boldsymbol{k} \wedge \boldsymbol{v}_0) + \boldsymbol{D}\left(\frac{H_2}{\gamma}\right) = 0 \,. \tag{6.79b}
$$

An equivalent system of two equations, in place of (6.79), can be written for H_2/γ and the vorticity $\Omega_0 = \partial v_0/\partial x - \partial u_0/\partial y = \boldsymbol{D}^2\psi_0$ if we take (6.70b,c) into account:

$$
\boldsymbol{D}^2\left(\frac{H_2}{\gamma}\right) = Ki^{-1}\,\Omega + 2\frac{\partial^2\psi_0}{\partial x^2}\frac{\partial^2\psi_0}{\partial y^2} - 4\frac{\partial^2\psi_0}{\partial x\,\partial y} \,, \tag{6.80a}
$$

$$
\left(\frac{\partial}{\partial t} + \frac{\partial\psi_0}{\partial x}\frac{\partial}{\partial y} - \frac{\partial\psi_0}{\partial y}\frac{\partial}{\partial x}\right)\boldsymbol{D}^2\psi_0 = 0 \,. \tag{6.80b}
$$

We also take into account the relations $u_0 = -\partial\psi_0/\partial y$ and $v_0 = \partial\psi_0/\partial x$, since $D_0 = 0$, as a consequence of $\partial\varpi_0/\partial p = 0$ in the continuity equation (6.69a). Equation (6.80a) is the 'balance' equation, and (6.80b) is an evolution (conservation) equation for $\boldsymbol{D}^2\psi_0$ on an isobaric surface. This Monin–Charney model system (6.79)/(6.80) is of the first order in the time t and necessitates one initial condition for $\boldsymbol{v}_0$ (or ψ_0): $\boldsymbol{v}_{0|t=0} = \boldsymbol{V}^0$. The question is: how can $\boldsymbol{V}^0$ be found?

The answer lies in posing a problem of unsteady adjustment (considered in Sect. 6.5.3) by introducing a short time $\tau = t/M$, significant close to the initial time $t = 0$, in order to take into account the internal gravity waves (these waves exist at the level of the Kibel primitive equations (6.69), the acoustic waves being filtered from the full Euler atmospheric equations as a consequence of the hydrostatic approximation when ε tends to zero) which were filtered out during the limiting process (6.77a), with (6.77b). In the quasi-non-divergent model (6.79)/(6.80), why is there no derivation with respect to the 'altitude' coordinate p? In fact, this quasi-non-divergent model describes only a plane flow of an incompressible atmosphere along each (isobaric) surface p = const, in a thin sheet spread over a plane tangential to the Earth's surface. The reader can find a pertinent discusion of the quasi-non-divergent (referred to as 'quasi-solenoidal') approximation in the book by Monin (1972, Sect. 8). The balance equation (6.80a) was presented in Charney (1955), and a justification (rather ad hoc) of the quasi-non-divergent/solenoidal approximation by means of an asymptotic expansion in powers of $M \ll 1$ was presented in a paper by Monin (1961) and in a similar form by Charney (1962); Gavrilin (1965) derived the equations of the quasi-solenoidal approximation for non-adiabatic synoptic processes on a spherical Earth. This model has quite a strong decoupling between the atmospheric motions in two surfaces at different values of p. At this level of approximation, the only way in which the motions in two isobaric surfaces may be coupled is through lateral or initial conditions. This peculiar feature is, of course, an artefact of the approximation procedure. As a consequence, our model of flow at low Mach number, at least at the lowest level of approximation, is only able to deal with the evolution of vertical vorticity once it has been created. Below, we consider the next, second order to see whether or not some coupling can be recovered.

6.5.2 The Second-Order Low-Mach-Number Model

We assume that the scalar α in the similarity rule (6.72) is greater than 2. With the expansion (6.77b) and (6.76), we find from (6.69d) and (6.69c) that

$$T_2 = -p\frac{\partial H_2}{\partial p}\,, \tag{6.81a}$$

$$\varpi_2 = \frac{p}{K_0(p)}\left[\frac{\partial T_2}{\partial t} + \boldsymbol{v}_0 \,.\, \boldsymbol{D}T_2\right]\,, \tag{6.81b}$$

and, considering that H_2 is computable from the above zeroth-order model (6.80), we may use (6.81) to compute ϖ_2. Going back to (6.77b), we obtain from (6.69a,b)

$$\boldsymbol{D} \,.\, \boldsymbol{v}_0 + \frac{\partial \varpi_2}{\partial p} = 0\,, \tag{6.82a}$$

$$\begin{aligned}\frac{\partial \boldsymbol{v}_2}{\partial t} &+ (\boldsymbol{v}_0 \,.\, \boldsymbol{D})\boldsymbol{v}_2 + \frac{1}{Ki}(\boldsymbol{k} \wedge \boldsymbol{v}_2) + (\boldsymbol{v}_2 \,.\, \boldsymbol{D})\boldsymbol{v}_0 \\ &+ \varpi_2\frac{\partial \boldsymbol{v}_0}{\partial p} + \boldsymbol{D}\left(\frac{H_4}{\gamma}\right) = 0\,.\end{aligned} \tag{6.82b}$$

It is obvious that some vertical coupling is recovered through the terms $\varpi_2\, \partial \boldsymbol{v}_0/\partial p$ and $\partial \varpi_2/\partial p$, so that the model looks sound. Recall that, at the lowest level of approximation, we have $H_0(P_{\mathrm{g}0}) = 0$ (see (6.78)). We use the notation $F_{\mathrm{g}} = F(t, x, y, P_{\mathrm{g}0})$, and find, for the second order,

$$\left(\frac{\partial H_0}{\partial p}\right)_{\mathrm{g}} P_{\mathrm{g}2} + H_{2\mathrm{g}} = 0 \quad \Rightarrow P_{\mathrm{g}2} = -\left[\frac{H_2}{(\partial H_0/\partial p)}\right]_{\mathrm{g}}\,. \tag{6.83}$$

The reader can check that this is compatible with the relation

$$\frac{\partial P_{\mathrm{g}2}}{\partial t} + \boldsymbol{v}_0 \,.\, \boldsymbol{D}P_{\mathrm{g}2} = \varpi_{2g}\,, \tag{6.84}$$

which may be derived from (6.71b). This compatibility is a consequence of the following identity (see (6.71a)):

$$\frac{\partial H}{\partial t} + \boldsymbol{v} \cdot \boldsymbol{D}H + \varpi \frac{\partial H}{\partial p} = 0 \quad \text{for } p \doteq P_{\mathrm{g}} \,. \tag{6.85}$$

6.5.3 Adjustment to Quasi-Non-Divergent Model Equations

With (6.77b), the low-Mach-number quasi-non-divergent flow (6.79) is highly constrained to glide on iso-p surfaces in such a way that ϖ_0, the limiting value of ϖ, is zero, and, consequently, the horizontal velocity is divergence-free (this is the reason why the model obtained from the starting Kibel primitive equations (6.69) is called quasi-non-divergent). Our purpose below is to investigate how the limiting state is reached if we start from an initial (inviscid and adiabatic) state $(\boldsymbol{v}^0, H^0)$ which does not agree with the constraint $\varpi = O(M)$. We find easily that the appropriate rescaling (short) time is

$$\tau = t/M \,, \tag{6.86}$$

and we write f^τ for any quantity f when it is considered as a function of τ rather than of t. We rewrite the starting Kibel primitive equations (6.69) as follows:

$$\boldsymbol{D} \cdot \boldsymbol{v}^\tau + \frac{\partial \varpi^\tau}{\partial p} = 0 \,,$$

$$\frac{1}{\gamma M} \boldsymbol{D}H^\tau + \frac{\partial \boldsymbol{v}^\tau}{\partial \tau} + M \left[(\boldsymbol{v}^\tau \cdot \boldsymbol{D})\boldsymbol{v}^\tau + \varpi^\tau \frac{\partial \boldsymbol{v}^\tau}{\partial p} + \frac{1}{Ki}(\boldsymbol{k} \wedge \boldsymbol{v}^\tau) \right] = 0 \,,$$

$$\frac{\partial T^\tau}{\partial \tau} + v^\tau \cdot DT^\tau + \varpi^\tau K(T^\tau) = 0, \quad T^\tau = -p \frac{\partial H^\tau}{\partial p} \,. \tag{6.87}$$

Next, in place of (6.77a), we write

$$M \downarrow 0 \quad \text{with } \tau, x, y \text{ and } p \text{ fixed} \,, \tag{6.88a}$$

together with the local expansion close to the intial time (in place of (6.77b)):

$$(\boldsymbol{v}^\tau, \varpi^\tau, H^\tau, T^\tau) = (\boldsymbol{v}_0^\tau, T_0(p), H_0(p), \varpi_0^\tau) + M(\boldsymbol{v}_1^\tau, \varpi_1^\tau, H_1^\tau, T_1^\tau) + O(M^2) \,, \tag{6.88b}$$

since

$$\mathrm{D}H_0^\tau = 0, \quad \frac{\partial}{\partial \tau}\left(\frac{\partial H_0^\tau}{\partial p} \right) = 0, \quad T_0^\tau = -p \frac{\mathrm{d}H_0^\tau}{\mathrm{d}p} \,.$$

For $\boldsymbol{v}_0^\tau, \varpi_0^\tau, H_1^\tau$ and T_1^τ, we find the following unsteady adjustment equations:

$$\frac{\partial \boldsymbol{v}_0^\tau}{\partial \tau} + \frac{1}{\gamma} \boldsymbol{D}H_1^\tau = 0 \,, \tag{6.89a}$$

$$\frac{\partial T_1^{\mathcal{T}}}{\partial \tau} - \frac{\varpi_0^{\mathcal{T}}}{p} K_0(p) = 0 \,, \tag{6.89b}$$

$$\boldsymbol{D} \,.\, \boldsymbol{v}_0^{\mathcal{T}} + \frac{\partial \varpi_0^{\mathcal{T}}}{\partial p} = 0 \,, \tag{6.89c}$$

$$T_1^{\mathcal{T}} = -p \frac{\partial H_1^{\mathcal{T}}}{\partial p} \,. \tag{6.89d}$$

From the above acoustic-type equations (6.89), we may derive a single equation for $H_1^{\mathcal{T}}$ alone through a straightforward computation, namely,

$$\gamma \frac{\partial^2}{\partial \tau^2} \left\{ \frac{\partial}{\partial p} \left[\frac{p^2}{K_0(p)} \frac{\partial H_1^{\mathcal{T}}}{\partial p} \right] \right\} + \boldsymbol{D}^2 H_1^{\mathcal{T}} = 0 \,. \tag{6.90}$$

In the case where we assume that $K_0(p) = K_{00} = \text{const}$, the above equation (6.90), with

$$t^* = \sqrt{\frac{K_{00}}{\gamma}} \tau \quad \text{and}\, \zeta = \log\left(\frac{1}{p}\right) \,, \tag{6.91a}$$

gives for the function

$$\Phi(t^*, x, y, \zeta) = \exp\left(-\frac{\zeta}{2}\right) H_1^{\mathcal{T}} \tag{6.91b}$$

the following equation with constant coefficients:

$$\frac{\partial^2}{\partial t^{*2}} \left(\frac{\partial^2 \Phi}{\partial \zeta^2} - \frac{1}{4} \Phi \right) + \boldsymbol{D}^2 \Phi = 0 \,. \tag{6.92}$$

A general solution of (6.92), valid in the whole space of x, y, ζ, is given by

$$\Phi = \int\!\!\int\!\!\int_{-\infty}^{+\infty} \exp[\mathrm{i}(k\zeta + lx + my)] \left[A \cos(\sigma t^*) + \frac{B}{\sigma} \sin(\sigma t^*) \right] \mathrm{d}k \, \mathrm{d}l \, \mathrm{d}m \,, \tag{6.93}$$

where $\sigma = [(l^2+m^2)/(k^2+1/4)]^{1/2}$, and A and B are functions of k, l, m which depend on the initial conditions. The only important issue with unsteady adjustment is whether or not there is decay or persistence when t^* tends to infinity. This question may be answered directly from the solution (6.93), without sophisticated computations. We have only to rewrite the quantity under the integration in (6.93) as a sum of exponentials $E = \exp[i(k\zeta + lx + my) + i\sigma t^*]$ and apply a method of stationary phase in order to obtain the asymptotic behaviour of Φ when $t^* \to \infty$. The stationary phase is reached at k^*, l^*, m^* such that

$$\zeta \pm t^* \left(\frac{\partial \sigma}{\partial k} \right)^* = x \pm t^* \left(\frac{\partial \sigma}{\partial l} \right)^* = y \pm t^* \left(\frac{\partial \sigma}{\partial m} \right)^* = 0 \,. \tag{6.94}$$

In this way, we obtain, asymptotically,

$$\Phi \approx \frac{\exp[\mathrm{i}(k^*\zeta + l^*x + m^*y)]}{t^{*3/2}}|\chi|, \quad \text{when } t^* \to \infty \,, \tag{6.95}$$

where $|\chi| = O(1)$. The decay of H_1^τ to zero when $t^* \to \infty$ becomes obvious, and this ensures also the decay of ϖ_0^τ, T_1^τ and $\partial \boldsymbol{v}_0^\tau/\partial\tau$ to zero according to the adjustment equations (6.89d,b,a). On the other hand, the matching between the expansions (6.88b) and (6.77b) for the horizontal velocities,

$$\lim_{\tau\to\infty} \boldsymbol{v}_0^\tau(\tau, x, y, p) = \lim_{t\to 0} \boldsymbol{v}_0(t, x, y, p) \,, \tag{6.96}$$

shows that, at $t = 0$, we can write

$$\boldsymbol{D} \,.\, \boldsymbol{V}^0 = 0 \,, \tag{6.97}$$

where $\boldsymbol{V}^0$ is the (unknown!) initial value (at $t = 0$) for the quasi-non-divergent horizontal velocity $\boldsymbol{v}_0(t, x, y, p)$, the solution of (6.79b).

Finally, the main conclusion is that the local-in-time expansion (6.88b) does not have any influence on the quasi-divergent model (6.79). Concerning the initial condition $\boldsymbol{V}^0$ for the model quasi-non-divergent equation (6.79b), more information is necessary, but to obtain this it seems that complementary investigations are necessary.

6.5.4 The Investigations of Guiraud and Zeytounian

The author's investigations with J.-P. Guiraud (briefly mentioned in Zeytounian and Guiraud 1984; see also Zeytounian 1990, Chaps. 12 and 13) show that a significant connection between different isobaric surfaces is obtained in the far field. Below, if the Coriolis term in (6.69b) may be neglected ($Ki \equiv \infty$), we have the possibility to consider a specific far-field ($r = x^2 + y^2 \to \infty$) behaviour,

$$\lim_{r\to\infty}(\boldsymbol{v}, \varpi, H, T) = (\boldsymbol{V}_\infty(p), \varpi_\infty = 0, H_\infty = H_0(p), T_\infty = T_0(p)) \,, \tag{6.98}$$

which is an exact solution of the Kibel primitive equations (6.69), where the term $(1/K_i)(\boldsymbol{k} \wedge \boldsymbol{v})$ is absent in (6.69b). We observe that if $\Omega_0 = 0$, as we have shown that then $D_0 = 0$, the condition at infinity (when $r \to \infty$) ensures that the conditions $\boldsymbol{v}_0 = \boldsymbol{V}_\infty(p), H_2 = 0$ and $T_2 = 0$ hold throughout. Now, since (6.98) is an exact solution of (6.69) with the Coriolis term neglected, we see that our quasi-non-divergent model is unable to generate a solution different from $\boldsymbol{V}_\infty(p)$, $\varpi_\infty = 0$, $H_\infty = H_0(p), T_\infty = T_0(p)$. The only thing that it may allow is to predict the evolution of some disturbance once it has been created. If we assume now that some distribution of vertical vorticity, localized in space, has been detected at some initial value of the time, then we may use the quasi-non-divergent model (6.79) to compute the later evolution of the flow, and (6.80) shows that $\Omega_0 = \boldsymbol{D}^2\psi_0$ remains localized in space at any finite value of the time.

Behaviour of v_0 and H_2 at Infinity When $r \to \infty$

Now we write, for the horizontal velocity $\boldsymbol{v}_0$, the following expression (which obviously is a solution of (6.79), since $D_0 = 0, u_0 = \partial\phi_0^H/\partial x - \partial\psi_0/\partial y$ and $v_0 = \partial\phi_0^H/\partial y + \partial\psi_0/\partial x$, where $\phi_0^{\mathrm{H}} = \boldsymbol{V}_\infty(p) \,.\, \boldsymbol{x}$ is a harmonic function and $\boldsymbol{x} = (x, y)$):

$$\boldsymbol{v}_0 = \boldsymbol{V}_\infty(p) + \boldsymbol{k} \wedge \boldsymbol{D}\psi_0 \,, \tag{6.99}$$

where ψ_0 (harmonic near infinity) may be computed from

$$\psi_0 = (2\pi)^{-1} \int\!\!\int_{R^2} \Omega_0(t, x', y', p) \log[(x - x')^2 + (y - y')^2]^{1/2} \, \mathrm{d}x' \, \mathrm{d}y' \,, \tag{6.100}$$

once Ω_0 is known. Here we are interested in only the behaviour at infinity, so that we need not know Ω_0 precisely. The only thing that matters is that Ω_0 vanishes outside a bounded domain. A straightforward computation (according to Guiraud 1983) shows that

$$\boldsymbol{v}_0 \approx \boldsymbol{V}_\infty(p) + \frac{\langle\Omega_0\rangle}{2\pi} \frac{\boldsymbol{k} \wedge \boldsymbol{x}}{|\boldsymbol{x}|^2}$$

$$- \frac{1}{2\pi|\boldsymbol{x}|^2} \left\{ (\boldsymbol{k} \wedge \boldsymbol{M}_0) - \frac{2}{|\boldsymbol{x}|^2} [(\boldsymbol{M}_0 \,.\, \boldsymbol{x})(\boldsymbol{k} \wedge \boldsymbol{x})] \right\} + O\left(\frac{1}{|\boldsymbol{x}|^3}\right) \,, \tag{6.101a}$$

$$\langle\Omega_0\rangle = \int\!\!\int_{R^2} \Omega_0(t', x', y', p) \, \mathrm{d}x' \, \mathrm{d}y' \text{ and } \frac{\partial \boldsymbol{M}_0}{\partial t} = \boldsymbol{V}_\infty(p) \langle\Omega_0\rangle \,, \tag{6.101b}$$

when we take into account the following asymptotic behaviour of ψ_0 :

$$\psi_0 = \frac{\langle\Omega_0\rangle}{2\pi} \log |\boldsymbol{x}| - \frac{(\boldsymbol{M}_0 \,.\, \boldsymbol{x})}{2\pi|\boldsymbol{x}|^2} + \frac{(\boldsymbol{x} \,.\, \boldsymbol{N}_0 \,.\, \boldsymbol{x})}{2\pi|\boldsymbol{x}|^4} + O\left(\frac{1}{|\boldsymbol{x}|^3}\right) \,, \tag{6.101c}$$

where

$$\boldsymbol{N}_0 = \int\!\!\int_{R^2} \Omega_0(t, x', y', p)[|\boldsymbol{x}|^2 \mathbf{1} - 2(\boldsymbol{x} \otimes \boldsymbol{x})] \, \mathrm{d}x' \, \mathrm{d}y' \,. \tag{6.101d}$$

In (6.101d), $\mathbf{1}$ is the second-order unit tensor and $\boldsymbol{x} \otimes \boldsymbol{x}$ is the tensor product of the vector $\boldsymbol{x}$ with the vector $\boldsymbol{x}$. Then the behaviour at infinity of H_2, the solution (together with $\boldsymbol{v}_0$) of (6.79), can be found from the following equation (since, in place of (6.79b), we have $\partial \boldsymbol{v}_0/\partial t + (\boldsymbol{v}_0 \,.\, D)\boldsymbol{v}_0 + D(H_2/\gamma) = 0$):

$$\boldsymbol{D}\left\{ \frac{H_2}{\gamma} + \frac{\partial \chi_0}{\partial t} + \left(\boldsymbol{V}_\infty(p) + \frac{1}{2}\boldsymbol{D}\chi_0 \right) \,.\, \boldsymbol{D}\chi_0 \right\} \approx 0 \,. \tag{6.101e}$$

This equation is significant near infinity; the function χ_0 is the harmonic conjugate of ψ_0, such that

$$\boldsymbol{k} \wedge \boldsymbol{D}\psi_0 = \boldsymbol{D}\chi_0 \,. \tag{6.101f}$$

From (6.101c) and (6.101f), we find the asymptotic behaviour of χ_0 at infinity, and then, from (6.101e), we compute the asymptotic behaviour of H_2 at infinity. If we go to a higher order it is found that H_4 is unbounded at infinity, and this is an indication that the expansion (6.77b), which leads to (6.79), (6.81) and also (6.82), is certainly not valid at infinity with respect to the horizontal variables.

Far-Field Expansion

When, in place of the horizontal coordinates x and y, we consider the far-field local horizontal coordinates

$$\xi = Mx, \quad \eta = My \,, \tag{6.102a}$$

and, in place of (6.77a) or (6.88a), we write

$$M \downarrow 0 \quad \text{with}\, t, \xi, \eta \,\text{and}\, p \,\text{fixed} \,, \tag{6.102b}$$

we must consider the following far-field representation:

$$\boldsymbol{v} = \boldsymbol{V}_\infty(p) + M^\beta \boldsymbol{v}^{\mathrm{ff}}, \quad \varpi = M^{\beta+1}\varpi^{\mathrm{ff}}, \quad H = H_0(p) + M^{\beta+1}H^{\mathrm{ff}} \,, \tag{6.102c}$$

$$T = T_0(p) + M^{\beta+1}T^{\mathrm{ff}}, \quad P_{\mathrm{g}} = P_{\mathrm{g}0} + M^{\beta+1}P_{\mathrm{g}}^{\mathrm{ff}} \,. \tag{6.102d}$$

We assume that the scalar α in (6.72) is sufficiently large that dissipation does not enter in to the far-field investigations up to the order considered. Concerning $P_{\mathrm{g}}^{\mathrm{ff}}$, we have the following condition:

$$\frac{\partial P_{\mathrm{g}}^{\mathrm{ff}}}{\partial t} - \varpi^{\mathrm{ff}} + M[\boldsymbol{V}_\infty(p) + M^\beta \boldsymbol{v}^{\mathrm{ff}}]\,.\,\boldsymbol{D}^{\mathrm{ff}} P_{\mathrm{g}}^{\mathrm{ff}} = 0 \quad \text{on}\, p = P_{\mathrm{g}0} + M^{\beta+1}P_{\mathrm{g}}^{\mathrm{ff}} \,, \tag{6.103}$$

where $\boldsymbol{D}^{\mathrm{ff}} = (\partial/\partial\xi, \partial/\partial\eta)$. Concerning the value of β, substitution of (6.102a) into (6.101), which describes the behaviour of $\boldsymbol{v}_0$, shows that (for matching relative to $|\boldsymbol{\xi}| \to 0$ with $|\boldsymbol{x}| \to \infty$)

$$\boldsymbol{v}_0 = \boldsymbol{V}_\infty(p) + M\frac{\langle \Omega_0 \rangle}{2\pi}\frac{\boldsymbol{k} \wedge \boldsymbol{\xi}}{|\boldsymbol{\xi}|^2} + O(M^2) \quad \Rightarrow \beta = 1 \,, \tag{6.104a}$$

where $\boldsymbol{\xi} = (\xi, \eta)$, and this is confirmed by

$$H_0(p) + M^2 H_2 = H_0(p) + M^4 O\left(\frac{1}{|\boldsymbol{\xi}|^2}\right) + \cdots \,. \tag{6.104b}$$

With $\beta = 1$, we write a far-field expansion

$$(\boldsymbol{v}^{\mathrm{ff}}, \varpi^{\mathrm{ff}}, H^{\mathrm{ff}}, T^{\mathrm{ff}}) = (\boldsymbol{v}_0^{\mathrm{ff}}, \varpi_0^{\mathrm{ff}}, H_0^{\mathrm{ff}}, T_0^{\mathrm{ff}}) + M(\boldsymbol{v}_1^{\mathrm{ff}}, \varpi_1^{\mathrm{ff}}, H_1^{\mathrm{ff}}, T_1^{\mathrm{ff}}) + \cdots \,. \tag{6.105}$$

and matching when $|\boldsymbol{\xi}| \to 0$ gives, for the leading order,

$$\boldsymbol{v}_0^{\mathrm{ff}} = \frac{\langle \Omega_0 \rangle}{2\pi} (\boldsymbol{k} \wedge \boldsymbol{\xi}) \frac{1}{|\boldsymbol{\xi}|^2}, \quad \left(\varpi_0^{\mathrm{ff}}, H_0^{\mathrm{ff}}, T_0^{\mathrm{ff}}\right) = 0 , \tag{6.106}$$

since $(\boldsymbol{v}_0^{\mathrm{ff}}, \varpi_0^{\mathrm{ff}}, H_0^{\mathrm{ff}}, T_0^{\mathrm{ff}}) \to 0$ when $|\boldsymbol{\xi}| \to \infty$. The reader will observe that the correctness of the above solution (6.106) is tied to the independence of $\langle \Omega_0 \rangle$ with respect to the time t. More precisely, $\partial \langle \Omega_0 \rangle / \partial t = 0$, and $\langle \Omega_0 \rangle (p)$, a function only of p, is the total eddy intensity contained within the isobaric surface of level p. At the order M, the solution is

$$\boldsymbol{v}_1^{\mathrm{ff}} = -\frac{1}{2\pi} (\boldsymbol{k} \wedge \boldsymbol{M}_0) \frac{1}{|\boldsymbol{\xi}|^2} + \frac{1}{\pi} (\boldsymbol{M}_0 \,.\, \boldsymbol{\xi})(\boldsymbol{k} \wedge \boldsymbol{\xi}) \frac{1}{|\boldsymbol{\xi}|^4} , \tag{6.107a}$$

$$\left(\varpi_1^{\mathrm{ff}}, H_1^{\mathrm{ff}}, T_1^{\mathrm{ff}}\right) = 0 . \tag{6.107b}$$

The reader can easily check that $\boldsymbol{D}^{\mathrm{ff}} \,.\, \boldsymbol{v}_1^{\mathrm{ff}} = 0$ and also that

$$\frac{\partial \boldsymbol{v}_1^{\mathrm{ff}}}{\partial t} + \boldsymbol{V}_\infty(p) \,.\, \boldsymbol{D} \boldsymbol{v}_0^{\mathrm{ff}} = 0.$$

We notice also that the first two terms of the outer (far-field) expansion (6.105) are identical to a reordering of the inner expansion. The reader might feel that this is a quite favourable circumstance, in that there is no back influence from the outer field on the inner field, but this is misleading because, owing to this happy feature, we lose the main effect of the outer field, which is, through a non-trivial solution, to bring in a coupling between the flows in various p-levels. Looking at the equation governing H_1^{ff},

$$\boldsymbol{D}^{\mathrm{ff}\,2} H_1^{\mathrm{ff}} + \gamma \frac{\partial^2}{\partial t^2} \left(\frac{p^2}{K_0(p)} \frac{\partial H_1^{\mathrm{ff}}}{\partial p} \right) = 0 , \tag{6.108}$$

we see that the reason for this feature is that the behaviour of H_1^{ff} near $|\boldsymbol{\xi}| = 0$ does not depend on the time t. If we examine (6.104), we see from (6.106) that we have a dependence on time in the terms $O(M^4)$ in (6.104b) through $\boldsymbol{M}_0$, but this dependency is quite peculiar because it is linear, and it will not work for our purpose, owing to the second-order time derivative in (6.108).

Unboundness of H_4

We need to compute H_4, and we try to do this now. The first step is to compute ϖ_2, Ω_2 and v_2; we find (see (6.81))

$$\varpi_2 = \frac{p^2}{K_0(p)} \left[\frac{\partial}{\partial t} + \boldsymbol{v}_0 \,.\, \boldsymbol{D} \right] \frac{\partial H_2}{\partial p} , \tag{6.109}$$

and from the equation (6.82b) for $\boldsymbol{v}_2$, once the Coriolis terms have been neglected ($Ki \equiv \infty$), we obtain near infinity (where Ω_0 is identically zero) an equation for $\Omega_2 = k.(\boldsymbol{D} \wedge \boldsymbol{v}_2)$:

$$\frac{\partial \Omega_2}{\partial t} + \boldsymbol{v}_0 \,.\, \boldsymbol{D}\Omega_2 + \boldsymbol{k} \,.\, \left(\boldsymbol{D}\varpi_2 \wedge \frac{\partial \boldsymbol{v}_0}{\partial p} \right) = 0\,. \tag{6.110}$$

This shows a mechanism by which vertical vorticity is created at order $O(M^2)$ – a statement which does not contradict the statement after (6.98), as a little reflection shows. If we use (6.109), provided Ω_2 tends to zero at infinity, we obtain

$$\Omega_2 \approx \frac{p^2}{K_0(p)} \left(\frac{\partial \boldsymbol{V}_\infty}{\partial p} \wedge \boldsymbol{k} \right) . \boldsymbol{D} \left(\frac{\partial H_2}{\partial p} \right) \quad \Rightarrow \Omega_2 \approx O\left(\frac{1}{|\boldsymbol{x}|^3} \right) . \tag{6.111}$$

We observe that it is not possible that Ω_2 does not tend to zero when $|\boldsymbol{x}| \to \infty$! Now, in order to compute $\boldsymbol{v}_2$, we put

$$\boldsymbol{v}_2 = \boldsymbol{D}\phi_2 + \boldsymbol{k} \wedge \boldsymbol{D}\psi_2 \quad \text{with}\, \boldsymbol{D}^2\phi_2 = -\frac{\partial \varpi_2}{\partial p} \,\text{and}\, \boldsymbol{D}^2\psi_2 = \Omega_2\,, \tag{6.112}$$

and we obtain near infinity

$$H_4 = -\gamma \frac{\partial \phi_2}{\partial t} + O\left(\frac{1}{|\boldsymbol{x}|} \right) , \tag{6.113}$$

when we use (6.82b) with $Ki \equiv \infty$, and (6.109). In (6.113), we have

$$\phi_2 = \phi_2^p - (2\pi)^{-1} \frac{\partial \langle \varpi_2 \rangle}{\partial p} \log|\boldsymbol{x}| + O\left(\frac{1}{|\boldsymbol{x}|} \right) , \tag{6.114}$$

where ϕ_2^{P} is a particular solution of the equation $\boldsymbol{D}^2\phi_2 = -\partial\varpi_2/\partial p$ near infinity in $|\boldsymbol{x}|$, and

$$\langle \varpi_2 \rangle (t,p) = \lim_{R\to\infty} \int\!\!\int_{r<R^2} \varpi_2(t,x',y',p)\, \mathrm{d}x'\, \mathrm{d}y' \,. \tag{6.115}$$

With (6.113) and (6.114), we confirm the statement that we made earlier about the unboundness of H_4. According to a process that will be explained below, matching with the outer solution requires that the expansion (6.77b) be modified by adding terms in $M^4 \log M$.

Matching

As a matter of fact, the modification of the expansion (6.77b) is very slight indeed: we need not change anything in $\boldsymbol{v}_2$, and $(M^4 \log M)(\gamma/2\pi)\partial^2\langle\varpi_2\rangle/\partial t\,\partial p$ has to be added to the expansion of H. More precisely (to avoid confusion), we must write the following outer expansions for $\boldsymbol{v}, H, \varpi$ and T in place of (6.77b):

$$\boldsymbol{v} = \boldsymbol{v}_0 + M^2 \boldsymbol{v}_2 + \cdots ,$$

$$H = H_0(p) + M^2 H_2 + (M^4 \log M) \frac{\gamma}{2\pi} \frac{\partial^2 \langle \varpi_2 \rangle}{\partial t \, \partial p} + M^4 H_4 + \cdots , \tag{6.116a}$$

$$\varpi = M^2 \varpi_2 + (M^4 \log M) \left\{ \frac{\gamma}{2\pi} \frac{\partial^2}{\partial t^2} \left[\frac{\partial}{\partial p} \left(\frac{p^2}{K_0(p)} \frac{\partial \langle \varpi_2 \rangle}{\partial p} \right) \right] \right\}$$

$$+ M^4 \varpi_4 + \ldots , \tag{6.116b}$$

$$T = T_0(p) + M^2 T_2 + (M^4 \log M) \left(-\frac{\gamma}{2\pi} p \frac{\partial^3 \langle \varpi_2 \rangle}{\partial t \, \partial p^2} \right)$$

$$+ M^4 T_4 + \ldots , \tag{6.116c}$$

Of course, T_4 and ϖ_4 may readily be computed once we know H_4, which, in (6.116), is exactly the same as the H_4 already considered, for example in (6.113). Concerning P_{g}, we have

$$P_{\mathrm{g}} = 1 + M^2 P_{\mathrm{g}2} + (M^4 \log M) \left\{ -\frac{\gamma}{2\pi} \left[\frac{\partial^2 \langle \varpi_2 \rangle}{\partial t \, \partial p} \right]_{\mathrm{g}} \left[\left(\frac{dH}{\partial p} \right)_{\mathrm{g}} \right]^{-1} \right\} + M^4 P_{\mathrm{g}4} + \ldots . \tag{6.117}$$

Now we take the expansion of H in (6.116), truncated just after $M^4 H_4$, put again $\boldsymbol{\xi} = (\xi, \eta) \equiv M\boldsymbol{x}$ with (6.102a) into it and re-expand according to (6.102b). We find

$$H = H_0(p) + M^4 \Bigg\{ - \left[\frac{\gamma}{2} \left[\frac{\langle \Omega_0 \rangle}{2\pi} \right]^2 + \frac{\gamma}{2\pi} \boldsymbol{V}_\infty \boldsymbol{.} (\boldsymbol{k} \wedge \boldsymbol{M}_0) \right] \frac{1}{|\boldsymbol{\xi}|^2}$$

$$+ \frac{\gamma}{2\pi} \frac{\partial^2 \langle \varpi_2 \rangle}{\partial t \, \partial p} \log |\boldsymbol{\xi}| - \frac{\gamma^2}{8\pi} \frac{\partial}{\partial p} \left(\frac{[\boldsymbol{\xi} \boldsymbol{.} \partial^3 \boldsymbol{N}_0 / \partial t^3 \boldsymbol{.} (\boldsymbol{k} \wedge \boldsymbol{\xi})]}{|\boldsymbol{\xi}|^2} \right)$$

$$+ O\left(\frac{1}{|\boldsymbol{\xi}|^4} \right) + \ldots \Bigg\} . \tag{6.118}$$

This allow us to match (6.116) with the expansion obtained from (6.102) and (6.105), which we state again:

$$\boldsymbol{v} = \boldsymbol{V}_\infty(p) + M \boldsymbol{v}_0^{\mathrm{ff}} + M^2 \boldsymbol{v}_1^{\mathrm{ff}} + M^3 \boldsymbol{v}_2^{\mathrm{ff}} + \ldots ,$$
$$H = H_0(p) + M^2 H_0^{\mathrm{ff}} + M^4 H_2^{\mathrm{ff}} + \ldots ,$$
$$T = T_0(p) + M^2 T_0^{\mathrm{ff}} + M^4 T_2^{\mathrm{ff}} + \ldots ,$$
$$\varpi = M^4 \varpi_2^{\mathrm{ff}} + \ldots , \tag{6.119}$$

Matching requires that when $|\boldsymbol{\xi}| \to 0$, H_2^{ff} behaves according to (see (6.118))

$$H_2^{\mathrm{ff}} \approx - \left[\frac{\gamma}{2} \left[\frac{\langle \Omega_0 \rangle}{2\pi} \right]^2 + \frac{\gamma}{2\pi} \boldsymbol{V}_\infty \boldsymbol{.} (\boldsymbol{k} \wedge \boldsymbol{M}_0) \right] \frac{1}{|\boldsymbol{\xi}|^2} + \frac{\gamma}{2\pi} \frac{\partial^2 \langle \varpi_2 \rangle}{\partial t \, \partial p} \log |\boldsymbol{\xi}|$$

$$- \frac{\gamma^2}{8\pi} \frac{\partial}{\partial p} \left(\frac{[\xi \, \partial^3 \mathbf{N}_0 / \partial t^3 \boldsymbol{.} (k \wedge \boldsymbol{\xi})]}{|\boldsymbol{\xi}|^2} \right) + O\left(\frac{1}{|\boldsymbol{\xi}|^4} \right) + \ldots , \tag{6.120}$$

However, we need not consider matching for the other quantities; we need simply to check, at the end, that it holds true.

Equation for H_2^{ff}

We shall avoid considering the above technicalities and state only that, by substituting (6.119) into the starting Kibel primitive equations (6.69), written in terms of the far-field coordinates (6.102a) in place of x and y, for the functions (6.102c,d) with superscripts 'ff', and using (6.106) and (6.107), we obtain the following far-field system of equations:

$$\frac{\partial \boldsymbol{v}_2^{\mathrm{ff}}}{\partial t} + \boldsymbol{D}^{\mathrm{ff}}\left(\frac{H_2^{\mathrm{ff}}}{\gamma}\right) + \boldsymbol{V}_\infty(p)\,.\,\boldsymbol{D}^{\mathrm{ff}}\boldsymbol{v}_1^{\mathrm{ff}} + \boldsymbol{v}_0^{\mathrm{ff}}\,.\,\boldsymbol{D}^{\mathrm{ff}}\boldsymbol{v}_0^{\mathrm{ff}} = 0\,,$$

$$\boldsymbol{D}^{\mathrm{ff}}\,.\,\boldsymbol{v}_2^{\mathrm{ff}} = 0, \quad \frac{\partial T_2^{\mathrm{ff}}}{\partial t} - \frac{K_0(p)}{p}\varpi_2^{\mathrm{ff}} = 0\,,$$

$$T_2^{\mathrm{ff}} + p\frac{\partial H_2^{\mathrm{ff}}}{\partial p} = 0\,. \tag{6.121}$$

It is straightforward to extract a single equation for H_2^{ff} from (6.121), namely,

$$\left\{\frac{\partial^2}{\partial \xi^2} + \frac{\partial^2}{\partial \eta^2} + \gamma\frac{\partial^2}{\partial t^2}\left[\frac{\partial}{\partial p}\left(\frac{p^2}{K_0(p)}\frac{\partial}{\partial p}\right)\right]\right\}H_2^{\mathrm{ff}} + \frac{2\gamma}{|\boldsymbol{\xi}|^4}\left[\frac{\langle \Omega_0\rangle}{2\pi}\right]^2 = 0\,. \tag{6.122}$$

This equation (6.122) for H_2^{ff} plays a role with respect to the Kibel primitive equations analogous to the one played by the acoustic equation with respect to the Euler equations. As far as external aerodynamics is concerned, acoustics govern the phenomenon of adaptation to incompressible flow; above, (6.122) governs the phenomenon of adaptation to the quasi-non-divergent approximation. We observe that $\langle \Omega_0\rangle(p)$ is the total amount of vertical vorticity which is contained in the isobaric surface $p = \mathrm{const}$ and which drives the potential vortex $\boldsymbol{v}_0^{\mathrm{ff}}$ given by (6.106). The homogeneous equation corresponding to (6.122) (with $\langle \Omega_0\rangle = 0$) is identical to (6.108) for H_1^{ff}. One the other hand, (6.108) is identical to (6.90) for H_1^{τ}, which describes the unsteady adjustment to quasi-non-divergent flow – a little reflection suggests that this should indeed be the case.

Now for H_1^{τ} we consider the derivation of the boundary conditions for the above equation (6.122). We examine first the case of the condition on the ground, which is located at (recall (6.103))

$$P_{\mathrm{g}}(t,\xi,\eta) = 1 + M^2\left[P_{\mathrm{g}0}^{\mathrm{ff}} + MP_{\mathrm{g}1}^{\mathrm{ff}} + M^2P_{\mathrm{g}2}^{\mathrm{ff}} + \ldots\right]\,, \tag{6.123}$$

and we have, thanks to the expansion of H in (6.119), written for P_{g}, and also (6.106) and (6.107b).

$$P_{\mathrm{g}0}^{\mathrm{ff}} = P_{\mathrm{g}1}^{\mathrm{ff}} = 0, \quad P_{\mathrm{g}2}^{\mathrm{ff}} = -H_2^{\mathrm{ff}} \left[\left(\frac{\partial H_0}{\partial p} \right)_{\mathrm{g}} \right]^{-1}, \tag{6.124}$$

where the subscript 'g' means (except for P_{g}) that the quantity so labelled is evaluated at $p = 1$. Now, going back to (6.103) and expanding, we find $\partial P_{\mathrm{g}2}^{\mathrm{ff}}/\partial t = \varpi_{2\mathrm{g}}^{\mathrm{ff}}$. However, from (6.121), we obtain

$$\varpi_2^{\mathrm{ff}} = -\left[\frac{p^2}{K_0(p)} \right] \frac{\partial^2 H_2^{\mathrm{ff}}}{\partial t\, \partial p},$$

so that, together with (6.123), (6.124) and the definition of P_{g} as p_{ground}, this leads to the ground condition

$$\frac{\partial}{\partial t} \left\{ H_{2\mathrm{g}}^{\mathrm{ff}} - \frac{1}{K_{0\mathrm{g}}} \left(\frac{\partial H_0}{\partial p} \right)_{\mathrm{g}} \left(\frac{\partial H_2^{\mathrm{ff}}}{\partial p} \right)_{\mathrm{g}} \right\} = 0\,. \tag{6.125}$$

In order to integrate out the time derivative in the above ground condition (6.125), an initial condition is needed. However, such a consideration belongs to the problem of unsteady adjustment, and we avoid it by putting

$$H_{2\mathrm{g}}^{\mathrm{ff}} - \frac{1}{K_{0\mathrm{g}}} \left(\frac{\partial H_0}{\partial p} \right)_{\mathrm{g}} \left(\frac{\partial H_2^{\mathrm{ff}}}{\partial p} \right)_{\mathrm{g}} = 0\,. \tag{6.126}$$

This is just the boundary condition on the ground for (6.122). The upper boundary condition relative to p, at $p = 0$, has been discussed by El Mabrouk and Zeytounian (1984), and we state only the result here:

$$\lim_{p \to 0} \left[\frac{p^2}{K_0(p)} H_2^{\mathrm{ff}} \frac{\partial H_2^{\mathrm{ff}}}{\partial p} \right] = 0\,. \tag{6.127}$$

Concerning infinity in relation to $|\boldsymbol{\xi}|$, we have

$$|\boldsymbol{\xi}| \to \infty : \quad H_2^{\mathrm{ff}} \to 0\,. \tag{6.128}$$

Finally, at $| \vec{\boldsymbol{\xi}} | \to 0$, we have the matching requirement given by (6.120).

A Tentative Solution to the Equation (6.122) with (6.126)–(6.128) Governing H_2^{ff}

We make, first, a simple observation, namely that if we put

$$H_2^{\mathrm{ff}} = -\left\{ \frac{\gamma}{2} \left[\frac{\langle \Omega_0 \rangle}{2\pi} \right]^2 + \frac{\gamma}{2\pi} \boldsymbol{V}_\infty \cdot (\boldsymbol{k} \wedge \boldsymbol{M}_0) \right\} \frac{1}{|\boldsymbol{\xi}|^2}$$
$$+ \frac{\gamma}{\pi} \{[(\boldsymbol{k} \wedge \boldsymbol{M}_0) \cdot \boldsymbol{\xi}](\boldsymbol{V}_\infty \cdot \boldsymbol{\xi})\} \frac{1}{|\boldsymbol{\xi}|^4} + H_2^{\mathrm{ff}\,*}\,, \tag{6.129}$$

then $H_2^{\mathrm{ff}\,*}$ must satisfy the homogeneous equation corresponding to (6.122), but the conditions are changed. For instance, the behaviour (6.120) (when $|\boldsymbol{\xi}| \to 0$) must be replaced by

$$H_2^{\mathrm{ff}*} = -\frac{\gamma}{2\pi}\left[\boldsymbol{\xi}\cdot\frac{\partial \boldsymbol{N}_0}{\partial t}\cdot(\boldsymbol{k}\wedge\boldsymbol{\xi})\right]\frac{1}{|\boldsymbol{\xi}|^4} + \frac{\gamma}{2\pi}\frac{\partial^2\langle\varpi_2\rangle}{\partial t\,\partial p}\log|\boldsymbol{\xi}| + \dots \quad (6.130)$$

In order to check (6.129), we observe that $\partial\langle\Omega_0\rangle/\partial t = 0$ and $\partial^2\boldsymbol{M}_0/\partial t^2 = 0$. The reader may wonder why, in (6.130), we have deleted the term which involves $\partial^3\boldsymbol{N}_0/\partial t^3$. Strictly speaking, such a statement is wrong. As a matter of fact, we should not consider (6.130) as an expansion of $H_2^{\mathrm{ff}\,*}$ near $|\boldsymbol{\xi}| = 0$; rather, we should consider it as an indication that $H_2^{\mathrm{ff}\,*}$ is the sum of two contributions, the *leading behaviour* of each one near $|\boldsymbol{\xi}| = 0$ being that of the corresponding term in the right-hand side of (6.130). If, then, we takes the homogeneous part of (6.122) into account, we can check that the above contributions, when expanded near $|\boldsymbol{\xi}| = 0$, will contain the term in $\partial^3\boldsymbol{N}_0/\partial t^3$ which has been deleted in (6.130). We summarize by stating the problem which needs to be solved for $H_2^{\mathrm{ff}\,*} \equiv H_2^{\mathrm{ff}*(1)} + H_2^{\mathrm{ff}*(2)}$. Namely:

$$\left\{\frac{\partial^2}{\partial\xi^2} + \frac{\partial^2}{\partial\eta^2} + \gamma\frac{\partial^2}{\partial t^2}\left[\frac{\partial}{\partial p}\left(\frac{p^2}{K_0(p)}\frac{\partial}{\partial p}\right)\right]\right\} H_2^{\mathrm{ff}*(\mathrm{n})} = 0, n = 1, 2\,,$$

$$\lim_{p\to 0}\left[\frac{p^2}{K_0(p)} H_2^{\mathrm{ff}\,*(\mathrm{n})}\frac{\partial H_2^{\mathrm{ff}\,*\mathrm{n}}}{\partial p}\right] = 0\,,$$

$$\left[H_2^{\mathrm{ff}\,*(\mathrm{n})} - \frac{1}{K_0(p)}\frac{\mathrm{d}H_0}{\mathrm{d}p}\frac{\partial H_2^{\mathrm{ff}\,*(\mathrm{n})}}{\partial p}\right]_{p=1} = 0,$$

$$H_2^{\mathrm{ff}\,*(\mathrm{n})} \to 0 \quad \text{when} |\boldsymbol{\xi}| \to \infty\,,$$

$$H_2^{\mathrm{ff}*\,(1)} \approx -\frac{\gamma}{2\pi}\left[\boldsymbol{\xi}\cdot\frac{\partial\boldsymbol{N}_0}{\partial t}\cdot(\boldsymbol{k}\wedge\boldsymbol{\xi})\right]\frac{1}{|\boldsymbol{\xi}|^4},\ \text{when}|\boldsymbol{\xi}| \to 0\,,$$

$$H_2^{\mathrm{ff}\,*(2)} \approx \frac{\gamma}{2\pi}\frac{\partial^2}{\partial t\,\partial p}(\langle\varpi_2\rangle)\log\,|\boldsymbol{\xi}|, \text{when}\,|\boldsymbol{\xi}| \to 0\,. \quad (6.131)$$

This is not the whole matter, because if we solve the problem (6.131) for $H_2^{\mathrm{ff}\,*} = H_2^{\mathrm{ff}\,*(1)} + H_2^{\mathrm{ff}\,*(2)}$ and substitute the result into (6.129), then the resulting H_2^{ff} will satisfy neither (6.126) nor (6.127), unless $\langle\Omega_0\rangle, \boldsymbol{V}_\infty$ and $\boldsymbol{M}_0$, as functions of p, meet specific requirements near $p = 0$ and near $p = 1$. However, we ignore this question for the time being and consider the above problem (6.131). We use a result of El Mabrouk and Zeytounian (1984), who proved that the operator $L = \partial/\partial p[(p^2/K_0(p)\partial/\partial p]$, operating on functions of p alone, is self-adjoint when associated with the boundary conditions at $p = 0$ and $p = 1$ in (6.131). With this operator we may associate an enumerable set of eigenvalues $\{-a_n^2\}$, and to each of them there corresponds an eigenfunction $\phi_n(p)$. We have the orthonormality property

$$\int_0^1 \phi_n(p)\phi_m(p)\,\mathrm{d}p = \delta_{\mathrm{nm}}\,, \tag{6.132a}$$

and a completeness property: for any function $f(p)$ which is square-integrable over (0, 1) with respect to p, we have

$$f(p) = \Sigma(f, \phi_n)\phi_n(p), \quad n = 1 \text{ to } \infty, \quad (f, g) = \int_0^1 fg\,\mathrm{d}p, \quad ||f||^2 = (f, f)\,. \tag{6.132b}$$

Now we look for a solution to (6.131) by writing

$$H_2^{\mathrm{ff}\,*} = \Sigma H_n(t, \xi, \eta)\phi_n(p), \quad n = 1 \text{ to } \infty\,, \tag{6.133}$$

so that each $H_n(t, \xi, \eta\,)$ must be a solution to the following problem, according to (6.131):

$$\frac{\partial^2 H_n}{\partial \xi^2} + \frac{\partial^2 H_n}{\partial \eta^2} - \frac{1}{C_n^2}\frac{\partial^2 H_n}{\partial t^2} = 0\,, \tag{6.134a}$$

$$H_n \to 0, \quad \text{when } |\boldsymbol{\xi}| \to \infty\,, \tag{6.134b}$$

$$H_n \approx -\frac{\gamma}{2\pi}\left[\boldsymbol{\xi}\,.\left(\frac{\partial \boldsymbol{N}_0}{\partial t}, \phi_n\right).(\boldsymbol{k}\wedge\boldsymbol{\xi})\right]\frac{1}{|\boldsymbol{\xi}|^4}\,,$$
$$+\frac{\gamma}{2\pi}\left(\frac{\partial^2\langle\omega_2\rangle}{\partial t\,\partial p}, \phi_n\right)\log|\boldsymbol{\xi}|\,, \text{ when } |\boldsymbol{\xi}| \to 0\,, \tag{6.134c}$$

where $C_n = (1/a_n\sqrt{\gamma})$. The solution of the above (linear) problem (6.134) is the sum of the solution of the following two problems:

$$\frac{\partial^2 H_n^{(1)}}{\partial \xi^2} + \frac{\partial^2 H_n^{(1)}}{\partial \eta^2} - \frac{1}{C_n^2}\frac{\partial^2 H_n^{(1)}}{\partial t^2} = 0\,,$$

$$H_n^{(1)} \to 0 \text{ when } |\boldsymbol{\xi}| \to \infty, H_n^{(1)} \approx -\frac{1}{2}(\boldsymbol{k}\wedge\boldsymbol{D}^{ff}).[\boldsymbol{A}.\boldsymbol{D}^{ff}\log|\boldsymbol{\xi}|] \text{ when } \xi \to 0\,, \tag{6.135a}$$

and

$$\frac{\partial^2 H_n^{(2)}}{\partial \xi^2} + \frac{\partial^2 H_n^{(2)}}{\partial \eta^2} - \frac{1}{C_n^2}\frac{\partial^2 H_n^{(2)}}{\partial t^2} = 0,$$

$$H_n^{(2)} \to 0 \quad \text{when } |\boldsymbol{\xi}| \to \infty\,,$$

$$H_n^{(2)} \approx -Q(t)\log|\boldsymbol{\xi}| \quad \text{when } |\boldsymbol{\xi}| \to 0\,. \tag{6.135b}$$

Here

$$Q(t) = \frac{\gamma}{2}\left(\frac{\partial^2\langle\varpi_2\rangle}{\partial t\partial p}, \phi_n\right), \quad \boldsymbol{A} = -\frac{\gamma}{2\pi}\left(\frac{\partial \boldsymbol{N}_0}{\partial t}, \phi_n\right),$$

and

$$[\boldsymbol{\xi}\,.\,\boldsymbol{A}\,.\,(\boldsymbol{k}\wedge\boldsymbol{\xi})]\frac{1}{|\boldsymbol{\xi}|^4}=-\frac{1}{2}(\boldsymbol{k}\wedge\boldsymbol{D}^{\mathrm{ff}})\,.\,[\boldsymbol{A}\,.\,\boldsymbol{D}^{\mathrm{ff}}\log|\boldsymbol{\xi}|]\,.$$

The reader may check that this is true by direct computation, provided **A** is a symmetric second-order tensor. According to Ashley and Landahl (1965, Chap. 6), the solution of the second problem, (6.135b), for $H_n^{(2)}$, is:

$$H_n^{(2)}=\int_{-\infty}^{t-\frac{|\xi|}{C_n}}\left\{\frac{Q(t')}{(t-t')^2-\left(\frac{|\xi|}{C_n}\right)^2}\right\}^{1/2}\mathrm{d}t'\,, \tag{6.136a}$$

and a moment's reflection will convince the reader that the solution of the first problem (6.135a), for $H_n^{(1)}$, is simply:

$$H_n^{(1)}=\frac{1}{2|\boldsymbol{\xi}|^2}(\boldsymbol{k}\wedge\boldsymbol{\xi})\,.\left\{\frac{\partial^2}{\partial|\boldsymbol{\xi}|^2}\int_{-\infty}^{t-|\boldsymbol{\xi}|/C_n}\left[\frac{\mathbf{A}(t')}{(t-t')^2-(|\boldsymbol{\xi}|/C_n)^2}\right]^{1/2}\mathrm{d}t'\right\}\,. \tag{6.136b}$$

In order to complete the solution, it is not sufficient to substitute the sum of the above two solutions (6.136) into (6.129), because we run into the difficulty mentioned earlier, after (6.131) in relation to $H_2^{\mathrm{ff}\,*}$. We shall deal with this matter, but before doing so, we observe that in the context of Fourier series we have a curious phenomenon (referred to as the Gibbs phenomenon; see Zygmund 1959), which is illustrated by the following simple example:

$$\sum_{n=1}^{\infty}\frac{\sin(nx)}{n}=\frac{1}{2}(\pi-x),\quad 0<x<2\pi\,,$$

where $(1/2)(\pi-x)$ does not satisfy the condition of vanishing at $x=2\pi$ which is met by any partial sum. We start from some function $f(p)$ which is space-integrable with respect to the variable p over (0, 1), and then we consider $\Phi\{f\}(p)=\sum_{n=1}^{\infty}(f,\phi_n)\phi_n(p)$; the convergence of this series is not guaranted pointwise if f is only square integrable. We may only state that $\lim\,||f-\sum_{n=1}^{N}(f,\phi_n)\phi_n(p)||=0$ when $N\to\infty$. Any partial sum $\Phi^N\{f\}(p)=\sum_{n=1}^{N}(f,\phi_n)\phi_n(p)$ is an infinitely differentiable function of p which satisfies (6.126) and (6.127), namely

$$\left\{\Phi^N\{f\}(p)-\frac{1}{K_0(p)}\frac{\partial H_0}{\partial p}\frac{\partial[\Phi^N\{f\}(p)]}{\partial p}\right\}_{p=1}=0\,,$$

$$\lim_{p\to 0}\left[\frac{p^2}{K_0(p)}\Phi^N\{f\}(p)\frac{\partial[\Phi^N\{f\}(p)]}{\partial p}\right]=0\,. \tag{6.137}$$

Even if the function $f(p)$ introduced above does not satisfy both relations in (6.137), we consider that $\Phi^N\{f\}(p)$ does! As a consequence, in order to complete the solution, we need only use (6.129) with the above two solutions (6.136) and with (6.137), provided we use the convention that the closed part of H_2^{ff} in (6.129) is interpreted as being modified by the operator Φ near $p=0$ and $p=1$.

6.5.5 Some Remarks

We conclude that, up to the order considered, a fully consistent inner–outer expansion scheme with respect to the small Mach number may be set up. Namely, we have obtained three expansions: for instance, for the function $H(t,x,p)$, we can write

$$H(t = M\tau, \boldsymbol{x}, p) = H^{\tau}(\tau, \boldsymbol{x}, p) = H_0(p) + MH_1^{\tau} + O(M^2)\,,$$

$$H(t, \boldsymbol{x}, p) = H_0(p) + M^2 H_2 + (M^4 \log M)\,\frac{\gamma}{2\pi}\,\frac{\partial^2\langle\varpi_2\rangle}{\partial t\,\partial p} + M^4 H_4 + \dots\,,$$

$$H\left(t, \boldsymbol{x} = \frac{\boldsymbol{\xi}}{M}, p\right) = H^{\mathrm{ff}}(t, \boldsymbol{\xi}, p) = H_0(p) + M^4 H_2^{\mathrm{ff}} + \dots\,. \tag{6.138}$$

More careful consideration shows that the inclusion of the term proportional to $(M^4 \log M)$ in the *main* 'Monin–Charney' expansion (with $t, \boldsymbol{x}, p$ fixed when $M \downarrow 0$) is indeed the only possibility for taking this term into account within the framework of a consistent asymptotic theory at low Mach number. In fact, if such a term is included in the *local* far-field Guiraud–Zeytounian expansion (with $t, \boldsymbol{\xi}, p$ fixed when $M \downarrow 0$), namely:

$$H\left(t, \boldsymbol{x} = \frac{\xi}{M}, p\right) = H^{\mathrm{ff}}(t, \boldsymbol{\xi}, p) = H_0(p) + (M^4 \log M) H_2^{\mathrm{ff}'} + M^4 H_2^{\mathrm{ff}} + \dots\,, \tag{6.139}$$

then we can prove that the corresponding initial–boundary-value problem for $H_2^{\mathrm{ff}'}$ has only a zero solution! We observe that, as a consequence of the matching ($\tau \to \infty \Leftrightarrow t = 0$) between the local unsteady-adjustment expansion (with $\tau, \boldsymbol{x}, p$ fixed when $M \downarrow 0$) and the *local* far-field Guiraud–Zeytounian expansion, we have proved that the term containing H_1 is really absent in the main 'Monin–Charney' expansion. However, for the determination of the initial value (at $t = 0$) of the term H_2, it is necessary to consider a second-order adjustment problem.

Of course, the main interest and conclusion of the investigations of Guiraud and Zeytounian is the necessity to take into account precisely the expansion (6.116) as a complement of the Monin–Charney expansion, which tells us that there is a back-influence from the expansion at infinity, expressed through the terms propotional to $M^4 \log M$. Except for that influence, the expansion near infinity may be ignored. One remarkable consequence of the low-Mach-number expansion (see, for instance, the expansion (6.75a), with the term $M^{\alpha} U_{\alpha}$ is the emergence of the 'standard atmosphere', characterized by $(H_0(p), T_0(p))$ such that the relation (6.75c) is satisfied, which we should identify with the relation referred in Zeytounian (1976), namely (see(6.68c$'$)):

$$\frac{\partial}{\partial p}\left[\frac{\mu_0}{Pr}\,\rho_0\,\frac{\partial T_0}{\partial p}\right] - \sigma\frac{\mathrm{dR_s}}{\mathrm{d}p} = 0\,. \tag{6.140}$$

Of course, this standard atmosphere is not at all arbitrary but is determined by the very process of expansion. This may have far-reaching consequences in 'theories of climate' (see, for instance, Monin 1972, Chap. 3, for a pertinent discussion of long-term weather processes). Obviously our analysis is still very crude and should be taken further. Two basic improvements are needed. The first concernes the energy balance: the term $\mathrm{dR_s}/\mathrm{d}p$ in (6.140) should be re-examined carefully. The second improvement concerns the absence of the Coriolis terms from the outset. Indeed, in order to extract the standard atmosphere from a low-Mach-number expansion, it should be necessary to work on the whole sphere. Further research is obviously necessary. We observe that the dependence (in the quasi-non-divergent expansion) on the vertical (pressure) coordinate p is realized via a local (inner–far-field) approximation which is valid horizontally near infinity in $\boldsymbol{x}$.

However, for the time being, it is not known whether or not a local approximation must also be considered in the vicinity of $p = 1$, which simulates the flat ground surface in the low-Mach-number model equations derived in Sects. 6.5.1 and 6.5.2 when $M \to 0$. In fact, from (6.85), we can derive the following value for ϖ_2 on the ground (as a consequence of the slip condition), simulated by $p = 1$:

$$\varpi_2 = \frac{1}{T_0(1)}\left[\frac{\partial H_2}{\partial t} + \boldsymbol{v}_0 \,.\, \boldsymbol{D} H_2\right] \quad \text{on } p = 1 \,. \tag{6.141a}$$

On the other hand, from the relation (6.81b), for ϖ_2, we obtain:

$$\lim\nolimits_{p\to 1}\varpi_2 = -\frac{1}{K_0(1)}\left[\frac{\partial}{\partial t}\left(\frac{\partial H_2}{\partial p}\right)_{p=1} + \boldsymbol{v}_{0\,p=1} \,.\, \boldsymbol{D}\left(\frac{\partial H_2}{\partial p}\right)_{p=1}\right] \,. \tag{6.141b}$$

It seems that we can conclude from (6.141) that

$$\lim\nolimits_{p\to 1}\varpi_2 \neq (\varpi_2)_{p=1} \,. \tag{6.142}$$

As a consequence, it can be stated that the vicinity of $p = 1$ is *singular* for the main asymptotic expansion (6.77b) which leads to the Monin–Charney quasi-non-divergent model (6.79) and then to the equations (6.81) and the second order system (6.82).

If it is true that we have (only!) explained how blocking is released thanks to the consideration of the far field (where $\boldsymbol{\xi} = O(1)$), nevertheless the blocking effect remains a mystery.

Related to this question, we would like to understand how waves are generated in the lee of mountains, in the context of a low-Mach-number approximation near the top of the mountain (see Hunt and Snyder 1980 for some considerations on this topic). Also, we would like to understand how the low-Mach-number approximation works at high altitude when $p \to 0$.

Clearly, further research is necessary. In Sect. 7.3, low-Mach-number flow over relief is considered and, in particular, a “primitive model” is considered, which is related with the investigations of the above Sect. 6.5.

6.6 Low-Mach-Number ($M \ll 1$) Asymptotics of the Hydrostatic Model Equations, with $Ki \ll 1$

In this section, it is useful to start with the following complete hydrostatic, dissipative system of equations with $\delta = O(1)$. In this case, we obtain the following system of equations:

$$\begin{aligned} &\frac{\mathrm{d}\boldsymbol{v}}{\mathrm{d}t} + \left[\frac{1}{Ki}\frac{\sin\phi}{\sin\phi^0} + \delta\tan\phi\, u\right](\boldsymbol{k}\wedge\boldsymbol{v}) \\ &\quad + \frac{1}{\gamma M^2}\boldsymbol{D}H = \frac{1}{Re_\perp}\frac{\partial}{\partial p}\left(\rho\mu\frac{\partial\boldsymbol{v}}{\partial p}\right), \\ &\frac{\mathrm{d}T}{\mathrm{d}t} - \frac{\gamma-1}{\gamma}\frac{T}{p}\omega = \frac{1}{Re_\perp}\left\{\frac{1}{Pr}\frac{\partial}{\partial p}\left(\rho k\frac{\partial T}{\partial p}\right)\right. \\ &\quad \left. + \mu(\gamma-1)M^2\rho\left|\frac{\partial\boldsymbol{v}}{\partial p}\right|^2 - \sigma\frac{\mathrm{dR_s}}{\mathrm{d}p}\right\}, \\ &\frac{\partial\omega}{\partial p} + \boldsymbol{D}\,.\,\boldsymbol{v} - \delta\tan\phi\,\boldsymbol{v} = 0, \quad \frac{\partial H}{\partial p} + \frac{T}{p} = 0\,, \end{aligned} \tag{6.143}$$

where

$$\omega = \frac{p}{T}\left[\frac{\partial H}{\partial t} + \boldsymbol{v}\,.\,\boldsymbol{D}H - w\right]. \tag{6.144}$$

In the above equations (6.143), written in the p-system, the operator $\boldsymbol{D}$ is the horizontal gradient on the isobaric, surface $p = \text{const}$, with two components $[(\cos\phi^0/\cos\phi)\,\partial/\partial x, \partial/\partial y)]$, and the material-derivative operator is $\mathrm{d}/\mathrm{d}t = \partial/\partial t + \boldsymbol{v}\,.\,\boldsymbol{D} + \omega\,\partial/\partial p$. In this case we assume that μ, k and R_s are known functions of p. One disadvantage of the p-system is that the lower boundary conditions at the flat ground surface become a condition at the unknown isobaric surface $H = 0$! Namely, it is necessary to impose the following boundary conditions:

$$\boldsymbol{v} = 0, \quad \omega = \frac{p}{T}\frac{\partial H}{\partial t} \quad \text{and} \quad kp\frac{\partial\log T}{\partial p} = \sigma\mathsf{R_s} \quad \text{on} \quad H = 0\,. \tag{6.145a}$$

As a boundary condition 'at infinite altitude', we can assume that at the upper end of the atmosphere,

$$\text{when } p = 0, \text{the total energy density decays sufficiently rapidly}\,. \tag{6.145b}$$

The set of equations (6.143) in the p-system, with the boundary conditions (6.145a,b) and the initial conditions

$$\text{at} \quad t \le 0, \quad \boldsymbol{v} = \boldsymbol{v}^0, \quad T = T^0\,, \tag{6.145c}$$

can be used as a theoretical basis for various investigations of features of atmospheric dissipative hydrostatic motions depending on the parameters

$Ki, Re_\perp, \delta, Pr$ and M^2. For instance, when $\tan\phi \neq \infty, Pr = O(1), Ki$ is fixed and $(1/Re_\perp) \to 0$, we can derive from the above equations (6.143) a non-dissipative (inviscid, adiabatic) set of equations which is, in fact, a generalized system of Kibel primitive equations, very suitable for dealing with large-scale motions ($\delta = O(1)$). In particular, our low-Mach-number ($M \ll 1$) theory with $Ki = O(1)$ described in Sect. 6.5 can be generalized on the basis of these inviscid, adiabatic primitive equations. However, among other things, we observe again that the smallness of the Mach number for the usual slow atmospheric motions poses many unresolved, hard problems, as has been demonstrated in the Sect. 6.5.4, where we considered some results of Guiraud and Zeytounian related to a deeper analysis of the non-divergent (quasi-solenoidal) approximation, derived in leading order when $M \downarrow 0$ with Ki fixed and $O(1)$. Below, we simplify the above problem (6.143)–(6.145) under the assumption $\delta \ll 1$ and also by writing a simple similarity rule between M and Ki, both being assumed to be small parameters. Namely, in the Kibel limiting process (2.15), we assume simply, as the similarity rule,

$$Ki = M \, , \tag{6.146}$$

This case gives us the possibility to avoid some complications in the presentation of the quasi-geostrophic approximation below. Obviously, other choices are possible, such that the estimates of the characteristic length scale L_c and of the characteristic velocity u_c are in better agreement with short-range weather predictions (for instance, such that the scale L_c is comparable to the Obukhov synoptic scale $L_B = \sqrt{gH_s}/f^0 \Rightarrow L_c/L_B = \gamma^{1/2}(M/Ki)$ – see Monin 1972, p. 33).

6.6.1 The Hydrostatic Dissipative Model Equations with the β-Effect

Now, in the first equation of the system (6.143), we assume that the sphericity parameter δ is much less than 1, such that

$$\frac{\sin\phi}{\sin\phi^0} = 1 + \frac{\delta}{\tan\phi^0} y + O(\delta^2), \quad \tan\phi = \tan\phi^0[1 + O(\delta)] \, , \tag{6.147}$$

since $\phi = \phi^0 + \delta y$, or

$$\frac{1}{Ki}\frac{\sin\phi}{\sin\phi^0} \approx \frac{1}{Ki} + \beta y \, , \tag{6.148}$$

with an error of $O(\delta^2)$, where (β-effect)

$$\beta = \frac{\delta}{Ki \tan\phi^0} \, , \tag{6.149}$$

which is our second similarity rule betwen the two small parameters $Ki = M$ and δ. We observe that $\tan\phi^0 \approx 1$ (for $\phi^0 \approx 45°$). On the other hand, in place of $Re_\perp$, assumed to be large ($Re_\perp \gg 1$), we introduce the (small) Ekman number

$$Ek_\perp = \frac{Ki}{Re_\perp} \tag{6.150}$$

and consider our third similarity rule,

$$Ek_\perp = \kappa^0 M^2, \quad \text{with} \quad \kappa^0 = O(1) \,, \tag{6.151}$$

where we assume that $Re_\perp$ is of the order $O(1/M)$. Finally, in (6.143), we neglect the terms containing $\delta \tan \phi^0 \, u$ and $\delta \tan \phi^0 \, v$, but take into account the term containing β and the above three similarity rules (6.146), (6.149) and (6.151). In this case, in place of (6.143), we can write the following simplified system of three equations for the three functions $\boldsymbol{v}, \omega$ and H:

$$M\frac{\mathrm{d}\boldsymbol{v}}{\mathrm{d}t} + [1 + M\beta y](\boldsymbol{k} \wedge \boldsymbol{v}) + \frac{1}{\gamma M}\boldsymbol{D}H = \kappa^0 M^2 \frac{\partial}{\partial p}\left(\rho\mu\frac{\partial \boldsymbol{v}}{\partial p}\right) \,, \tag{6.152a}$$

$$M\left[\frac{\mathrm{d}(p\,\partial H/\partial p)}{\mathrm{d}t} - \frac{\gamma-1}{\gamma}\omega\frac{\partial H}{\partial p}\right]$$
$$= \kappa^0 M^2 \left\{\frac{1}{Pr}\frac{\partial}{\partial p}\left(\rho k \frac{\partial(p\,\partial H/\partial p)}{\partial p}\right) - M^2 \mu\rho\frac{\gamma-1}{\gamma}\left|\frac{\partial \boldsymbol{v}}{\partial p}\right|^2 + \sigma\frac{\mathrm{dR_s}}{\mathrm{d}p}\right\} \,, \tag{6.152b}$$

$$\frac{\partial \omega}{\partial p} + \boldsymbol{D} \,.\, \boldsymbol{v} = 0 \,, \tag{6.152c}$$

where $T = -p\,\partial H/\partial p$, and the main small parameter is the Mach number M. We have, as boundary conditions on the ground for $\boldsymbol{v}, \omega$ and H,

$$\boldsymbol{v} = 0, \quad \omega = \frac{p}{T}\frac{\partial H}{\partial t} \qquad \text{and} \; -kp\frac{\partial \log(p\,\partial H/\partial p)}{\partial p} = \sigma \mathrm{R_s} \;\text{on}\; H = 0 \,, \tag{6.153a}$$

and, as initial conditions for $\boldsymbol{v}$ and H (with $T^0 = -p\,\partial H^0/\partial p$),

$$\text{at}\; t \le 0 : \quad \boldsymbol{v} = \boldsymbol{v}^0, H = H^0 \,. \tag{6.153b}$$

We observe that the above hydrostatic dissipative equations (6.152) including the β-effect, (where, again, $d/dt = \partial/\partial t + \boldsymbol{v} \,.\, \boldsymbol{D} + \omega\,\partial/\partial p$) are significant only when $\varepsilon \ll 1, \delta \ll 1, Ki \ll 1$, and $M \ll 1$, with the three similarity rules (6.146), (6.149) and (6.151); β and κ^0 are fixed and $O(1)$, and the terms containing $\delta \tan\phi^0 u$ and $\delta \tan\phi^0 v$ are neglected. The above model problem (6.152) with (6.153), when $M \downarrow 0$, is investigated below.

6.6.2 The Quasi-Geostrophic Approximation

After the unsuccessful attempt of Richardson (1922), the most fundamental step was taken made by Kibel, who, in the well-known paper Kibel (1940), first proposed a basic principle for simplifying the equations governing atmospheric motions and derived a hydrodynamic quasi-geostrophic (QG) working model for short-range weather prediction, just before the Second World War (see also Kibel 1970, 1984). The main equation of the QG model contains only one unknown function and is first-order with respect to time, since the low-Mach-number asymptotics of the Kibel equations (6.152) filter out the family of short gravitational waves in these third-order-in-time hydrostatic, dissipative equations. A systematic matched (inner (close to initial time) $\rightarrow$ outer $\leftarrow$ inner (near the ground)) asymptotic expansion with respect to $M \ll 1$ leads first to the classical QG (outer) model and then to a correction to it, which we call the ageostrophic (AG) model. It is then found that four regions have to be studied separately: main region I away from the ground and from the initial time; an initial region II close to the initial time; an Ekman-layer region III near the ground; and a corner region IV, simultaneously close to the initial time and near the ground, which is a region of unsteady adjustment to the steady Ekman layer. In region I (of the QG and AG model equations), t and p are both $O(1)$; region II (of adjustment problems) corresponds to $t = O(Ki)$ and $p = O(1)$; and region III (of Ackerblom problems)corresponds to $t = O(1)$ and $|p-1| = O(Ki)$. Finally, in the corner region IV, we have, simultaneously, $t = O(Ki)$ and $|p-1| = O(Ki)$. Concerning the dissipative coefficients μ and k in (6.152a) and (6.152b) and the radiative heat transfer R_s in (6.152b) and in the boundary conditions (6.153a), we assume that they are dependent only on p. Obviously, we can consider another case, where they are dependent on the vertical structure of the Ekman layer, characterized by $\pi = (p-1)/Ki$.

Below, we give only brief information for the AG model. The reader can find in Guiraud and Zeytounian (1980) a detailed asymptotic (low-Kibel-number) theory. Concerning, more specifically, the AG asymptotic model, see Zeytounian (1990, Sect. 11.4). In Bourgeois and Beale (1994), the reader can find some mathematical results concerning the justification of the QG asymptotics for large-scale flow in the atmosphere and in the ocean, and in Lions and Temam (1994), the QG asymptotics of the primitive equations for the atmosphere are considered from the point of view of an applied mathematician. Unfortunately, because the (well-justified) rigorous results in the latter two papers are obtained on the basis of an atmospheric problem formulated in a rather 'ad hoc' way, the present author is very critical concerning the real value of these two papers.

QG (Outer) Model Equation, Singular Close to $t = 0$ and Near $p = 1$

Here we expand $H, \boldsymbol{v}, \omega$ and T in the model problem (6.152) with (6.153), written in the p-system, according to the following scheme (where the time t and the pressure coordinate p are both fixed when M tends to zero):

$$(H,T) = (H_0,T_0) + M(H_{\mathrm{QG}},\ T_{\mathrm{QG}}) + M^2(H_{\mathrm{AG}},\ T_{\mathrm{AG}})$$
$$+ M^3(H^*_{\mathrm{AG}}, T^*_{\mathrm{AG}}) + \dots\ , \tag{6.154a}$$
$$(\boldsymbol{v},\omega) = (\boldsymbol{v}_{\mathrm{QG}},\ \omega_0) + M(\boldsymbol{v}_{\mathrm{AG}},\ \omega_{\mathrm{QG}})$$
$$+ M^2(\boldsymbol{v}^*_{\mathrm{AG}},\ \omega_{\mathrm{AG}}) + \dots\ , \tag{6.154b}$$

From (6.152) and (6.153), we find that (H_0, T_0) do not depend on the horizontal variables, and then we assume that they do not depend on the time t either. Although this does not follow directly from the equations (6.152), it will be found to be consistent with the constancy (with respect to t) of

$$\frac{\mathrm{d}}{\mathrm{d}p}\left(kp\frac{\mathrm{d}\,\log T_0}{\mathrm{d}p}\right) = \sigma\frac{\mathrm{dR_s}}{\mathrm{d}p}\ , \tag{6.155a}$$

which is a consequence of (6.153a). Of course, we have

$$T_0(p) = -\mathrm{p}\frac{\mathrm{d}H_0}{\mathrm{d}p} \quad \text{and} \quad \rho_0(p) = \frac{p}{T_0}\ , \tag{6.155b}$$

but we do not know yet how $T_0(p)$ depends on p. On the other hand, from the boundary condition (6.153a), written for $T_0 = -p\,\mathrm{d}H_0/\mathrm{d}p$, we can write also

$$kp\frac{\mathrm{d}\,\log T_0}{\mathrm{d}p} = \sigma\mathrm{R_s} \quad \text{on } p = 1\ , \tag{6.155c}$$

where $p = 1$ is the (dimensionless) solution of the equation $H_0(p) = 0$; it is found that the relation (6.155c) holds throughout. From our point of view, we suppose that $T_0(p)$ is given (as the temperature of the standard atmosphere), and we assume that the relation

$$\mathrm{R_s} = \frac{k}{\sigma}p\frac{\mathrm{d}\,\log T_0}{\mathrm{d}p}\ , \tag{6.155d}$$

allows us to compute $\mathrm{R_s}(p)$. Then, assuming again (see (6.76)) that

$$\frac{\gamma-1}{\gamma}T_0 - p\frac{\mathrm{d}T_0}{\mathrm{d}p} \equiv K_0(p) \neq 0\ ,$$

we find, from (6.152b), that $\omega_0 = 0$, and this same equation (6.152b) leads to the following equation for T_{QG}:

$$\frac{\partial T_{\mathrm{QG}}}{\partial t} + \boldsymbol{v}_{\mathrm{QG}}\,.\,\boldsymbol{D}T_{\mathrm{QG}} - \frac{K_0(p)}{p}\omega_{\mathrm{QG}} = 0\ . \tag{6.156}$$

Of course, from the first equation of (6.152a), we derive the well-known geostrophic balance: for a divergenceless geostrophic wind $\boldsymbol{v}_{QG}$,

$$\gamma(\boldsymbol{k}\wedge\boldsymbol{v}_{\mathrm{QG}}) + \boldsymbol{D}H_{\mathrm{QG}} = 0 \quad \Leftrightarrow \boldsymbol{v}_{\mathrm{QG}} = \frac{1}{\gamma}(\boldsymbol{K}\wedge\boldsymbol{D}H_{\mathrm{QG}}), \quad \text{with } \boldsymbol{D}\,.\,\boldsymbol{v}_{\mathrm{QG}} = 0\ . \tag{6.157}$$

Going to higher order, we find from (6.152a,c,d) the following three equations:

$$\boldsymbol{v}_{\mathrm{AG}} = \boldsymbol{k} \wedge \left[\frac{\partial \boldsymbol{v}_{\mathrm{QG}}}{\partial t} + (\boldsymbol{v}_{\mathrm{QG}} \cdot \boldsymbol{D})\boldsymbol{v}_{\mathrm{QG}} + \frac{1}{\gamma} \boldsymbol{D} H_{AG} \right] - \beta y \boldsymbol{v}_{\mathrm{QG}} \,, \tag{6.158a}$$

$$\frac{\partial \omega_{\mathrm{QG}}}{\partial p} + \boldsymbol{D} \cdot \boldsymbol{v}_{\mathrm{AG}} = 0 \,, \tag{6.158b}$$

$$T_{\mathrm{QG}} = -p \frac{\partial H_{\mathrm{QG}}}{\partial p} \,. \tag{6.158c}$$

From the (continuity) equation (6.158b), together with the equation (6.158a) and the expression for ω_{QG} (obtained through the elimination of T_{QG} and $\boldsymbol{v}_{\mathrm{QG}}$ in (6.156) from the equation (6.158c) and second equation of (6.157)), we obtain the following single QG model equation for the potential vorticity as a function of H_{QG}:

$$\frac{\partial \mathsf{L} H_{\mathrm{QG}}}{\partial t} + \frac{1}{\gamma} \left[\frac{\partial H_{\mathrm{QG}}}{\partial x} \frac{\partial \mathsf{L} H_{\mathrm{QG}}}{\partial y} - \frac{\partial H_{\mathrm{QG}}}{\partial y} \frac{\partial \mathsf{L} H_{\mathrm{QG}}}{\partial x} \right] + \frac{1}{\gamma} \beta \frac{\partial H_{\mathrm{QG}}}{\partial x} = 0 \,, \tag{6.159}$$

where

$$\mathsf{L} H_{\mathrm{QG}} = \frac{1}{\gamma} \boldsymbol{D}^2 H_{\mathrm{QG}} + \frac{\partial}{\partial p} \left[\frac{p^2}{K_0(p)} \frac{\partial H_{\mathrm{QG}}}{\partial p} \right] . \tag{6.160}$$

We observe that the Q-G model equation (6.159) contains one derivation with respect to the time t and, as a consequence, only one initial condition must be supplied for H_{QG}, via an unsteady adjustment problem, which we shall derive below. The boundary condition that must be supplied on the ground at $p = 1$ will be derived subsequently. Finally, concerning the boundary conditions that must be applied at the upper end of the atmosphere at $p = 0$ and far off in the horizontal plane, we can assume again that the total energy density

$$\frac{p^2}{K_0(p)} \left[\frac{\partial H_{\mathrm{QG}}}{\partial p} \right]^2 + \frac{1}{\gamma} \left| \boldsymbol{D}^2 H_{\mathrm{QG}} \right|^2 \text{decays sufficiently rapidly} \,. \tag{6.161}$$

Derivation of the Initial Condition at $t = 0$ for $\mathsf{L} H_{\mathrm{QG}}$ via Adjustment to Geostrophic Balance

It is not difficult to verify that the inner-in-time equations of unsteady adjustment to geostrophic balance (6.157) can be derived by setting $\tau = t/M$ and applying the initial limiting process $M \downarrow 0$, with τ fixed. Concerning, more precisely, the introduction of a short adjustment time (such as τ) in the case of the unsteady adjustment problems (and matching) of meteorological fields, I would like to bring in a typical fact that shows the present author's difficulty (in 1957, as a young mathematician) in understanding the subtlety of the arguments related to the adjustment to geostrophic balance. Namely, I very well remember, how as a graduate (PhD) student in 1957 in Kibel's Dynamic Meteorology Department (in what is now the Oboukhov Institute of Atmospheric Physics) and reading the just-edited monograph of Kibel (1957; Chap. 4), I was in no way capable of understanding the simultaneous appearance of an (adjustment) short time and an evolution prediction time – both times being denoted (in Kibel 1957) by t. Only after 1967, at ONERA, in France, while working in the Aerodynamics Department, did I have the possibility–thanks to the book of Van Dyke (1964) – to really understand the profound significance of these two (inner and outer) times in asymptotic outer–inner expansion theory with a matching ($\tau \to \infty \Leftrightarrow t \to 0$) procedure Il'ya Afanas'yevich Kibel' was able (in 1955) to settle the main issue of the adjustment to geostrophic balance and, in fact, derive, the single equation (6.171) below.

Let us write f^τ for any quantity f considered as a function of $\tau = t/M$ instead of t. First, we rewrite the equations (6.152) with $(1/M)\,\partial/\partial\tau$ in place of $\partial/\partial t$, for the new functions $\boldsymbol{v}^\tau, \omega^\tau, H^\tau$ and T^τ, and then expand:

$$(\boldsymbol{v}^\tau, \omega^\tau, H^\tau, T^\tau) = (\boldsymbol{v}_0^\tau, \omega_0^\tau, H_0^\tau, T_0^\tau) + M(\boldsymbol{v}_1^\tau, \omega_1^\tau, H_1^\tau, T_1^\tau)$$
$$+ M^2(\boldsymbol{v}_2^\tau, \omega_2^\tau, H_2^\tau, T_2^\tau) + \dots . \tag{6.162}$$

We find, first,

$$(H_0^\tau, T_0^\tau) = (H_0(p), T_0(p)), \quad \text{with } T_0 = -p\frac{\partial H_0}{\partial p} . \tag{6.163}$$

In order to find equations for $(\boldsymbol{v}_0^\tau, \omega_0^\tau, H_1^\tau, T_1^\tau)$, we have to go to higher order:

$$\frac{\partial \boldsymbol{v}_0^\tau}{\partial \tau} + \boldsymbol{k} \wedge \boldsymbol{v}_0^\tau + \frac{1}{\gamma}\boldsymbol{D}H_1^\tau = 0 ,$$
$$\frac{\partial \omega_0^\tau}{\partial p} + \boldsymbol{D} \cdot \boldsymbol{v}_0^\tau = 0, \quad T_1^\tau = -p\frac{\partial H_1^\tau}{\partial p} ,$$
$$\frac{\partial T_1^\tau}{\partial \tau} - \frac{K_0(p)}{p}\omega_0^\tau = 0 . \tag{6.164}$$

The system of equations (6.164) is the system governing the unsteady process of adjustment to the geostrophic balalance (6.157). We observe that the derivation of the unsteady-adjustment local equations (6.164), significant close to the initial time, in the case of low-Kibel/Mach-number asymptotics, is not a consequence of the linearization of the primitive Kibel inviscid equations, as was stated in Kibel (1957, p. 83). Here, these inner-in-time (linear) equations of unsteady adjustment to geostrophic balance are rationally derived, in the general case, from (6.152) as a significant limit when $M \to 0$ with the short time τ fixed as this is place in t being fixed, as this is the case in the derivation of the Q-G equation (6.159). From the last two equations of (6.164), we derive

$$\omega_0^\tau = -\frac{p^2}{K_0(p)}\frac{\partial}{\partial\tau}\left[\frac{\partial H_1^\tau}{\partial p}\right], \tag{6.165}$$

and, going back to the first two equations of (6.164), we find two equations for $\boldsymbol{v}_0^\tau$ and H_1^τ, Namely

$$\frac{\partial \boldsymbol{v}_0^\tau}{\partial\tau} + \boldsymbol{k}\wedge\boldsymbol{v}_0^\tau + \frac{1}{\gamma}\boldsymbol{D}H_1^\tau = 0\,, \tag{6.166}$$

$$\boldsymbol{D}\,.\,\boldsymbol{v}_0^\tau - \frac{\partial}{\partial p}\left[\frac{p^2}{K_0(p)}\frac{\partial H_1^\tau}{\partial\tau\,\partial p}\right] = 0\,. \tag{6.167}$$

For the two evolution equations (6.166) and (6.167) for $\boldsymbol{v}_0^\tau$ and H_1^τ (describing evolution in the time τ), it is necessary to give initial conditions for $\boldsymbol{v}_0^\tau$ and H_1^τ at $\tau = 0$. Concerning $\boldsymbol{v}_0^\tau$, we may use the initial data $\boldsymbol{v}_0^0$ at $\tau = 0$ in $\boldsymbol{v}^0 = \boldsymbol{v}_0^0 + M\boldsymbol{v}_1^0 + \ldots$. For H_1^τ, we may use the initial data H_1^0 at $\tau = 0$ in $H^0 = H_0(p)) + MH_1^0 + M^2H_2^0 + \ldots$. In this case (and only in this case) we obtain, as initial conditions for (6.166) and (6.167),

$$\tau = 0: \quad \boldsymbol{v}_0^\tau = \boldsymbol{v}_0^0, \quad H_1^\tau = H_1^0\,. \tag{6.168}$$

Whenever the assumed data H^0 in (6.153b) cannot be put into the above form, we must expect that another adjustment process will hold. For the inviscid adjustment equations (6.166) and (6.167), it is necessary to write only a slip condition on the flat ground surface (rigorously speaking, the above slip condition (6.169) is, in fact, a matching condition with the region IV considered in the AG theory). Namely, as a consequence of the boundary condition (6.153a) for ω , with (6.165) for ω_0^τ, we obtain (at the leading order) the following consistent slip boundary condition:

$$\frac{\partial}{\partial\tau}\left[H_1^\tau + p\frac{T_0(p)}{K_0(p)}\frac{\partial H_1^\tau}{\partial p}\right] = 0 \text{ on } p = 1\,. \tag{6.169}$$

We stress again that $p = 1$ is the dimensionless solution of the equation $H_0(p) = 0$, which simulates, at the leading order (M^0), the flat ground surface $H^\tau = 0$. As a matter of fact, we assume that we have the following matching condition:

$$\lim{}_{\tau\uparrow\infty}(\boldsymbol{v}_0^\tau, H_1^\tau) = [\boldsymbol{v}_{\mathrm{QG}}(t=0,x,y,p), H_{\mathrm{QG}}(t=0,x,y,p)]\,, \tag{6.170a}$$

with

$$(\boldsymbol{k}\wedge\boldsymbol{v}_{\mathrm{QG}}) + \frac{1}{\gamma}\boldsymbol{D}H_{\mathrm{QG}} = 0 \quad \text{at } t = 0\,. \tag{6.170b}$$

There is, also, an important observation, which was known to Kibel and which concerns the way in which H_1^τ is related to the initial data $\boldsymbol{v}_0^0$ and H_1^0

in (6.168) when $\tau \uparrow \infty$. From the linear adjustment system (6.164), we can obtain easily the following single equation between $\boldsymbol{v}_0^\tau$ and H_1^τ:

$$\frac{\partial}{\partial \tau}\left\{\boldsymbol{k}\,.\,(\boldsymbol{D}\wedge \boldsymbol{v}_0^\tau) + \frac{\partial}{\partial p}\left[\frac{p^2}{K_0(p)}\frac{\partial H_1^\tau}{\partial p}\right]\right\} = 0\,. \tag{6.171}$$

If we integrate this equation between $\tau = 0$ and $\tau = \infty$ and use the geostrophic balance for the limiting values of $\boldsymbol{v}_0^\tau$ and H_1^τ when $\tau \uparrow \infty$, we obtain, by matching, the following initial condition:

$$\mathsf{L}H_{\mathrm{QG}} = \boldsymbol{k}\,.\,\{\boldsymbol{D}\wedge \boldsymbol{v}_0^0\} + \frac{\partial}{\partial p}\left[\frac{p^2}{K_0(p)}\frac{\partial H_1^0}{\partial p}\right] \quad \text{at } t = 0\,, \tag{6.172}$$

where $\mathsf{L}H_{\mathrm{QG}}$ is the same operator as that which appears in the QG main outer model equation (6.159) and is given by (6.160). We observe that to predict the field H_{QG} in the QG approximation, it is sufficient to know only the initial value of H_{QG}. However, unfortunately, this single initial value is related, according to (6.172), to the initial data (6.168), which are obtained from (a priori prescribed) data (6.153b) for the hydrostatic dissipative equations (6.152a,b), under the constraints mentioned before (6.168). Finally, we observe also that by integrating the slip boundary condition (6.169) with respect to τ and taking the initial conditions (6.168) into account, as well as the matching $(H_1^\tau)_{\tau\uparrow\infty} = (H_{\mathrm{QG}})_{t=0}$, we obtain the following condition:

$$\left\{\left[H_{\mathrm{QG}} + p\frac{T_0(p)}{K_0(p)}\frac{\partial H_{\mathrm{QG}}}{\partial p}\right]_{p=1}\right\}_{t=0} = \left[H_1^0 + p\frac{T_0(p)}{K_0(p)}\frac{\partial H_1^0}{\partial p}\right]_{p=1}\,. \tag{6.173}$$

The role of this 'curious' condition (6.173) is to serve as an initial condition for the boundary condition on $p = 1$ – namely (6.187), derived below – that must be supplied for the main QG outer equation (6.159), written for H_{QG}. More precisely, (6.187) is derived via the formulation of the Ackerblom problem in a steady Ekman layer (region III) near the ground, and contains a time derivative. As a consequence, (6.187) may be considered as a boundary condition for our main QG outer equation (6.159) only if it is complemented with an initial condition, which is just the above condition (6.173).

The Ackerblom Problem in a Steady Ekman Layer and the Boundary Condition at $p = 1$ for the QG Outer Model Equation (6.159)

For the main outer QG single equation (6.159) for H_{QG} derived above, with the initial condition (6.172) at time $t = 0$, and the behaviour condition (sufficiently rapid decay)(6.161) at $p = 0$ and at infinity in the horizontal plane, it is necessary also to specify a boundary condition on the (flat) ground, simulated (for low Mach/Kibel number) by $p = 1$. For this purpose, within the region III (the steady Ekman layer), we introduce a new vertical coordinate $\pi = (p-1)/M$, and in this region we consider the following inner limiting process:

$$Ki \to 0 \quad \text{with } t, x, y \text{ and } \pi \text{ fixed} . \tag{6.174a}$$

Together with (6.174a), we consider the inner expansion

$$\begin{aligned}(\boldsymbol{v}, \omega, H, T, \rho) &= (\boldsymbol{v}_0^{\mathrm{Ek}}, 0, 0, T_0(1), \rho_0(1)) + M(\boldsymbol{v}_1^{\mathrm{Ek}}, \omega_1^{\mathrm{Ek}}, H_1^{\mathrm{Ek}}, T_1^{\mathrm{Ek}}, \rho_1^{\mathrm{Ek}}) \\ &\quad + M^2(\boldsymbol{v}_2^{\mathrm{Ek}}, \omega_2^{\mathrm{Ek}}, H_2^{\mathrm{Ek}}, T_2^{\mathrm{Ek}}, \rho_2^{\mathrm{Ek}}) + \dots ,\end{aligned} \tag{6.174b}$$

and we obtain from the starting equations (6.152) the following set of equations for $\boldsymbol{v}_0^{\mathrm{Ek}}, \omega_1^{\mathrm{Ek}}, H_1^{\mathrm{Ek}}$ and T_1^{Ek}:

$$\begin{aligned}(\boldsymbol{k} \wedge \boldsymbol{v}_0^{\mathrm{Ek}}) + \frac{1}{\gamma}\boldsymbol{D}H_1^{\mathrm{Ek}} - \kappa_0 \frac{\partial}{\partial \pi}\left(\frac{\partial \boldsymbol{v}_0^{\mathrm{Ek}}}{\partial \pi}\right) &= 0 , \\ \frac{\partial \omega_1^{Ek}}{\partial \pi} + \boldsymbol{D} \,.\, \boldsymbol{v}_0^{Ek} &= 0 , \\ \frac{1}{Pr}\kappa_0 \frac{\partial}{\partial \pi}\left(\frac{\partial T_1^{\mathrm{Ek}}}{\partial \pi}\right) &= 0 ,\end{aligned} \tag{6.175}$$

with $T_0(1) = -\partial H_1^{\mathrm{Ek}}/\partial \pi$, and we assume that $\rho_0(1) = 1, k(1) = 1$ and $\mu(1) = 1$. From the boundary conditions (6.153a) on the flat ground surface, we obtain (at $\pi = \Pi_{\mathrm{g}}^{\mathrm{Ek}}$)

$$H_1^{\mathrm{Ek}} = 0, \quad \boldsymbol{v}_0^{\mathrm{Ek}} = 0, \quad \omega_1^{\mathrm{Ek}} = \frac{\partial H_1^{\mathrm{Ek}}}{\partial t}, \quad -\frac{\partial T_1^{\mathrm{Ek}}}{\partial \pi} = \sigma \mathsf{R}_{\mathrm{s}}(1) , \tag{6.176}$$

where we assume that, in region III, the ground surface for low Mach/Kibel number is located at

$$\Pi \mathrm{g}^{\mathrm{Ek}} = \Pi_{\mathrm{g}0}^{\mathrm{Ek}} + M\Pi_{\mathrm{g}1}^{\mathrm{Ek}} + \dots , \tag{6.177}$$

and that $T_0(1) = 1$; the main radiative transfer (as a function of p only) does not have an Ekman structure. In the whole region III, using the third equation of (6.175) for T_1^{Ek} with the last condition in(6.176), and matching with region I, we derive the following expressions for T_1^{Ek} and H_1^{Ek}:

$$T_1^{\mathrm{Ek}} = \left(\frac{\mathrm{d}T_0}{\mathrm{d}p}\right)_{p=1} \pi - \left(\frac{\partial H_{\mathrm{QG}}}{\partial p}\right)_{p=1} \quad \text{and} \quad H_1^{\mathrm{Ek}} = (H_{\mathrm{QG}})_{p=1} - \pi . \tag{6.178}$$

As a consequence, we derive, for $\Pi_{\mathrm{g},0}$ in (6.177), the expression

$$\Pi_{\mathrm{g}0}^{\mathrm{Ek}} = (H_{\mathrm{QG}})_{p=1} . \tag{6.179}$$

Now, in the first equation of (6.175), we set $\boldsymbol{v}_0^{\mathrm{Ek}} = (\boldsymbol{v}_{\mathrm{QG}})_{p=1} + \boldsymbol{V}_0^{\mathrm{Ek}}$, and from matching with the main quasi-geostrophic region, when π tends to infinity, we obtain by matching

$$\boldsymbol{V}_0^{\mathrm{Ek}} = 0 \quad \text{when } \pi \uparrow -\infty, \quad \text{and} \quad \left[\boldsymbol{k} \wedge \boldsymbol{v}_{\mathrm{QG}} + \frac{1}{\gamma}\boldsymbol{D}H_{\mathrm{QG}}\right]_{p=1} = 0\,. \qquad (6.180)$$

As a consequence, we recover in a consistent way the following classical Ackerblom problem for $\boldsymbol{V}_0^{\mathrm{Ek}}$:

$$\kappa_0 \frac{\partial^2 \boldsymbol{V}_0^{\mathrm{Ek}}}{\partial \pi^2} - \boldsymbol{k} \wedge \boldsymbol{V}_0^{\mathrm{Ek}} = 0\,,$$
$$\boldsymbol{V}_0^{\mathrm{Ek}} = -(\boldsymbol{v}_{\mathrm{QG}})_{p=1} \quad \text{on } \pi = (H_{\mathrm{QG}})_{p=1}\,,$$
$$\boldsymbol{V}_0^{\mathrm{Ek}} \to 0 \quad \text{when } \pi \to -\infty. \qquad (6.181)$$

We observe that, according to (6.178) for H_1^{Ek}, it is obvious that the external boundary of the steady Ekman layer must be at $\pi = -\infty$. The solution of the Ackerblom problem (6.181) is obtained in a standard way, namely

$$\boldsymbol{V}_0^{\mathrm{Ek}} - \mathrm{i}\boldsymbol{k} \wedge \boldsymbol{V}_0^{\mathrm{Ek}} = -[(\boldsymbol{v}_{\mathrm{QG}})_{p=1} + \mathrm{i}\boldsymbol{k} \wedge (\boldsymbol{v}_{\mathrm{QG}})_{p=1}](1-E)\,, \qquad (6.182)$$

where

$$E = \exp\left\{\frac{(1+\mathrm{i})}{(2\kappa_0)^{1/2}}[\pi - (H_{\mathrm{QG}})_{p=1}]\right\}. \qquad (6.183)$$

From the second equation of (6.175), with the condition

$$\omega_1^{\mathrm{Ek}} = \left(\frac{\partial H_{\mathrm{QG}}}{\partial t}\right)_{p=1} \qquad \text{on } \pi = (H_{\mathrm{QG}})_{p=1}\,, \qquad (6.184)$$

according to the second relation of (6.178), thanks to (6.182), and integrating with respect to π and then matching the terms $M\omega_1^{\mathrm{Ek}}$ with $M\omega_{\mathrm{QG}}$, we obtain the main relation (at $p = 1$)

$$(\omega_{\mathrm{QG}})_{p=1} = \left(\frac{\partial H_{\mathrm{QG}}}{\partial t}\right)_{p=1} - \left(\frac{\kappa_0}{2\gamma}\right)^{1/2} \boldsymbol{D}^2(H_{\mathrm{QG}})_{p=1}\,, \qquad (6.185)$$

when we observe that a straightforward computation gives

$$\boldsymbol{D}\,.\int_{\Pi_{\mathrm{g}0}^{\mathrm{Ek}}}^{+\infty} \boldsymbol{V}_0^{\mathrm{Ek}}\,dp = -\left(\frac{\kappa_0}{2\gamma}\right)^{1/2} \boldsymbol{D}^2(H_{\mathrm{QG}})_{p=1}\,. \qquad (6.186)$$

Finally, we can express $(\omega_{\mathrm{QG}})_{p=1}$ through $(H_{\mathrm{QG}})_{p=1}$, according to (6.156) and the equation (6.158c). Then, we derive from (6.185) the boundary condition which must be applied to the main QG outer equation (6.159) at the ground surface $p = 1$. Namely, we obtain the following boundary condition:

$$\frac{\partial}{\partial t}\left[H_{\mathrm{QG}} + \frac{1}{K_0(1)}\frac{\partial H_{\mathrm{QG}}}{\partial p}\right] + \frac{1}{K_0(1)}\left[\boldsymbol{v}_{\mathrm{QG}}\,.\,\boldsymbol{D}\left(\frac{\partial H_{\mathrm{QG}}}{\partial p}\right)\right]$$
$$-\left(\frac{\kappa_0}{2\gamma}\right)^{1/2} \boldsymbol{D}^2 H_{\mathrm{QG}} = 0 \quad \text{on } p = 1\,. \qquad (6.187)$$

This last boundary condition (6.187), on the flat ground surface $p = 1$, takes into account the influence of the viscous Ekman steady boundary layer on the main quasi-geostrophic flow governed by the QG model equation (6.159).

Conclusion

The main result of the above analysis is that we may formulate a definite problem for the function H_{QG} in region I, consisting of : the QG equation (6.159), with (6.160), the initial conditions at $t = 0$, (6.172) and (6.173), the condition (6.187), on the ground at $p = 1$, and a suitable condition (for instance (6.161)) at infinity, with respect to the horizontal coordinates, and when $p \uparrow 0$. From H_{QG}, we may obtain $\boldsymbol{v}_{QG}$ using geostrophy (6.157), T_{QG} using the relation (6.158c), and also ω_{QG} using (6.156). Finally, we observe that $T_0(p)$ and $H_0(p)$ are assumed to be known beforehand, and we have said that $T_0(p)$ is related to the thermal balance (see (6.155c,d)) and that $H_0(p)$ can be deduced from the condition of hydrostatic equilibrium, namely

$$H_0(p) = -\int_1^p \frac{T_0(q)}{q} \mathrm{d}q ,$$

if use is made of the condition $H_0(1) = 0$ on the ground. However, it is necessary to observe that the similarity rule (6.146b) between Ki and M is not completely satisfactory, because it leads to a value of the characteristic velocity U_c that is too large. Obviously, it is possible to change our above asymptotic scheme by using more suitable similarity rules between M, Ki and $Ek_\perp$, by choosing, for instance, a suitable function $F(M)$ in (2.15). Here, we do not consider this problem.

6.6.3 The Second-Order Ageostrophic Model of Guiraud and Zeytounian

For the derivation of the AG model, we need to consider three local (inner) expansions in addition to the main one. Two of them are higher approximations relative to the ones considered previously in the framework of the QG model derived above. From the outer approximation, we derive the main AG, second-order model equation for H_{AG}. Then, from the first local (close to the initial time) approximation, via the problem of unsteady adjustment to the AG model equation, we obtain the initial condition at $t = 0$ for the AG equation by matching (between regions I and II). Afterwards, from the second local (near the ground) approximation, via a second-order Ackerblom problem in the steady Ekman layer, we have the possibility to derive a boundary condition on the ground $p = 1$, for this AG main equation by matching (between regions I and III). However, in the framework of the AG second-order model, the consideration of a new, third local (simultaneously close to the initial time and near the ground) approximation is necessary. Indeed, in the corner region IV, corresponding to an unsteady Ekman layer, we must consider a problem of unsteady adjustment to the classical Ackerblom model problem (6.181) – this will provide a boundary condition on the ground for region II, in the second approximation, which itself provides the initial condition for the AG model equation. We observe that the analysis of the corner region IV raises many difficult problems, and some questions related to its analysis remain open.

AG (Outer) Model Equation

First, it is necessary to derive the AG outer (region I) main model equation, which governs the evolution of H_{AG} in the low-Kibel/Mach-number expansion (6.154a). By expanding the three equations (6.152) to order M^2, we obtain a system of equations that we need not write out. In fact, we may compute $\boldsymbol{v}_{\mathrm{AG}}$ from this system of equations, and then we obtain $T_{\mathrm{AG}} = -p\,\partial H_{\mathrm{AG}}/\partial p$ and also find ω_{AG}. As a result, we obtain from the continuity equation

$$\frac{\partial \omega_{\mathrm{AG}}}{\partial p} + \boldsymbol{D} \cdot \boldsymbol{v}_{\mathrm{AG}} = 0, \tag{6.188}$$

written at the order M^2, the following AG outer model equation for H_{AG}:

$$\left[\frac{\partial}{\partial t} + \boldsymbol{v}_{\mathrm{QG}} \cdot \boldsymbol{D}\right] (\mathsf{L}H_{\mathrm{AG}}) + \frac{1}{\gamma}[\boldsymbol{k} \wedge \boldsymbol{D}H_{\mathrm{AG}}] \cdot \boldsymbol{D}\mathsf{L}H_{\mathrm{QG}} + \frac{1}{\gamma}\beta\frac{\partial H_{\mathrm{AG}}}{\partial x}$$
$$= \boldsymbol{D} \cdot \boldsymbol{U}(H_{\mathrm{QG}}) + \frac{\partial \Omega(H_{\mathrm{QG}})}{\partial p}, \tag{6.189}$$

where $\boldsymbol{U}$ and Ω are complicated, lengthy known expressions, that is, konwn when the QG model has been solved. The above equation (6.189) for H_{AG} is the second-order equation corresponding to (6.159), considered as the first-order equation for H_{QG}, both being components of the asymptotic expansion (6.154a) for the function H. We observe that the horizontal wind is given by (with an error of order M^2)

$$\boldsymbol{v} = \boldsymbol{v}_{\mathrm{QG}} + M\left\{\frac{1}{\gamma}(\boldsymbol{k} \wedge \ \boldsymbol{D}H_{\mathrm{AG}})\right.$$
$$\left. + \boldsymbol{k} \wedge \ \left[\frac{\partial \boldsymbol{v}_{\mathrm{QG}}}{\partial t} + (\boldsymbol{v}_{\mathrm{QG}} \cdot \boldsymbol{D})\boldsymbol{v}_{\mathrm{QG}} + \beta y(\boldsymbol{k} \wedge \boldsymbol{v}_{\mathrm{QG}})\right]\right\} + O(M^2)\,, \tag{6.190}$$

where $\boldsymbol{v}_{\mathrm{QG}} = (1/\gamma)(\boldsymbol{k} \wedge \boldsymbol{D}H_{\mathrm{QG}})$.

Initial Conditions for the AG Equation (6.189)

To use the above AG model equation (6.189) for weather prediction, we must again derive an initial condition for $t = 0$. For this purpose it is necessary, again, to consider the initial region II and its matching with region I. First, taking into account the remark made before (6.166) and (6.167), we write the initial conditions as

$$\tau = 0: \quad \boldsymbol{v}_1^{\tau} = \boldsymbol{v}_1^0\,, \mathsf{H}_2^{\tau} = H_2^0\,. \tag{6.191}$$

However, in order to formulate completely the problem of the adjustment to the AG main model equation (6.189), it is necessary to examine, in the corner region IV, a problem for an unsteady Ekman layer and the adjustment of its solution to the solution of the Ackerblom problem (6.175) when $\tau \to \infty$. The first interesting point of the analysis of region IV is that the matching between regions IV and II gives again the slip condition (6.169), which emerges naturally for the problem (6.164) of the adjustment to the geostrophic balance (6.157) and shows that the process of adjustment to the geostrophic balance (6.157) is really a purely non-viscous, adiabatic process. On the other hand, a detailed analysis of the initial region II, in the framework of the second-order problem of adjustment to the AG model equation (6.189), shows convincingly that the unsteady adjustment in the low-Mach/Kibel-number asymptotics is of decaying type and, as a consequence, that matching between regions II and I is always realizable. However, for the problem of unsteady adjustment to the AG model equation (6.189), we must add, again, a boundary condition at $p = 1$, which must be derived from matching with the expansion in the corner region IV. Now, by analogy with the analysis performed for the adjustment to the geostrophic balance, we derive a system of four equations for $(\boldsymbol{v}_1^\tau, \omega_1^\tau, H_2^\tau, T_2^\tau)$ in place of (6.164), where, in the right-hand sides of the equations for $\boldsymbol{v}_1^\tau$ and T_2^τ we have various terms containing $\boldsymbol{v}_0^\tau$, ω_0^τ and $\boldsymbol{v}_0^\tau$, T_1^τ, ω_0^τ respectively in place of zero. As a consequence, in place of (6.171), we derive

$$\frac{\partial}{\partial \tau}\left\{\boldsymbol{k} \cdot (\boldsymbol{D} \wedge \boldsymbol{v}_1^\tau) + \frac{\partial}{\partial p}\left[\frac{p^2}{K_0(p)}\frac{\partial H_2^\tau}{\partial p}\right]\right\} = \Phi^\tau , \tag{6.192}$$

and Φ^τ is a known function of $\boldsymbol{v}_0^\tau$ and H_1^τ when we take into account the last two equations of (6.164). Equation (6.192) may again be integrated from $\tau = 0$ to $\tau = \infty$, taking the above initial conditions (6.191) into account and also the matching between the expansions (6.154) in region I (at $t = 0$) with (6.162) in region II (when $\tau \to \infty$). As result we derive the following initial condition for (6.189):

$$\begin{aligned} \mathsf{L} H_{\mathrm{AG}} &= \frac{1}{\gamma}\boldsymbol{k} \cdot \left\{\boldsymbol{D} \wedge \left[\frac{\partial}{\partial t} + \boldsymbol{v}_{\mathrm{QG}} \cdot \boldsymbol{D}\right](\boldsymbol{D} H_{\mathrm{QG}})\right\} \\ &+ \int_0^\infty [\Phi^\tau - \Phi^{\tau\infty}]\, \mathrm{d}\tau. \quad \text{at} \quad t = 0 . \end{aligned} \tag{6.193}$$

We observe that $\Phi^{\tau\infty}$ is obtained by replacing $\boldsymbol{v}_0^\tau$ and H_1^τ in the expression for Φ^τ by $\boldsymbol{v}_{\mathrm{QG}}$ and H_{QG} and then taking the values of the resulting quantity at $t = 0$ (in accordance with the matching between regions I and II). It is also important to note here that, from the rate of approach of $\boldsymbol{v}_0^\tau$ and H_1^τ to their limiting values, which is governed by a sum of terms, $\tau^{-1/2}\cos\tau$ (see, for instance, Kibel 1957, Sect. 4.2), we can see that the integration of (6.192) from $\tau = 0$ to limiting values for $\tau = \infty$ is justified.

When one examines the AG model equation (6.189), it seems that the above initial condition (6.193) will provide all that is necessary at $t = 0$. However, as in the case of the QG model (see (6.159)), this is not quite true, because the boundary condition for the AG model equation (6.189) at $p = 1$ (which is mentioned below – see the condition (6.196) in page 222) contains a time derivative and must itself be supplied with an initial condition, which must be derived in the framework of an analysis of the (rather strange) corner region IV. The present author has shown, with J.-P. Guiraud, that the initial condition (6.195) (written below) at $t=0$ on $p=1$, for the boundary condition (6.196)(on $p=1$), can be derived by integrating the following condition (which is an analogue of (6.169), but for H_2^τ and with a right-hand side Ψ^τ in place of zero):

$$\frac{\partial}{\partial \tau}\left[H_2^\tau + p\frac{T_0(p)}{K_0(p)}\frac{\partial H_2^\tau}{\partial p}\right] = \Psi^\tau \quad \text{on } p=1\,. \tag{6.194}$$

This gives just

$$\left\{\left[H_{\mathrm{AG}} + p\frac{T_0(p)}{K_0(p)}\frac{\partial H_{\mathrm{AG}}}{\partial p}\right]_{t=0}\right\}_{p=1} = \int_0^\infty \left[\Psi^\tau - \Psi^{\tau\infty}\right] d\tau. \tag{6.195}$$

In (6.195), $\Psi^{\tau\infty}$ is the value at $\tau=\infty$ of Ψ^τ, which is itself obtained as result of an analysis of the corner region IV. The expression for Ψ^τ in (6.195) is very complicated and has been derived by Guiraud (1983).

The Second-Order Ackerblom Problem

In the Ekman-layer region III, the derivation of the boundary condition at $p=1$ for the above AG main, outer equation (6.189), through the consideration of a second-order Ackerblom problem, is technically rather complicated, but the various logical steps of this derivation are in principle clear.

The left-hand side of this boundary condition (6.196) for H_{AG} on $p=1$ is, in fact, the linearized form of the condition (6.187), but its right-hand side, $M(H_{\mathrm{QG}})$, is again a very complicated, lengthy expression, known when the QG model has been solved. The condition is

$$\frac{\partial H_{\mathrm{AG}}}{\partial t} + \frac{p^2}{K_0(p)}\left[\frac{\partial}{\partial t} + \boldsymbol{v}_{\mathrm{QG}}\,.\,\boldsymbol{D}\right]\frac{\partial H_{\mathrm{AG}}}{\partial p} - \left(\frac{\kappa_0}{2\gamma}\right)^{1/2}\boldsymbol{D}^2 H_{\mathrm{AG}}$$
$$-\frac{1}{\gamma}\frac{p^2}{K_0(p)}\left[\boldsymbol{k}\wedge\boldsymbol{D}\left(\frac{\partial H_{\mathrm{QG}}}{\partial p}\right)\right]\,.\,\boldsymbol{D}H_{\mathrm{AG}} = M(H_{\mathrm{QG}}) \quad \text{on } p=1\,. \tag{6.196}$$

As was observed above, the condition (6.196) contains a time derivative, and an initial condition must be applied to this boundary condition (6.196), namely (6.195).

But here the situation is more complicated than for the boundary condition (6.187), for which the initial condition is (6.173), in view of the fact that the initial condition in the present case cannot be obtained rationally without an analysis of the corner region IV. Fortunately, the full solution of the unsteady Ekman-layer problem in this corner region IV is not necessary if one is interested only in the derivation of (6.196). More precisely, the basic relation which through, a very lengthy set of calculations, gives us the possibility to derive the higher-order part associated with the linearized form of (6.187), is

$$(\omega_2)_{\mathrm{g}} = \left(\omega_2^{\mathrm{Ek}}\right)_{\mathrm{g}} - \Pi_{\mathrm{g0}}^{\mathrm{Ek}}\left[\boldsymbol{D}\,.\,(\boldsymbol{v}_1)_{\mathrm{g}}\right] + \left(\boldsymbol{V}_1^{\mathrm{Ek}}\right)_{\mathrm{g}}\,.\,\boldsymbol{D}\Pi_{\mathrm{g0}}^{\mathrm{Ek}} + \boldsymbol{D}\,.\int_{\Pi_{\mathrm{g0}}^{\mathrm{Ek}}}^{+\infty} \boldsymbol{V}_1^{\mathrm{Ek}}\,\mathrm{d}\pi\,. \tag{6.197}$$

In (6.197), $\Pi_{\mathrm{g0}}^{\mathrm{Ek}}$ is the first term in (6.177), and $\boldsymbol{V}_1^{\mathrm{Ek}}$ (which vanishes at infinity) is the solution of the non-homogeneons equation (associated with the equation for $\boldsymbol{V}_0^{\mathrm{Ek}}$ in the Ackerblom problem (6.81))

$$\kappa_0 \frac{\partial^2 \boldsymbol{V}_1^{\mathrm{Ek}}}{\partial \pi^2} - \boldsymbol{k} \wedge \boldsymbol{V}_1^{\mathrm{Ek}} = \boldsymbol{V}, \tag{6.198}$$

where $\boldsymbol{V}$ is known from the solution (6.182), with (6.183), of the classical Ackerblom problem (6.81) for $\boldsymbol{V}_0^{\mathrm{Ek}}$. Finally, for $(\boldsymbol{V}_1^{\mathrm{Ek}})_{\mathrm{g}}$ in (6.197), we have the relation

$$(\boldsymbol{V}_1^{\mathrm{Ek}})_{\mathrm{g}} = \Pi_{\mathrm{g0}}^{\mathrm{Ek}}\left(\frac{\partial \boldsymbol{v}_0}{\partial p}\right)_{\mathrm{g}} - (\boldsymbol{v}_1)_{\mathrm{g}} - \Pi_{\mathrm{g1}}^{\mathrm{Ek}}\left(\frac{\partial \boldsymbol{v}_0^{\mathrm{Ek}}}{\partial \pi}\right)_{\mathrm{g}}. \tag{6.199}$$

Unfortunately, obtaining an explicit expression for the integral in the last term of (6.197), as a functional of H_{QG}, is associated with a lot of very tedious algebra.

The Problem of the Unsteady Ekman Boundary Layer and the Adjustment to the Ackerblom Model

Here, we stress, first, that the subscript 'g' stands for any quantity evalued on the flat ground surface, but the reader should be cautious when using this notation. We stress that the notation $(f^{\tau})_{\mathrm{g}}$ used in region II must be distinguished from the $(f^{\tau,\,\mathrm{Ek}})_{\mathrm{g}}$ used in region IV. As a matter of fact, when dealing with the unsteady-adjustment region II, we do not see the ground, but rather the location of the matching, which is at $p = 1$, so that we put $(f^{\tau})_{\mathrm{g}} = f^{\tau}|_{p=1}$. On the other hand, when dealing with the unsteady Ekman boundary-layer region IV, we see properly the ground, which is located at

$$\pi = \Pi_{\mathrm{g}}^{\tau,\,\mathrm{Ek}}\ \left(= \Pi_{\mathrm{g0}}^{\tau,\mathrm{Ek}} + M\Pi_{\mathrm{g1}}^{\tau\mathrm{Ek}} + \ldots \right), \tag{6.200}$$

and we write $(f^{\tau,\,\mathrm{Ek}})_{\mathrm{g}}$ for $f^{\tau,\,\mathrm{Ek}}$ on $\pi = \Pi_{\mathrm{g0}}^{\tau,\,\mathrm{Ek}}$. In region IV, where $\tau = t/Ki$ and $\pi = (p-1)/Ki$, we assume that

$$\begin{aligned}(\boldsymbol{v}, \omega, H, T, \rho) = {} & (\boldsymbol{v}_0^{\tau,\,\mathrm{Ek}}, \omega_0^{\tau,\,\mathrm{Ek}}, 0, 1, 1) \\ & + M(\boldsymbol{v}_1^{\tau,\,\mathrm{Ek}}, \omega_1^{\tau,\,\mathrm{Ek}}, H_1^{\tau,\,\mathrm{Ek}}, T_1^{\tau,\,\mathrm{Ek}},\ \rho_1^{\tau,\,\mathrm{Ek}}) + \ldots . \end{aligned} \tag{6.201}$$

The first interesting point of the analysis of this region IV is the derivation of the initial–boundary problem for the corresponding horizontal velocity,

$$\boldsymbol{V}_0^{\tau,\,\mathrm{Ek}} = \boldsymbol{v}_0^{\tau,\,\mathrm{Ek}} - (\boldsymbol{v}_0^{\tau})_{p=1},$$

where, according to the condition for matching between regions IV and II, when $\pi \to \infty$,

$$\boldsymbol{v}_0^{\tau,\,\mathrm{Ek}} \to (\boldsymbol{v}_0^{\tau})_{p=1}.$$

We extract from the starting equations (6.152), rewritten for the region IV with τ and π, through a Laplace transform, a classical problem which is considered in Greenspan (1968, Sect. 2.3). However, the solution of this problem is not simple, and its inverse Laplace transform is even more complicated. Here we discuss only the result, since we are not interested in the process of adjustment itself. The only point that really matters is: whether or not $\boldsymbol{v}_0^{\tau,\,\mathrm{Ek}}$ tends to a definite limit when $\tau \to \infty$. Matching between regions IV and III requires that

$$\text{when } \tau \to \infty, \quad \boldsymbol{v}_0^{\tau,\,\mathrm{Ek}} = \left(\boldsymbol{v}_0^{\mathrm{Ek}}\right)_{t=0} = \left[(\boldsymbol{v}_{\mathrm{QG}})_{p=1} + \boldsymbol{V}_0^{\mathrm{Ek}}\right]_{t=0}. \tag{6.202a}$$

Matching between regions I and II (with respect to the time τ), for $p = 1$ shows that (6.202a) may be rewritten as the following condition (which is, in fact, a matching condition between regions IV and III with respect to the velocities $\boldsymbol{V}_0^{\tau,\mathrm{Ek}}$ and $\boldsymbol{V}_0^{\mathrm{Ek}}$):

$$\text{when } \tau \to \infty, \quad \boldsymbol{V}_0^{\tau,\,\mathrm{Ek}} = \left[\boldsymbol{V}_0^{\mathrm{Ek}}\right]_{t=0}. \tag{6.202b}$$

The interesting point that is worth discussing is the rate at which the limit (6.202b) is reached. According to the adjustment to the geostrophic balance, the rate at which $\boldsymbol{v}_0^{\tau}$, tends to $(\boldsymbol{v}_{\mathrm{QG}})_{t=0}$, when $\tau \to \infty$ (at $p = 1$) is $O(\tau^{-1/2})$, a result obtained by Kibel in the case when $K_0(p) = \text{const}$. On the other hand, the solution of the pure initial-value problem associated with the initial–boundary-value problem for the velocity $\boldsymbol{V}_0^{\tau,\,\mathrm{Ek}}$ decays to zero (according to Greenspan 1968) as $O(\tau^{-1/2})$. Now, if we take into account the fact that when $\tau \to \infty, \omega_0^{\tau,\,\mathrm{Ek}} \to 0$, we can write for the function $\boldsymbol{V}_{0\infty}^{\tau,\,\mathrm{Ek}}$ (equal to the limiting value of $\boldsymbol{V}_0^{\tau,\,\mathrm{Ek}}$ at $\tau \to \infty$) a boundary-value problem very similar to (6.181), but with a boundary condition at $\tau \to \infty$, for $\Pi_{\mathrm{g}0}^{\tau,\,\mathrm{Ek}} = (H_{\mathrm{QG}}|_{p=1})_{t=0}$ with respect to π. Then, from a comparison, we can see that the matching (6.202b) holds, because, from the relation (6.178) for H_1^{Ek}, we have just $\Pi_{\mathrm{g}0}^{\mathrm{Ek}}|_{t=0} = (H_{\mathrm{QG}}|_{p=1})_{t=0}$. Finally, we observe that the adjustment process decays through some fractional negative power of $\tau = t/M$, and this inevitably raises the question of fractional powers of M in the asymptotic expansion for region II. Whether or not such fractional powers occur before the level of the ageostrophic field, in region I, remains an open question.

6.6.4 Conclusion

As is often the case with higher-order models derived from limiting processes and asymptotic expansions, one runs into inconsistencies if one does not rely on a systematic, logical step-by-step expansion.

The need for approximations that are more suitable than the QG one requires one to be very cautious in their derivation, and one of the most reliable ways to obtain the desired model is to look carefully at solutions of (6.152) with the conditions (6.153), which may be obtained from a systematically constructed expansion emerging from the low-Mach/Kibel-number asymptotics of the hydrostatic, dissipative model. This was precisely the purpose of Sect. 6.6, of the present Chap. 6. Now, there is a second reason for carrying out a careful study of the behaviour of the solutions of (6.152) with (6.153) as M tends to zero; this arises from the fact that the equations (6.152) are rather intricate and that one may learn a good deal about them by examining the behaviour of their solutions when M tends to zero. We observe that we have worked throughout with the model (6.152) with (6.153), in which hydrostatic balance is used. This means that we have considered no correction with respect to the small parameter ε. We conjecture that under the reasonable hypothesis $\varepsilon = O(M^2)(\alpha = 2$ in (2.16)), the first correction with respect to the quasi-static parameter ε is of a higher order than the order considered here, except for one term that will appear in $\Omega(H_{\mathrm{QG}})$ in the AG main model equation (6.189), which comes from a slight change in the equation of hydrostatic balance (6.157).

At this point, we must add a comment concerning the 'balance equation' derived by, for instance, Monin (1958, 1961). This balance equation cannot be derived in the framework of the low-Mach/Kibel-number asymptotics of the hydrostatic equations (even as a second correction with respect to Ki or M, with (6.146)). Within our rational asymptotic theory, referring to (6.190), with $\beta = 0$, we can easily derive the relation

$$\boldsymbol{k}\,.\,(\boldsymbol{D}\wedge\boldsymbol{v}_{\mathrm{AG}}) = \frac{1}{\gamma}\boldsymbol{D}^2 H_{\mathrm{AG}} + 2\left(\frac{1}{\gamma}\right)^2\left[\left(\frac{\partial^2 H_{\mathrm{QG}}}{\partial x\,\partial y}\right)^2 - \frac{\partial^2 H_{\mathrm{QG}}}{\partial x^2}\frac{\partial^2 H_{\mathrm{QG}}}{\partial y^2}\right], \tag{6.203}$$

which may be interpreted in different ways. In Sect. 6.6, it is considered as an explicit expression for the vertical component of the AG vorticity, once H_{QG} and H_{AG} have been computed from the AG (6.189) and QG (6.159) main model equations. One the other, in an 'ad hoc' non-rational theory, if one confuses H_{QG} and H_{AG} with H^* , it may be viewed as a Monge–Ampere equation for computing H^*, when the left-hand side of (6.203) is known, and in this case it is just called the 'balance equation' (see, for instance, (6.80a)). This last, improper interpretation is obviously a consequence of the absence of a rigorous, logical, step-by-step, approach to the derivation of sequential approximations. A final comment concerning the book by Monin (1988), where the reader can find a modern approach to geophysical hydrodynamics: curiously, in this very complete and pertinent book, the reader cannot find any reference to an asymptotic approach! Concerning (modern) analytical methods and numerical models for large-scale atmospheric–ocean dynamics, the reader can find various pieces of useful information in two recent volumes edited by Norbury and Roulstone (2002), especially in the survey papers by

Cullen and by White. In two recent papers (Zeytounian 2003, 2004), both devoted to an asymptotic approach to atmospheric motions (the first being related to the foundation of the Boussinesq approximation and the second being in honour of the 100th birthday of I. A. Kibel), the reader can also find various information and references on the validity of the Boussinesq equations and on the derivation of the system of hydrostatic, dissipative equations.

References

H. Ashley and M. T. Landahl, *Aerodynamics of Wings and Bodies.* Addison-Wesley, 1965.

G. K. Batchelor. Q. J. Roy. Meteorol. Soc., **79** (1953), 224–235.

H. Bénard. Rev. Gén. Sci. Pures Appl., **11** (1900), 1261–1271.

P. A. Bois. J. de Mécanique, **15** (1976), 781–811.

P. A. Bois. J. de Mécanique, **18** (1979), 395–417.

P. A. Bois. Geophys. Astrophys. Fluid Dyn., **29** (1984), 267–303.

A. J. Bourgeois and J. T. Beale. SIAM J. Math. Anal., **25**(4) (1994), 1023–1068.

J. Boussinesq, *Théorie Analytique de la Chaleur,* Vol. 2. Gauthier-Villars, Paris, 1903.

S. Chandrasekhar, *Hydrodynamic and Hydromagnetic Stability.* Clarendon press, Oxford, 1961.

J. G. Charney. Tellus, **7** (1955), 22.

J. G. Charney. In: *Procedings of the International Symposium on Numerical Weather Prediction,* Tokyo, Novemver 1960, edited by. Meteorological Society of Japan, Tokyo, 1962 p. 131.

P. G. Drazin. Tellus, **13** (1961), 239–251.

P. G. Drazin and W. H. Reid, *Hydrodynamic Stability.* Cambridge University Press, Cambridge, 1981.

J. A. Dutton and G. H. Fichtl. J. Atmosph. Sci., **26** (1969), 241.

C. Eckart and H. G. Ferris. Rev. Mod. Phys. **28**(1) (1956), 48–52.

M. El Mabrouk and R. Kh. Zeytounian, Sur les conditions au bord pour equations primitives. Rev. Roumaine Math. Pures Appl. **29**(3) (1984), 235–238.

B. L. Gavrilin. Izv. Akad. Nauk SSSR, Ser. Fiz. Atmos. Okean, **1** (1965), 557.

P. Germain, *Mécanique*, vol. 2. Ellipses, Palaiseau, 1986.

H. P. Greenspan, *The Theory of Rotating Fluids.* Cambridge University Press, Cambridge, 1968.

J. P. Guiraud, *Examples of Applications of Asymptotic Techniques to the Derivation of Models for Atmospheric Flows.* Advanced School coordinated by R. Kh. Zeytounian, CISM, Von Karman Session, 3–5 October 1983, Udine, Italy, unpublished notes, LMM, University of Paris VI.

J. P. Guiraud and R. Kh. Zeytounian, A note on the asymptotic interpretation of the classical theory of lee waves in the troposphere and the role of the upper boundary condition. Geophys. Astrophys. Fluid Dyn. **12**(1/2) (1979), 61–72.

J. P. Guiraud and R. Kh. Zeytounian, Asymptotic derivation of the ageostrophic model for atmospheric hydrostatic flows. Geophys. Astrophys. Fluid Dyn. **15**(3+4) (1980), 283–295.

J. P. Guiraud and R. Kh. Zeytounian, Note on the adjustment to hydrostatic balance. Tellus, **34**(1) (1982), 50–54.

J. C. R. Hunt and W. H. Snyder. J. Fluid Mech., **96**(4) (1980), 671.

D. D. Joseph, *Stability of Fluid Motions*, Vol. 2. Springer, Heidelberg, 1976.

I. A. Kibel. Izv. Akad. Nauk. SSSR, Ser. Geograph. Geophys., No. 5 (1940), 627–638.

I. A. Kibel, *Vevedeniyé v guidrodinamitcheskiye metodey kratkosrochnovo prognoza pogodey.* Gos Isd Tekh-Teor Literatouré, Moscow, 1957. (English edition): *An Introduction to the Hydrodynamical Methods of Short Period Weather Forecasting.* Macmillan, London, 1963.)

I. A. Kibel, Hydrodynamical (numerical) short-range weather prediction. In: *Mechanics in the USSR During the Last 50 Years (1917–1967)*, Vol. 2: *Fluids and Gases, Mechanics*, edited by Nauka, Moscow, 1970, pp. 661–683.

I. A. Kibel, *Selected Works on Dynamical Meteorology.* Hydrometeo-Izdat, Leningrad, 1984.

N. E. Kotchin, I. A. Kibel and N. V. Roze, *Teoretitcheskaya Guidromekhanika*, Part 1, 6th edition. FM, Moscow, 1963.

V. N. Kozhevnikov. Izv. Akad. Nauk SSSR, Ser. Geofiz. **7** (1963), 1108–1116.

V. N. Kozhevnikov, Izv. Akad. Nauk SSSR, Atmosph. Ocean. Phys. **4**(1) (1968), 16–27.

L. D. Landau and E. M. Lifshitz, *Guidrodinamika*, 4th edition. Teoretitcheskaya Fizika, Vol. 6, Nauka, Moscow, 1988.

J. L. Lions, R. Temam and S. Wang. Topol. Methods Nonlinear Anal., **4** (1994), 253–287.

R. Long. Tellus. **5** (1953), 42–57.

W. V. R. Malkus, Boussinesq equations. In: *Geophysical Fluid Dynamics*, Vol. 1, edited by Woods Hole Oceanographic Institution, Woods Hole, MA, 1964, pp. 1–12.

L. Mahrt. J. Atmosph. Sci. **43**(10) (1986), 1036–1044.

J. Mihaljan. Astrophys. J., **136** (1962), 1126.

J. W. Miles. J. Fluid Mech., **33**(4) (1968), 803–814.

A. S. Monin. Izv. Akad. Nauk SSSR, Ser. Geofiz., **4** (1958), 497.

A. S. Monin. Izv. Akad. Nauk SSSR, Ser. Geofiz., **7** (1961), 602.

A. S. Monin, *Weather Forecasting as a Problem in Physics.* MIT Press, Cambridge, MA, 1972.

A. S. Monin, *Fundamentals of Geophysical Fluid Dynamics.* Hydrometeo-Izdat, Leningrad, 1988.

J. Norbury and I. Roulstone (eds.), *Large-Scale Amospheric–Ocean Dynamics.* Vol. 1, *Analytical Methods and Numerical Models*; Vol. 2, *Geometric Methods and Models.* Cambridge University Press, Cambridge, 2002.

A. Oberbeck. Ann. Phys. Chem., Neue Folge, **7** (1879), 271–292.

Y. Ogura and N. A. Phillips, J. Atmosph. Sci., **19** (1962), 173–179.

Ye. M. Pekelis. Bull. (Izv.) Acad. Sci. USSR, Atmosph. Ocean. Phys. **12**(5) (1976), 470–477.

R. Perez Gordon and M. G. Velarde, J. de Physique, **36**(7–8) (1979), 591–601.

R. Perez Gordon and M. G. Velarde, J. de Physique, **37**(3) (1976), 177–182.

N. A. Phillips, Models for weather prediction. Annu. Rev. Fluid Mech., **2** (1970), 251–292.

O. M. Phillips, *The Dynamics of the Upper Ocean.* 2nd edition. Cambridge University Press, Cambridge, 1977.

Lord Rayleigh. Philos. Mag., **32** (1916), 529–546.

L. F. Richardson, *Weather Prediction by Numerical Processes.* Cambridge University Press, Cambridge, 1922 (reprinted by Dover, New York, 1966).

E. A. Spiegel and G. Veronis. Astrophys. J., **131**(1960), 442–447. (Correction, Astrophys. J., **135** (1960), 655–656).

M. Van Dyke, *Perturbation Methods in Fluid Mechanics.* Academic Press, New York, 1964.

M. G. Velarde and R. Kh. Zeytounian, *Interfacial Phenomena and the Marangoni Effect.* CISM Courses and Lectures, No. 428, Springer, Vienna, 2002.

C. Wilcox, *Scattering Theory for the d'Alembert Equation in Exterior Domain.* Lecture Notes in Mathematics, No. 442, Springer, Berlin, 1975.

C.-S. Yih, *Stratified Flows.* Academic Press, London, 1980.

R. Kh. Zeytounian, *Etude des phénomènes d'ondes dans les écoulement stationnaires d'un fluide stratifié non visqueux.* ONERA, Chatillon, Publication No. 126, Febuary 1969. (English translation edited by A. McPherson, Royal Aircraft Establishment, Farnborough, Library Translation 1404, December 1969). See also R. Kh. Zeytounian, J. de Mécanique, **8**(1969), 239–263, and R. Kh. Zeytounian, J. de Mécanique **8**(1969), 335–355.

R. Kh. Zeytounian. C. R. Acad. Sci. Paris, **274A** (1972), 1056. R. Kh. Zeytounion. C. R. Acad. Sci. Paris, **274A** (1972), 1413.

R. Kh. Zeytounian. Arch. Mech. (Archiwum Mechaniki Stosowanej), **26**(3) (1974), 499–509.

R. Kh. Zeytounian, La Météorologie du point de vue du Mécanicien des Fluides. Fluid Dyn. Trans. Polish Acad. Sci. **8**(1976), 289–352.

R. Kh. Zeytounian, Izv. Acad. Sci. USSR, Atmosph. Ocean. Phys. **15**(5) (1979), 498–507.

R. Kh. Zeytounian, *Asymptotic Modeling of Atmospheric Flows.* Springer, Heidelberg, 1990.

R. Kh. Zeytounian, *Meteorological Fluid Dynamics.* Lecture Notes in Physics, m5, Springer, Heidelberg, 1991.

R. Kh. Zeytounian. Phys.-Uspekhi, **41**(3) (1998), 241–267.

R. Kh. Zeytounian, *Theory and Applications of Nonviscous Fluid Flows.* Springer, Heidelberg, 2002a.

R. Kh. Zeytounian, *Asymptotic Modelling of Fluid Flow Phenomena.* Fluid Mechanics and Its Applications, vol. 64, series Ed. R. Moreau, Kluwer Academic, Dordrecht, 2002b.

R. Kh. Zeytounian, On the foundations of the Boussinesq approximation applicable to atmospheric motions. Izv. Atmosph. Ocean. Phys., **39**, Suppl. (2003), S1–S14.

R. Kh. Zeytounian, Weather prediction as a problem in fluid dynamics (on the 100th birthday of I. A. Kibel). Izv. Atmosph. Ocean. Phys., **40**(5) (2004), 527–539.

R. Kh. Zeytounian and J. P. Guiraud, Asymptotic features of low Mach number flows in aerodynamics and in the atmosphere. In: *Advances in Computational Methods for Boundary and Interior Layers, BAIL III,* edited by J. J. H. Miller. Boole Press, Dublin, 1984, pp. 95–100.

A. Zygmund, *Trigonometric Series,* 2nd edition. Cambridge University Press, New York, 1958.

7

Miscellaneous: Various Low-Mach-Number Fluid Problems and Motions

In this last chapter, we consider, first, in Sect. 7.1, mainly the asymptotic derivation of the KZK equation of nonlinear acoustics, which generalizes the well-known Burgers' unsteady one-dimensional dissipative model equation (Burgers 1948) to an equation with a diffraction and parabolic effect. Then, in Sect. 7.2, from the Bénard thermal-convection problem, we derive consistently the RB problem governed by the shallow-convection (Boussinesq) equations, and we show that to the starting exact Bénard problem ('heated from below') there corresponds, in the low-Mach-number limit, only the rigid–free model RB problem (very well analysed in Chandrasekhar (1981) and Drazin and Reid (1981)). Section 7.3 is devoted to various aspects of the low-Mach-number flow over relief, in an inviscid adiabatic atmosphere. In particular, we extend the results of Sect. 6.5.4; we consider also the steady 2D Eulerian case and derive a low-Mach-number model for lee waves over a mountain. In Sect. 7.4, from the full system of NSF equations (3.48), we show that close to the initial time $t = 0$ (where the initial conditions are prescribed) and near a rigid wall (for instance, $z = 0$ in a Cartesian system of coordinates x, y, z, for simplicity), the model equations governing the low-Mach-number motion for a large Reynolds number ($Re \gg 1$) such that $M^2 Re = 1$ are necessarily 'unsteady, one-dimensional (t, z)'; in particular, this gives us the possibility to derive in a consistent asymptotic way the single equation of Howarth (1951) for the Rayleigh compressible problem. Finally, in Sect. 7.5, we discuss various aspects of the 'nearly-incompressible-flow' description, considered by Zank and Matthaeus (1990, 1991) and Ghosh and Matthaeus (1992).

7.1 An Asymptotic Derivation of the Kuznetsov–Zabolotskaya–Khokhlov (KZK) Model Equation

A fundamental model equation in classical nonlinear acoustics is the 'parabolic KZK equation', which was first derived by Zabolotskaya and Khokhlov (1969)

R. Kh. Zeytounian: *Topics in Hyposonic Flow Theory*, Lect. Notes Phys. **672**, 229–278 (2006)
www.springerlink.com

in the inviscid case and then generalized by Kuznetsov (1970) to the dissipative case as an approximation of the Kuznetsov equation. This Kuznetsov equation, mentioned in Sect. 2.4 (see (2.18)), is written (according to Coulouvrat 1992, p. 333) in a curious form, since we have the Mach number M in front of the nonlinear terms on the right-hand side. From this equation, with an additional assumption about the slow divergence of the acoustic beam, the nonlinear Lighthill–Westervelt equation (obtained by Westervelt 1963, from an equation considered by Lighthill 1954) is obtained. Namely (when the dimensionless equation of state is $p = \rho\, T$)

$$\frac{\partial^2 \Phi}{\partial t^2} - \Delta\Phi - 2S\frac{\partial^3 \Phi}{\partial t^3} = \frac{\gamma+1}{2}\, M\frac{\partial}{\partial t}\left(\frac{\partial \Phi}{\partial t}\right)^2 . \tag{7.1}$$

The reader can find a derivation of the Kuznetsov equation in Coulouvrat (1992, pp. 334–338). The KZK equation was derived for the first time using a multiple-scale asymptotic method by Tjøtta and Tjøtta (1981). The same method was used (though in a slightly different way) in Coulouvrat (1992). It would be interesting to reconsider the derivation of the various approximate equations of nonlinear acoustics from a fully consistent logical, asymptotic viewpoint. We do not claim that these model equations, derived by a 'pseudo-asymptotic' method, are not interesting for various applications related to nonlinear acoustics: for instance, for the sound attenuation in seawater, thermoviscous and nonlinear attenuation in the near field and far field, and analysis of dispersion relations; more examples are provided in Hamilton and Blackstock (1990) and, see also, the 'heavy' paper by Crighton (1981). Nevertheless, it appears necessary that these acoustic models should be seriously revisited to examine the consistency of the results and conclusions obtained from these model equations, taking into account, in particular, the associated similarity rules. This, obviously, is a somewhat difficult and challenging problem.

Below, we give (as an example) a tentative asymptotic derivation of the KZK equation for the far field. We start with the system of nonlinear acoustic equations (1.37). Namely, for the velocity vector $\boldsymbol{v}$ and the thermodynamic perturbations π, ω and θ, we consider, with an error of order $O(M^2)$, the following equations:

$$\frac{\partial \omega}{\partial t} + \nabla \cdot \boldsymbol{v} = -M\, \nabla \cdot (\omega \boldsymbol{v}) , \tag{7.2a}$$

$$\frac{\partial \boldsymbol{v}}{\partial t} + \frac{1}{\gamma}\nabla\pi = \frac{1}{Re_{\mathrm{ac}}}\left\{\nabla^2 \boldsymbol{v} + \left[\frac{1}{3} + \frac{\mu_{\mathrm{vc}}}{\mu_{\mathrm{c}}}\right]\nabla(\nabla \cdot \boldsymbol{v})\right\}$$
$$- M\left[\omega\frac{\partial \boldsymbol{v}}{\partial t} + (\boldsymbol{v} \cdot \nabla)\boldsymbol{v}\right] , \tag{7.2b}$$

$$\frac{\partial\theta}{\partial t} + (\gamma - 1)\nabla \cdot \boldsymbol{v} = \frac{\gamma}{Pr}\frac{1}{Re_{\mathrm{ac}}}\nabla^2\theta + \gamma(\gamma-1)\frac{M}{Re_{\mathrm{ac}}}\left\{\frac{1}{2}[\boldsymbol{D}(\boldsymbol{v}) \cdot \boldsymbol{D}(\boldsymbol{v})]\right.$$

$$\left. + \left[\frac{1}{3} + \frac{\mu_{\mathrm{vc}}}{\mu_{\mathrm{c}}}\right](\nabla \cdot \boldsymbol{v})^2\right\} - M\left[\omega\frac{\partial\theta}{\partial t} + \boldsymbol{v} \cdot \nabla\theta + (\gamma-1)\pi\nabla \cdot \boldsymbol{v}\right]\Bigg\} \,, \quad (7.2c)$$

$$\pi - (\theta + \omega) = M\theta\omega \,. \quad (7.2d)$$

The above 'dominant' equations (7.2) are the main starting point for the derivation of various model equations in nonlinear acoustics, considered as a branch of fluid mechanics Crighton (1981). For the frequencies and media commonly used in nonlinear acoustics, the acoustic Reynolds number Re_{ac} is always very large compared to unity, and this means that the (continuous) medium is weakly dissipative at the frequency chosen – in the most common experimental situations, the above nonlinear equations (7.2a–c), together with (7.2d), are accurate enough. As a consequence, the three equations (7.2a–c) turn out to be perturbation equations with two small parameters, M and $1/Re_{\mathrm{a}}$. Obviously, since we neglect the terms proportional to $O(M^2)$, the above equations (7.2b,c) for $\boldsymbol{v}$ and θ make sense only if

$$M^2 \ll \frac{1}{Re_{\mathrm{ac}}} \,. \quad (7.3)$$

7.1.1 Asymptotic Approach

More particularly, concerning the asymptotic derivation of the KZK equation, the main idea is, basically, that it is assumed that the 3D acoustic field is locally plane, such that the nonlinear wave propagates in the same way as a linear plane wave over a few wavelengths, the wave profile or amplitude being significantly altered only at large distances away from the source (in the far field). As a consequence, obviously, the parabolic approximation, which leads to the KZK model equation, may not be valid close to the source (in the near field). The condition for that approximation to be valid is that the width of the acoustic source d should be much larger than the wavelength $1/k$, so that transverse field variations are slow compared with longitudinal variations along the acoustic axis.

Close to the origin of the acoustic axis, near the acoustic source, we have a 'near field', and at a large distance from this acoustic source a far field – the KZK equation being significant (asymptotically) only for this acoustic far-field. From the above consideration, the purely linear acoustic homogeneous field, which is derived when the Mach number tends to zero ($1/Re_{\mathrm{ac}}$ being, in fact, of the order of M), depends mostly on the retarded time $\chi = t - x$, the dependence on the two other (transverse) space variables being slow.

Thus, if the dimensionless 3D position variable is $\boldsymbol{x} = (x, y, z)$, then two new dimensionless variables η and ζ are defined as

$$\eta = \alpha y \quad \text{and } \zeta = \alpha z \,, \quad (7.4a)$$

where the ratio $\alpha = (1/k)/d$ is much less than 1 and tends to zero as $M \downarrow 0$. However, the transverse components of the fluid velocity should be changed also: namely, if the dimensionless velocity vector is $\boldsymbol{v} = (u, v, w)$, then we set

$$V = \frac{v}{\alpha}, \quad W = \frac{w}{\alpha}, \quad \boldsymbol{v} = u\boldsymbol{i} + \alpha(V\boldsymbol{j} + W\boldsymbol{k}) \, . \tag{7.4b}$$

Using (7.4), we can write

$$\nabla = \frac{\partial}{\partial x}\boldsymbol{i} + \alpha\left(\frac{\partial}{\partial \eta}\boldsymbol{j} + \frac{\partial}{\partial \zeta}\boldsymbol{k}\right) \equiv \frac{\partial}{\partial x}\boldsymbol{i} + \alpha\boldsymbol{\nabla}_{\perp} \tag{7.5}$$

and

$$\nabla \,.\, \boldsymbol{v} = \frac{\partial u}{\partial x} + \alpha^2\left[\frac{\partial V}{\partial \eta} + \frac{\partial W}{\partial \zeta}\right] , \tag{7.6a}$$

$$(\boldsymbol{v} \,.\, \nabla) = u\frac{\partial}{\partial x} + \alpha^2\left[V\frac{\partial}{\partial \eta} + W\frac{\partial}{\partial \zeta}\right] . \tag{7.6b}$$

As a consequence of (7.4)–(7.6), it is necessary to assume that the functions $u, V, W, \pi, \omega, \theta$ are dependent on the variables t, x, η, ζ and also on M (at least when we take into account the two similarity rules (7.7) below).

However, on the other hand, for a consistent asymptotic derivation of the KZK model equation it is necessary also to assume the existence of two similarity relations between the three small parameters $1/Re_{\mathrm{ac}}$, M and α, namely

$$\frac{1}{Re_{\mathrm{ac}}} = \kappa M \quad \text{and} \quad \frac{\alpha^2}{M} = \beta, \quad \text{with } \kappa = O(1) \quad \text{and } \beta = O(1) \, . \tag{7.7}$$

The above relations (7.7) – the first similarity rule being compatible with (7.3) for $\kappa \approx 1$ – gives us the possibility to take into account, in the KZK model equation derived below, the influence of dissipative effects (via the scalar κ) and also of the diffraction related to transverse variations. Now, with the above relations (7.4)–(7.7), it is necessary, first, in place of the starting equations (7.2), to write the transformed equations for $u, V, W, \pi, \omega, \theta$ as dependent functions of t, x, η, ζ and M. We obtain the following system:

$$\frac{\partial \omega}{\partial t} + \frac{\partial u}{\partial x} + M\left[\frac{\partial(\omega u)}{\partial x} + \beta\left(\frac{\partial V}{\partial \eta} + \frac{\partial W}{\partial \zeta}\right)\right] = O(M^2) \, , \tag{7.8a}$$

$$\frac{\partial u}{\partial t} + \frac{1}{\gamma}\frac{\partial \pi}{\partial x} + M\left[\omega\frac{\partial u}{\partial t} + u\frac{\partial u}{\partial x} - \kappa\left(\frac{4}{3} + \frac{\mu_{\mathrm{vc}}}{\mu_{\mathrm{c}}}\right)\frac{\partial^2 u}{\partial x^2}\right] = O(M^2), \tag{7.8b}$$

$$\frac{\partial V}{\partial t} + \frac{1}{\gamma}\frac{\partial \pi}{\partial \eta} + M\left[\omega\frac{\partial V}{\partial t} + u\frac{\partial V}{\partial x} - \kappa\frac{\partial^2 V}{\partial x^2} - \kappa\left(\frac{1}{3} + \frac{\mu_{\mathrm{vc}}}{\mu_{\mathrm{c}}}\right)\frac{\partial^2 u}{\partial \eta \, \partial x}\right] = O(M^2) \, , \tag{7.8c}$$

$$\frac{\partial W}{\partial t}+\frac{1}{\gamma}\frac{\partial \pi}{\partial \zeta}+M\left[\omega\frac{\partial W}{\partial t}+u\frac{\partial W}{\partial x}-\kappa\frac{\partial^2 W}{\partial x^2}-\kappa\left(\frac{1}{3}+\frac{\mu_{\mathrm{vc}}}{\mu_{\mathrm{c}}}\right)\frac{\partial^2 u}{\partial \zeta\,\partial x}\right]=O(M^2)\,. \tag{7.8d}$$

$$\frac{\partial \theta}{\partial t}+(\gamma-1)\frac{\partial u}{\partial x}+M\left[\omega\frac{\partial \theta}{\partial t}+u\frac{\partial \theta}{\partial x}+(\gamma-1)\pi\frac{\partial u}{\partial x}\right.$$
$$\left.+(\gamma-1)\beta\left(\frac{\partial V}{\partial \eta}+\frac{\partial W}{\partial \zeta}\right)-\kappa\left(\frac{\gamma}{Pr}\right)\frac{\partial^2 \theta}{\partial x^2}\right]=O(M^2)\,, \tag{7.8e}$$

$$\pi-(\theta+\omega)-M\theta\omega=0\,. \tag{7.8f}$$

For low Mach numbers, we consider the following asymptotic expansion:

$$U=(u,V,W,\pi,\theta,\omega)=U_0+MU_1+\dots\,, \tag{7.9}$$

where M tends to zero with t, x, η, ζ and $\gamma, Pr, \mu_{\mathrm{vc}}/\mu_{\mathrm{c}}, \kappa, \beta$ fixed.

7.1.2 The Leading-Order System for U_0

With (7.9), from the above equations (7.8a,b,e,f) we derive, at the leading order, for u_0, π_0, θ_0 and ω_0, the following acoustic system:

$$\frac{\partial \omega_0}{\partial t}+\frac{\partial u_0}{\partial x}=0\,, \tag{7.10a}$$

$$\frac{\partial u_0}{\partial t}+\frac{1}{\gamma}\frac{\partial \pi_0}{\partial x}=0\,, \tag{7.10b}$$

$$\frac{\partial \theta_0}{\partial t}+(\gamma-1)\frac{\partial u_0}{\partial x}=0, \tag{7.10c}$$

$$\pi_0=\theta_0+\omega_0\,. \tag{7.10d}$$

On the other hand, the equations (7.8c,d) are reduced to

$$\frac{\partial V_0}{\partial t}+\frac{1}{\gamma}\frac{\partial \pi_0}{\partial \eta}=0\,, \tag{7.11a}$$

$$\frac{\partial W_0}{\partial t}+\frac{1}{\gamma}\frac{\partial \pi_0}{\partial \zeta}=0, \tag{7.11b}$$

or

$$\frac{\partial}{\partial t}\left[\frac{\partial V_0}{\partial \zeta}-\frac{\partial W_0}{\partial \eta}\right]=0\,, \tag{7.12}$$

and we observe that the relation (7.12) is a consistent consequence of our asymptotic approach. In an 'ad hoc' derivation, Coulouvrat (1992, p. 335) uses (in the dissipative case?) the potential equation $\boldsymbol{v}=-\nabla\phi$, assuming the motion to be irrotational. Obviously, this (physically unrealized) assumption is superfluous.

First, from (7.10), we derive two equations for u_0 and π_0,

$$\gamma\frac{\partial u_0}{\partial t}+\frac{\partial \pi_0}{\partial x}=0\,, \tag{7.13a}$$

$$\frac{\partial \pi_0}{\partial t}+\gamma\frac{\partial u_0}{\partial x}=0\,, \tag{7.13b}$$

and we obtain for u_0 the classical one-dimensional (plane wave) equation of linear acoustics,

$$\frac{\partial^2 u_0}{\partial t^2}-\frac{\partial^2 u_0}{\partial x^2}=0\,. \tag{7.14}$$

As a solution, when it is assumed that the fluid is unbounded, we write (for an outgoing wave propagating towards $x>0$) $u_0 = F(\chi,\eta,\zeta), \chi = t-x$. However, this acoustic solution is in fact a good solution *only* for the near field close to the acoustic source. On the other hand, the KZK equation is a far-field equation, and to avoid various cumulative effects it is necessary to consider (7.9) as a non-secular two-scale expansion relative to variations along the acoustic axis and to define the slow scale as

$$\xi = Mx, \quad \text{with}\ \frac{\partial}{\partial x}=-\frac{\partial}{\partial \chi}+M\frac{\partial}{\partial \xi},\quad \frac{\partial^2}{\partial x^2}=\frac{\partial^2}{\partial \chi^2}+O(M)\,. \tag{7.15}$$

In such a case, we consider the acoustic solution for u_0 as a function of both of the fast (local) scale x, via χ, and the slow scale ξ, namely

$$u_0 = F^*(\chi,\xi,\eta,\zeta)\,. \tag{7.16}$$

Obviously, using (7.16), we can write the solution of the acoustic equations (7.10) in the following form (where we observe that $\partial/\partial t = \partial/\partial\chi$):

$$\omega_0 = F^*(\chi,\xi,\eta,\zeta),\quad \pi_0=\gamma F^*(\chi,\xi,\eta,\zeta),\quad \theta_0=(\gamma-1)F^*(\chi,\xi,\eta,\zeta)\,. \tag{7.17}$$

One the other hand, from (7.11), we obtain also (because $\partial/\partial t = \partial/\partial\chi$)

$$\frac{\partial V_0}{\partial \chi}=-\frac{\partial F^*}{\partial \eta}\quad \text{and}\quad \frac{\partial W_0}{\partial \chi}=-\frac{\partial F^*}{\partial \zeta}\,. \tag{7.18}$$

7.1.3 The Second-Order System for U_1

Using (7.9), (7.15), (7.16) and (7.17), we derive from the above equations (7.8a,b,e,f) at the second order the following equations for u_1, π_1, θ_1 and ω_1

$$\frac{\partial \omega_1}{\partial t}+\frac{\partial u_1}{\partial x}=2F^*\frac{\partial F^*}{\partial \chi}-\frac{\partial F^*}{\partial \xi}-\beta\left(\frac{\partial V_0}{\partial \eta}+\frac{\partial W_0}{\partial \zeta}\right), \tag{7.19a}$$

$$\frac{\partial u_1}{\partial t}+\frac{1}{\gamma}\frac{\partial \pi_1}{\partial x}=\kappa\left(\frac{4}{3}+\frac{\mu_{\mathrm{vc}}}{\mu_{\mathrm{c}}}\right)\frac{\partial^2 F^*}{\partial \chi^2}-\frac{\partial F^*}{\partial \xi}\,, \tag{7.19b}$$

$$\frac{\partial \theta_1}{\partial t} + (\gamma - 1)\frac{\partial u_1}{\partial x} = (\gamma - 1)\left\{\kappa\left(\frac{\gamma}{Pr}\right)\frac{\partial^2 F^*}{\partial \chi^2} - \frac{\partial F^*}{\partial \xi} + \gamma F^* \frac{\partial F^*}{\partial \chi}\right\} \quad (7.19c)$$

$$\pi_1 - \theta_1 - \omega_1 = (\gamma - 1)(F^*)^2 . \quad (7.19d)$$

Finally, from (7.18) we obtain also the following relation:

$$-\frac{\partial}{\partial \chi}\left(\frac{\partial V_0}{\partial \eta} + \frac{\partial W_0}{\partial \zeta}\right) = \frac{\partial^2 F^*}{\partial \eta^2} + \frac{\partial^2 F^*}{\partial \zeta^2} \equiv \nabla_\perp^2 F^*. \quad (7.20)$$

We do not write the two inhomogeneous equations for V_1 and W_1, derived from (7.8c,d), since both these equations are related to the derivation of a second-order equation 'à la KZK'. From the above system of equations (7.19), together with (7.20), taking into account that the right-hand sides of these equations (7.19a–d) are functions (via F^*) of the variables $\chi = t - x$ and $\xi = Mx$, and not of t and x separately, such that $\partial/\partial t = +\partial/\partial\chi$ and $\partial/\partial x = -\partial/\partial\chi$, and also taking account of (7.15), we derive a single inhomogeneous acoustic equation for u_1 with a source term, namely

$$\frac{\partial^2 u_1}{\partial t^2} - \frac{\partial^2 u_1}{\partial x^2} = \frac{\partial \Phi(F^*)}{\partial \chi} + \frac{\beta}{\gamma}\nabla_\perp^2 F^* , \quad (7.21)$$

where $F^* = F^*(\chi, \xi, \eta, \zeta)$ and

$$\Phi(F^*) = \kappa\left[\frac{4}{3} + \frac{\mu_{vc}}{\mu_c} + \frac{1}{Pr}(\gamma - 1)\right]\frac{\partial^2 F^*}{\partial \chi^2} + (\gamma + 1)F^*\frac{\partial F^*}{\partial \chi} - 2\frac{\partial F^*}{\partial \xi} . \quad (7.22)$$

7.1.4 The KZK Model Equation as a Compatibility–Non-Secularity Condition

Since the right-hand-side source term in (7.21) gives (as a consequence of $F^* = F^*(\chi, \xi, \eta, \zeta)$, u_1 being assumed to be a function of t, x, η and ζ) a term in the solution for u_1 which will ultimately be greater than u_0, whatever the smallness of the Mach number, the only way for that expansion (7.9) to be non-secular is the validity of the following compatibility condition (according to the usual multiple-scale method of Kevorkian and Cole 1981):

$$\frac{\partial \Phi(F^*)}{\partial \chi} + \frac{\beta}{\gamma}\nabla_\perp^2 F^* = 0 . \quad (7.23)$$

Using (7.23), with (7.22) we derive the following KZK equation:

$$\frac{\partial^2 F^*}{\partial \chi\, \partial \xi} - \frac{\beta}{2\gamma}\vec{\nabla}_\perp^2 F^* - \frac{\gamma + 1}{2}\frac{\partial}{\partial \chi}\left[F^*\frac{\partial F^*}{\partial \chi}\right] = S^*\frac{\partial^3 F^*}{\partial \chi^3} , \quad (7.24)$$

where

$$S^* = \frac{\kappa}{2}\left[\left(\frac{4}{3} + \frac{\mu_{vc}}{\mu_c}\right) + \frac{\gamma - 1}{Pr}\right] . \quad (7.25)$$

The parameter β in the KZK equation (7.24) measures the relative orders of magnitude of diffraction and nonlinearity, and for $\beta = 0$, we derive again a Burgers equation valid for the far field (at large distances from the source) in the following form, after integration with respect to χ:

$$\frac{\partial F^*}{\partial \xi} - \frac{1}{2}(\gamma+1)F^*\frac{\partial F^*}{\partial \chi} = S^*\frac{\partial^2 F^*}{\partial \chi^2} \,. \tag{7.26}$$

The above KZK equation (7.24) is the simplest equation that takes into account, simultaneously, in an asymptotically consistent way, *nonlinearity* (the term proportional to $(1/2)(\gamma+1)$), *dissipation* (the term proportional to S^*) and *diffraction* (the term proportional to $\beta/2\gamma$). The KZK equation is by far the most useful equation, owing to its (relative) simplicity, and practically all numerical schemes dealing with strongly nonlinear propagation rely on it, even for rather wide-banded signals and for interaction within the near field of piston sources. Now, the question arises: how is it possible to do without the parabolic approximation? The above KZK equation can also be written in the 'Coulouvrat form'

$$\frac{\partial^2 F^*}{\partial \chi\, \partial \sigma} - \frac{N}{4}\nabla_\perp^2 F^* - \frac{1}{\Gamma}\frac{\partial^3 F^*}{\partial \chi^3} = \frac{\partial}{\partial \chi}\left(F^*\frac{\partial F^*}{\partial \chi}\right) , \tag{7.27}$$

where $\sigma = (1/2)(\gamma+1)\xi$, and in this case we have two main parameters,

$$\Gamma = \frac{2S^*}{\gamma+1} \quad \text{and} \quad N = \frac{4\beta}{\gamma(\gamma+1)} \,. \tag{7.28}$$

For the far field, when (case (i)) $\Gamma \ll 1$ but $N\Gamma \gg 1$, we have 'thermoviscous attenuation', and when (case (ii)) $\Gamma \gg 1$ and $N \gg 1$, we have 'nonlinear attenuation' as the leading effect. For a parametric acoustic source emission, Moffett and Mellen 1977, case (i), where the Gol'dberg number Γ is small, plays an important role. Finally, we observe that experimental values show that Γ and N remain roughly of order 1, which proves that the hypotheses that we made in deriving the KZK equation are consistent.

7.2 From the Bénard Problem to the Rayleigh–Bénard Rigid–Free Model Problem

Below, we consider the classical Bénard problem: *the onset of thermal instability in a horizontal one-layer of fluid (liquid) heated from below.* The liquid layer (d is the depth of this layer) is bounded below by a rigid plane and bounded above by a passive gas (having negligible density and viscosity, a constant pressure p_{A} and a constant temperature T_{A}), separated from the liquid layer by an interface. The rigid plane (a perfect heat conductor fixed at a temperature T_{w}) lies at $z = 0$, and the mean position of the interface lies at $z = d$. The liquid is a Newtonian fluid, assumed from the beginning to be compressible, viscous and heat-conducting, with constant values (at a temperature $T = T_0$, where T_0 is defined by (7.35) below) of the viscosities μ_0 and λ_0, of the reference density $\rho_0 = \rho(T_0)$, of the specific heat C_0, of the thermal conductivity k_0, and of the volume expansion coefficent

$$\alpha_0 = \alpha(T_0) \text{ with } \alpha(T) = -\frac{1}{\rho}\frac{\mathrm{d}\rho(T)}{\mathrm{d}T} , \tag{7.29a}$$

When we assume that the equation of state of the expansible liquid is simply

$$\rho = \rho(T) . \tag{7.29b}$$

As a consequence of the introduction of a perturbation of the pressure (see (7.44) below), this equation of state is asymptotically consistent at the leading order for the derivation of the model RB instability problem (the reader can easily verify this fact). The constant thermal diffusivity is $\kappa_0 = k_0/C_0\rho_0$, and $\nu_0 = \mu_0/\rho_0$ is the constant kinematic viscosity. We assume also that the surface tension σ of the interface is constant, equal to $\sigma_0 = \sigma(T_0)$, which implies that the interface is maintained at a constant temperature T_0 in the conduction motionless steady state and *also* in the convection flow regime, when the interface is deformed. This strong hypothesis is very natural in the framework of a rigorous asymptotic derivation of the RB model problem where, according to the alternatives set out by Zeytounian (1997):

> "Either the buoyancy is taken into account and in a such case the free surface deformation effect is negligible in the leading-order RB instability problem, or the free surface deformation effect is taken into account and in this case the buoyancy does not plays a significant role in the leading-order Bénard–Marangoni–Biot (BMB) problem with a deformable free surface and full thermocapillary and Biot effects".

7.2.1 Some Physical Aspects of the Bénard Problem

It is well known that the buoyancy effect is operative (in the RB problem) when the Grashof number,

$$Gr = \alpha(T_0)\frac{(T_{\mathrm{w}} - T_0)}{Fr^2} , \tag{7.30}$$

is fixed. However, for a weakly expansible/dilatable liquid, the expansibility parameter

$$\varepsilon = \alpha(T_0)\,\Delta T , \tag{7.31}$$

where $\Delta T = T_w - T_0$, is a small parameter and, consequently, it is necessary to consider the case of a small squared Froude number,

$$Fr^2 = \frac{(\nu_0/d)^2}{gd} \ll 1 \quad \Rightarrow d \gg \left(\frac{\nu_0^2}{g}\right)^{1/3} . \tag{7.32}$$

In other words, the RB model problem can be asymptotically derived in a consistent way only when the following similarity rule is taken into account:

$$\frac{\varepsilon}{Fr^2} = Gr \quad \text{fixed, when } \varepsilon \text{ and } Fr^2 \text{ both tend to zero} . \tag{7.33}$$

Concerning the constant temperature T_0, we observe that in the steady, motionless conduction state, the temperature $T_s(z)$ is the solution of

$$\frac{\mathrm{d}^2 T_s(z)}{\mathrm{d}z^2} = 0 \quad \text{with } T_s(0) = T_w , \tag{7.34}$$

and we obtain as solution,

$$T_s(z) = T_w - \beta_s z , \quad \text{with} \quad \beta_s = -\frac{\mathrm{d}T_s(z)}{\mathrm{d}z} \tag{7.35a}$$

and the constant temperature T_0 is defined by:

$$T_s(z = d) = T_w - \beta_s d \equiv T_0 . \tag{7.35b}$$

In (7.35a), β_s is the adverse *conduction* temperature gradient. In the steady, motionless conduction state, β_s is determined from the following Newton's law of heat transfer, written for $T_s(z)$:

$$k_0 \frac{\mathrm{d}T_s(z)}{\mathrm{d}z} + q_s[T_s(z) - T_A] = 0 \quad \text{at } z = d , \tag{7.36}$$

where the coefficient q_s is the constant conduction heat transfer coefficient which is assumed known. As a consequence of (7.36) with (7.35a) we obtain for β_s, the following relation (as in Takashima 1981):

$$\beta_s = \frac{Bi,s}{(1 + Bi,s)} \left[\frac{T_w - T_A}{d}\right] , \tag{7.37}$$

where

$$Bi,s = \frac{d q_s}{k_0} , \tag{7.38}$$

is the *conduction* constant Biot number.

In (7.37) and (7.38) we have assumed that the conduction heat transfer coefficient q_s is constant and, as a consequence, that Bi,s = const, mainly because T_0 = const is uniform over the flat free surface $z = d$ in the steady, motionless conduction regime linked to the temperature $T_s(z)$ in the motionless steady state; it should be realized that β_s is always different from zero in the framework of the Bénard thermal-instability problem, and as a consequence, in what follows, it is necessary to bear in mind that *always* $Bi,s \neq 0$. On the other hand, we observe that if, in the starting Bénard problem, the data are d, q_s, k_0 (or Bi,s), T_w and T_A, then, when β_s is defined by (7.37), we must determine also the reference temperature $T_s(z = d) = T_0$ and, according to (7.35b), we have the relation $\beta_s = \Delta T/d$, where $\Delta T = T_w - T_0$.

The conduction Biot number $Bi,s \neq 0$ defined above by (7.38), also plays a role in the mathematical formulation of the dimensionless RB instability model problem and appears as a consequence of a Bénard conduction effect. In convection problem it is necessary to take into account the fact that, a consequence of the small squared Froude number, the deformation of the interface characterized by the parameter δ in the dimensionless equation for the interface:

$$\frac{z}{d} = 1 + \delta h(t', x', y') \tag{7.39}$$

(The time–space coordinates t', x', y' here are dimensionless; see Sect. 7.2.2) must tend to zero as Fr^2 tends to zero, such that:

$$\frac{\delta}{Fr^2} = \delta^* = O(1)\ . \tag{7.40}$$

In conclusion, since the temperature is assumed constant over the interface in the convection regime ($T \equiv T_0$, no Marangoni effect) and, in fact, the interface is $z = d$ for the RB model problem, it seems that we have the possibility (and this is a very admissible *conjecture*) to write Newton's cooling law for the *convection* regime (at $z' \equiv z/d = 1$), using the conduction constant Biot number $Bi,s = dq_s/k_0$ given by (7.38), for the dimensionless perturbation of the temperature of the liquid

$$\theta = \frac{(T - T_0)}{\Delta T}\ . \tag{7.41}$$

Namely, we can write:

$$\boldsymbol{\nabla}'\theta.\boldsymbol{n}' + Bi,s\theta + 1 = O(\varepsilon) \quad \text{at } z' = 1 + O(\varepsilon)\ , \tag{7.42}$$

where $\boldsymbol{n}'$ is the dimensionless unit outward normal vector to the free surface and $\boldsymbol{\nabla}' = d\boldsymbol{\nabla}$.

We observe that in the steady, motionless conduction state, we have the dimensionless perturbation of the temperature

$$\theta_s(z') \equiv \frac{T_s(z/d) - T_0}{T_w - T_0} = 1 - z'\ . \tag{7.43}$$

Together with (7.41), for a coherent, consistent derivation of the RB model problem, it is necessary also to introduce the following perturbation of the pressure:

$$\pi = \frac{1}{Fr^2}\left(\frac{p - p_A}{gd\rho_0} + z' - 1\right)\ . \tag{7.44}$$

Usually, in the RB model equation for θ (in shallow-convection/Oberbeck–Boussinesq equations), the effect of viscous dissipation is neglected. Below (in Sect. 7.2.4), we show that this is true only if the parameter,

$$\chi = \frac{gd}{\Delta T C_0} \approx 1 \quad \Rightarrow d \approx \frac{C_0\,\Delta T}{g} \equiv d_{sh}\ , \tag{7.45}$$

which is a bound on the thickness of the liquid layer, d, in the RB model problem. On the other hand, since in this case $gd_{\text{sh}} \approx \Delta T\ C_0$, we can write (or identify) the squared Froude number with a *low* squared ('liquid') Mach number M_1^2 via

$$Fr^2 \approx \frac{(\nu/d_{\text{sh}})^2}{C_0\ \Delta T} = \left[\frac{\nu/d_{\text{sh}}}{\sqrt{\Delta T C_0}}\right]^2 \equiv M_1^2 \ll 1\ . \tag{7.46}$$

Finally, for our weakly expansible liquid, in place of Fr^2, we can use M_1^2, which is the ratio of the reference (intrinsic) velocity $U_1 = \nu/d_{\text{sh}}$ to the ('liquid') pseudo-sound speed $c_1 = \sqrt{C_0\ \Delta T}$.

7.2.2 A Mathematical Formulation of the Bénard Convection Leading-Order Problem

Assuming a constant surface tension $\sigma_0 = \sigma(T_0)$, we can define the following Weber number:

$$\text{We} = \sigma(T_0)\frac{d}{\rho_0\nu_0^2}\ . \tag{7.47}$$

However, physicists usually use, in place of the Weber number, the *crispation number*

$$Cr = \frac{1}{We\,Pr} = \frac{\rho_0\nu_0\kappa_0}{\sigma(T_0)d}\ , \tag{7.48}$$

where $Pr = \nu_0/\kappa_0$ is the Prandtl number. For most liquids in contact with air, Cr is very small (see, for instance, the range of values of Cr in Takashima 1981, p. 2748). On the other hand, if we use the equation of state (7.29b) and take into account (7.41) and (7.31), we obtain, with an error of $O(\varepsilon^2)$, the following linear relation between the density ρ of the liquid and the convective perturbation of the temperature, θ

$$\rho = \rho_0[1 - \varepsilon\theta]\ . \tag{7.49}$$

As a consequence of (7.49), the specific energy of our weakly expansible liquid is, in fact, a function only of the perturbation of the temperature θ, and we can write

$$\frac{\mathrm{D}E}{\mathrm{D}t} = \Delta T\ C_0\frac{\mathrm{D}\theta}{\mathrm{D}t} + O(\varepsilon)\ , \tag{7.50}$$

where $\mathrm{D}/\mathrm{D}t = \partial/\partial t + (\boldsymbol{u}\,.\,\nabla)$ is the material time derivative, and $\boldsymbol{u}$ is velocity vector and $\nabla = (\partial/\partial z, \partial/\partial x, \partial/\partial y)$ is the gradient operator.

Dimensionless Dominant Governing Equations

First, with the approximate equation of state (7.49) and with an error of $O(\varepsilon^2)$, we can write the following dimensionless equation of continuity:

$$\nabla'\,.\,\boldsymbol{u}' = \varepsilon\frac{\mathrm{D}\theta}{\mathrm{D}t'}\ , \tag{7.51a}$$

where $\boldsymbol{u}' = \boldsymbol{u}/(\nu_0/d)$ and $\mathrm{D}/\mathrm{D}t' = \partial/\partial t' + (\boldsymbol{u}' \,.\, \nabla')$, and $t' = t/(d^2/\nu_0)$. If we work (for simplicity) with constant viscosities μ_0 and λ_0, then, with the above definitions (7.41) and (7.44) of θ and π, we can derive the following dimensionless equation of motion for $\boldsymbol{u}'$:

$$\frac{\mathrm{D}\boldsymbol{u}'}{\mathrm{D}t'} + \nabla'\pi - Gr\,\theta\boldsymbol{k} - \nabla'^2\boldsymbol{u}' = O(\varepsilon)\,. \tag{7.51b}$$

Finally, from (7.50), we obtain the following dimensionless equation for the perturbation θ, of the temperature;

$$\frac{\mathrm{D}\theta}{\mathrm{D}t'} - \frac{1}{Pr}\nabla'^2\theta = O(\varepsilon)\,. \tag{7.51c}$$

We assume here, according to (7.45), that $gd/\Delta T\; C_0 \approx 1$. The above system of three dominant equations (7.51) for $\boldsymbol{u}', \pi$ and θ is very significant for the asymptotic modelling of the Bénard thermal problem, when we consider a weakly expansible ideal liquid and assume (7.49) and (7.33).

Dimensionless Dominant Boundary Conditions

Concerning the boundary conditions on the interface (a free surface simulated by the dimensionless equation $z' = 1 + \delta h(t, x', y')$, the relevant boundary condition for the pressure perturbation π is (see, for instance, Zeytounian 2002, pp. 378–380)

$$\pi = \delta^* h + 2d'_{ij}n'_i n'_j + We(\nabla'_{\|} \,.\, \boldsymbol{n}') + O(\varepsilon)\,, \tag{7.52a}$$

if we use (7.40); d'_{ij} are the dimensionless components of the strain rate tensor (deformation tensor; see (1.2)) $\boldsymbol{D}'(\boldsymbol{u}')$, n'_i are the dimensionless components $\boldsymbol{n}'$ and $\boldsymbol{\nabla}'_{\|}$ is the dimensionless gradient vector on the free surface. In addition to (7.52a), we have also three free boundary conditions at $z' = 1 + \delta h(t', x', y')$, namely:

$$d'_{ij}t'^{(s)}_i n'_j = O(\varepsilon) \quad (s = 1, 2)\,, \tag{7.52b}$$

$$\boldsymbol{u}' \,.\, \boldsymbol{k} = \delta\left[\frac{\partial h}{\partial t'} + \boldsymbol{u}' \,.\, \nabla' h\right]\,, \tag{7.52c}$$

and, together with the free-surface condition (7.42) for θ derived above, we have a complete set of free-surface boundary conditions. In the two ($s = 1$ and 2) dimensionless dominant boundary conditions (7.52b), $t'^{(1)}_i$ and $t'^{(2)}_i$ are the dimensionless components of two orthonormal tangent vectors $\boldsymbol{t}'^{(s)}$, $s = 1, 2$, to the free surface $z' = H'(= 1 + \delta h)$. Following Pavithran and Redeekopp (1994), we write

$$\boldsymbol{t}'^{(1)} = \left(\frac{1}{N_1'}\right)^{1/2} \left(1; 0; \delta \frac{\partial h}{\partial x'}\right) ,$$

$$\boldsymbol{t}'^{(2)} = \left(\frac{1}{N_1' N'}\right)^{1/2} \left(-\delta^2 \frac{\partial h}{\partial x'} \frac{\partial h}{\partial y'}; N_1'; \delta \frac{\partial h}{\partial y'}\right) ,$$

$$N_1' = 1 + \delta^2 \left(\frac{\partial h}{\partial x'}\right)^2 , \quad N' = N_1' + \delta^2 \left(\frac{\partial h}{\partial y'}\right)^2 . \tag{7.53}$$

In reality, the fact that, in the equation of the interface $z' = 1 + \delta h(t', x', y')$, the amplitude parameter δ vanishes with the square of the Froude number to zero, according to the similarity rule (7.40), simplifies strongly the above conditions (7.52). Namely, a straightforward examination gives the following leading-order interface conditions:

$$\pi_{z'=1} = \delta^* h - We^* \left[\frac{\partial^2 h}{\partial x'^2} + \frac{\partial^2 h}{\partial y'^2}\right] , \tag{7.54a}$$

$$\boldsymbol{u}' \cdot \boldsymbol{k} \equiv u_3 = 0, \quad \frac{\partial u_1'}{\partial z'} = 0, \quad \frac{\partial u_2'}{\partial z'} = 0, \quad \text{at } z' = 1 . \tag{7.54b}$$

In the above relation (7.54a,b), we take into account also the relation for $(\nabla'_{||} \cdot \boldsymbol{n}')$, for vanishing δ,

$$\nabla'_{||} \cdot \boldsymbol{n}' = -\delta \left[\frac{\partial^2 h}{\partial x'^2} + \frac{\partial^2 h}{\partial y'^2}\right] , \tag{7.55}$$

and the similarity rule for a large Weber number (which is the case in various physical situations)

$$\delta \, We = We^* . \tag{7.56}$$

The condition (7.42) for θ is unchanged, and at the lower rigid plate we have the dimensionless boundary conditions

$$\boldsymbol{u}' = 0 \quad \text{and } \theta = 1 \quad \text{at } z' = 0 . \tag{7.57}$$

The terms $O(\varepsilon)$ in (7.51b) and (7.51c) and the free-surface boundary conditions (7.42), (7.52a) and (7.52b) are given explicitly in Zeytounian (2002, pp. 376–380). As a consequence of the above asymptotic approach (if we take into account the terms proportional to ε), we have, in practice, the possibility to derive the associated 'second-order' linear problem (significant at the order ε, with various non-Boussinesq effects related with expansibility.). However, here we do not consider such a (certainly laborious but straightforward) derivation.

Initial Conditions

Obviously, it is necessary also to write initial conditions for u' and θ at $t' = 0$. These initial data characterize the physical nature of the dominant Bénard problem derived above. Strictly speaking, the given starting physical data for the density do not necessarily satisfy the approximate linear equation of state (7.49). Indeed, unfortunately, the problem of the initial data for this dominant Bénard problem has been very poorly investigated, and certainly this dominant Bénard problem is not significant close to the initial time, because the partial derivative of the density with respect to time is lost in this dominant problem. As a consequence, it is necessary to derive, close to the initial time, a *local* dominant Bénard problem (with a short time) and then to consider an adjustment (inner) problem. At the end of this adjustment process, when the short time tends to infinity, we have in principle the possibility to obtain by matching the well-defined data for the above dominant Bénard (outer) problem.

A pertinent initial–boundary-value problem related to the development of nonlinear waves on the surface of a horizontally rotating thin film (without Marangoni and Biot effects) has been considered by Needham and Merkin (1987) and, more recently, by Bailly (1995). In these two studies, an incompressible viscous liquid is injected onto a disc at a specified flow rate through a small gap of height a at the bottom of a cylindrical reservoir of radius l situated at the centre of the disc. With the assumption $a \ll l$, a long-wave unsteady theory is considered, with a thin 'inlet' region and also a region corresponding to a very small time, in which rapid adjustment to the initial conditions occurs. Through matching, these regions provide appropriate 'boundary' and 'initial' conditions for the leading-order (outer) problem in the main region. Without doubt, the approach of Needham and Merkin (1987) and Bailly (1995) can be used for various thermal-convection model problems which usually are not valid close to the initial time, and this obviously deserves further investigation.

7.2.3 The RB Rigid–Free Shallow-Convection Model Problem

First, we use the squared Mach number $M_1^2 \ll 1$ in place of ε and Fr^2, in accordance with (7.46), such that

$$\varepsilon = Gr\, M_1^2 \quad \text{and } Fr^2 = M_1^2\,. \tag{7.58a}$$

Assuming

$$M_1^2 \to 0, \quad \text{with} \quad t', x', y', z', Gr,\ We^*, Bi, s, \delta^* \quad \text{fixed}\,, \tag{7.58b}$$

we consider the following low-Mach-number expansions:

$$\boldsymbol{u}' = \boldsymbol{u}_0 + M_1^2 \boldsymbol{u}_2 + \dots, \quad \theta' = \theta_0 + M_1^2 \theta_2 + \dots\,,$$
$$\pi' = \pi_0 + M_1^2 \pi_2 + \dots\,, \tag{7.59a}$$

$$h = h_0 + M_1^2 h_2 + \cdots\,. \tag{7.59b}$$

Using (7.58) and (7.59), from the above equations (7.51) and the conditions (7.54), (7.57) and (7.42), we derive an RB model problem, governed by the following shallow (Oberbeck–Boussinesq) equations:

$$\nabla' \cdot \boldsymbol{u}_0 = 0 \, ,$$

$$\frac{\mathrm{D}\boldsymbol{u}_0}{\mathrm{D}t'} + \nabla' \pi_0 - Gr\, \theta_0 \boldsymbol{k} - \nabla'^2 \boldsymbol{u}_0 = 0 \, ,$$

$$\frac{D\theta_0}{\mathrm{D}t'} - \frac{1}{Pr} \nabla'^2 \theta_0 = 0 \, . \tag{7.60}$$

For the above system (7.60) of RB shallow-convection equations, we write the boundary conditions, in the framework of a rigid–free model problem, as

$$\boldsymbol{u}_0 = 0 \quad \text{and} \quad \theta_0 = 1, \quad \text{at } z' = 0 \, , \tag{7.61a}$$

$$\boldsymbol{u}_0 \cdot \boldsymbol{k} = 0, \quad \frac{\partial (\boldsymbol{u}_0 \cdot \boldsymbol{k})}{\partial z'} = 0, \quad \frac{\partial \theta_0}{\partial z'} + Bi{,}s\, \theta_0 + 1 = 0, \quad \text{at } z' = 1 \, . \tag{7.61b}$$

In (7.60), $\mathrm{D}/\mathrm{D}t' = \partial/\partial t' + \boldsymbol{u}_0 \cdot \nabla'$ and $\nabla' = (\partial/\partial x', \partial/\partial y', \partial/\partial z')$.

The RB model problem (7.60) and (7.61) is a consistent asymptotic, low-Mach-number approximation, via (7.58) and (7.59), of the starting 'exact' Bénard thermal-convection problem. This RB model problem is operative in a (Boussinesq) liquid layer of thickness d such that

$$\left(\frac{\nu_0^2}{g} \right)^{1/3} \ll d \approx C_0 \frac{\Delta T}{g} \equiv d_{\mathrm{sh}} \, , \tag{7.62}$$

as a consequence of (7.45), if we assume that the parameter $\chi = gd/C_0\, \Delta T \approx 1$, and this condition characterizes the case of the 'shallow convection'. The leading order of the deformation of the free surface $h_0(t', x', y')$ in the expansion (7.59b) is then determined, when the perturbation of the pressure $\pi_0(t', x', y', z')$ is known at $z = 1$, after the solution of the RB problem (7.60) and (7.61), by the equation

$$\frac{\partial^2 h_0}{\partial x'^2} + \frac{\partial^2 h_0}{\partial y'^2} - \frac{\delta^*}{We^*} h_0 = -\frac{1}{We^*} \pi_0(t', x', y', z' = 1) \, , \tag{7.63}$$

thanks to the similarity relation (7.56), for a large Weber number.

We observe also that, in our approximate low-Mach-number 'theory', the *Bond number* is of order 1, i.e.

$$Bo = \frac{Cr}{Fr^2} = O(1) \, , \tag{7.64}$$

since $Cr = \delta / PrWe^* \ll 1$ according to (7.48) and (7.56), and Bo is a ratio of two small parameters ($B_0 \approx \delta^*$) when $Pr = O(1)$ and $We^* = O(1)$. The above equation (7.63) for the deformation of the interface $h_0(t', x', y')$ seems not have been derived in the framework of the classical, 'ad hoc' theory (see, for instance, Dauby and Lebon 1996, where the Marangoni effect is also taken into account), emerges very naturally in our asymptotic approach.

7.2.4 Some Complementary Remarks

First, concerning the equation of state (7.29b), it is necessary to note that, from the relation (7.44) for the perturbation of the pressure π, we have

$$p = p_A + gd\rho_0[1 - z' + M_1^2\pi] . \tag{7.65}$$

It is clear that the presence of the pressure p in a full baroclinic equation of state $\rho = \rho(T, p)$, in place of (7.29b), does not (in the low-Mach-number limit (7.58b)) radically change the form of the Oberbeck–Boussinesq, shallow-convection, approximate RB equations (7.60) derived. From (7.45), we obtain also the result that our RB, rigid–free model problem (7.60) and (7.61) is operative when

$$\Delta T \equiv T_w - T_0 \approx \frac{gd_{sh}}{C_0} . \tag{7.66}$$

A second remark concerns the constraint (7.45), $\chi \approx 1$, related to the viscous dissipation in the dimensionless dominant equation for θ. Namely, when the parameter χ defined by (7.45) is large, such that (since $Fr^2 \ll 1$)

$$\chi\, Fr^2 \approx 1 , \tag{7.67}$$

then in this case

$$d = \frac{d_{sh}}{Fr^2} \equiv d_{deep} \gg d_{sh} \text{ since } Fr^2 \ll 1 . \tag{7.68}$$

Thermal convection in a layer of thickness $d \equiv d_{deep}$ is termed "deep convection". Obviously, when $\chi \gg 1$, such that the relation (7.67) is realized, we do not have the possibility to use above derived RB model problem (7.60), (7.61). For the deep convection problem we do not have the possibility to neglect, in the equation (7.51c) for θ, two terms which appear in the right hand side of this equation (in $O(\varepsilon)$). Namely, the terms:

$$\varepsilon\chi(z' - 1)\frac{\mathrm{D}\theta}{\mathrm{D}t'} \quad \text{and} \quad -\frac{\chi Fr^2}{2}\left(\frac{\partial u'_i}{\partial x_j} + \frac{\partial u'_j}{\partial x_i}\right) ,$$

respectively.

The deep convection limiting equation for θ_d in this case is obtained via the following limiting process:

$$\varepsilon \to 0 \text{ and } Fr^2 \to 0, \text{ such that } \frac{\varepsilon}{Fr^2} = Gr \text{ fixed,}$$
$$\text{and } \chi\varepsilon = O(1),\ \chi Fr^2 = O(1) . \tag{7.69}$$

The resulting leading-order deep convection equation for θ_d, from (7.69), is:

$$[1 + \Gamma(z' - 1)]\frac{\mathrm{D}\theta_d}{\mathrm{D}t'} = \frac{1}{Pr}\nabla'^2\theta_d + \frac{1}{2}\frac{\Gamma}{Gr}\left(\frac{\partial u'_i}{\partial x'_j} + \frac{\partial u'_j}{\partial x'_1}\right)^2 , \tag{7.70}$$

where

$$\Gamma \equiv \chi\varepsilon = \frac{g\alpha(T_0)d}{C_0} , \tag{7.71}$$

is the deep convection parameter, according to Zeytounian (1989).

The above deep-convection equation (7.70) has been written on the assumption that $p_A = 0$ and $p' = p/gd\rho_0$ in (7.65). Finally, from $\Gamma \approx 1$, we obtain the following relation for d_{deep}:

$$d \equiv d_{\text{deep}} = \frac{C_0}{g\alpha(T_0)} . \tag{7.72}$$

A third remark concerns the derivation of the second-order model equations for the RB problem. These equations are written for the terms proportional to M_l^2 in the expansion (7.59a), $\boldsymbol{u}_2$, θ_2 and π_2, when we take into account that, according to (7.33) and (7.46), we have $\varepsilon = M_l^2 Gr$. First, we derive a second-order equation for $\boldsymbol{u}_2$ that is not divergenceless:

$$Pr\ \nabla' . \boldsymbol{u}_2 = \nabla'^2\theta_0 . \tag{7.73a}$$

Then, we obtain the following second-order linear equation for the RB convective motion, for $\boldsymbol{u}_2$:

$$\begin{aligned}\frac{\partial \boldsymbol{u}_2}{\partial t'} &+ (\boldsymbol{u}_0 . \nabla')\boldsymbol{u}_2 + (\boldsymbol{u}_2 . \nabla')\boldsymbol{u}_0 + \nabla'\pi_2 - Gr\ \theta_2\boldsymbol{k} - \nabla'^2\boldsymbol{u}_2 \\ &= Gr\theta_0\left[\frac{\partial \boldsymbol{u}_0}{\partial t'} + (\boldsymbol{u}_0 . \nabla')\boldsymbol{u}_0\right] \\ &+ Gr\left[1 + \frac{\lambda_0}{\mu_0}\right]\nabla'\left(\frac{\partial\theta_0}{\partial t'} + \boldsymbol{u}_0 . \nabla'\theta_0\right) .\end{aligned} \tag{7.73b}$$

Finally, for θ_2, we derive the following second-order linear thermal equation:

$$\begin{aligned}&\frac{\partial\theta_2}{\partial t'} + \boldsymbol{u}_0 . \nabla'\theta_2 + \boldsymbol{u}_2 . \nabla'\theta_0 - \frac{1}{Pr}\nabla'^2\theta_2 \\ &= Gr\left\{[\chi(1 - z') + \theta_0]\left(\frac{\partial\theta_0}{\partial t'} + \boldsymbol{u}_0 . \nabla'\theta_0\right)\right\} + \frac{\chi}{2}\left[\frac{\partial u'_{i0}}{\partial x'_j} + \frac{\partial u'_{j0}}{\partial x'_i}\right]^2 .\end{aligned} \tag{7.73c}$$

The above system of equations (7.73a–c) for the RB model problem are the only possible asymptotically consistent second-order shallow-convection equations. We shall not go further in this direction, but the reader can easily derive for the equations (7.73) the associated boundary conditions at $z = 0$ and $z = 1$. We observe that the first terms in the right-hand sides of (7.73b,c) take an effect of compressibility into account.

We make a final, very short comment concerning the dissipation coefficients, λ, μ and k, assumed constant in our above theory. This hypothesis does not have any influence in the derivation of the RB model problem (7.60) and (7.61). For the second-order system of equations, however, this is obviously not the case. For instance, if we assume that the thermal conductivity coefficient is a function of temperature, i.e. $k = k(T)$, which is a physically consistent assumption having regard to the equation of state (7.29b), then, using (7.41), we obtain

$$k(T) = k(T_0 + \Delta T\theta) = k(T_0) + \left.\frac{\mathrm{d}k}{\mathrm{d}T}\right|_{T_0} \Delta T\theta \approx k_0[1 - \varepsilon\beta\theta] , \tag{7.74a}$$

where β is the following ratio;

$$\beta = \frac{\rho(T_0)}{k(T_0)} \left[\frac{\mathrm{d}k(T)}{\mathrm{d}T} \bigg/ \frac{\mathrm{d}\rho(T)}{\mathrm{d}T}\right]_{T0} . \tag{7.74b}$$

This ratio is assumed to be fixed and $O(1)$ when $\varepsilon(= M_l^2 Gr)$ tends to zero. With (7.74a), we have an additional term in the right-hand side of the second-order equation (7.73c) for θ_2, of the form

$$-\frac{\beta}{Pr}\nabla'[\theta_0\nabla'\theta_0] . \tag{7.74c}$$

7.3 Flow Over Relief When M Tends to Zero

Below, we consider first some aspects of a 'primitive model', which is related with the investigations of the Sect. 6.5. Then we examine a Boussinesq-like model, which was considered by Drazin (1961) and Brighton (1978). Finally, the steady 2D Eulerian case is considered.

7.3.1 The Primitive Model

Our purpose is to investigate the scheme of Sect. 6.5, when there is a relief defined by a function $h(x, y)$ such that $|h| = O(1)$. In this case, to leading order, we obtain (with the notation of Sect. 6.5)

$$H_0(P_{g0}(x, y)) = h(x, y) , \tag{7.75a}$$

since an obvious consequence of the fact that the topography does not depend on time is $\partial P_{g0}/\partial t = 0$. On the other hand, we can write

$$\boldsymbol{v}_0 \,.\, \boldsymbol{D}h = 0 \quad \text{on } p = P_{g0}(x, y) , \tag{7.75b}$$

and for the leading-order functions $\boldsymbol{v}_0$ and H_2, by analogy with (6.79), but without the effect of the Coriolis force ($Ki = \infty$), we obtain the following problem:

$$\frac{\partial \boldsymbol{v}_0}{\partial t} + (\boldsymbol{v}_0 \cdot \boldsymbol{D})\boldsymbol{v}_0 + \boldsymbol{D}\left(\frac{H_2}{\gamma}\right) = 0, \quad \boldsymbol{D} \cdot \boldsymbol{v}_0 = 0 , \tag{7.76a}$$

$$\boldsymbol{v}_0 \cdot \boldsymbol{D}h = 0, \quad \text{for } (x, y) \subset C_0(p) , \tag{7.76b}$$

$$\boldsymbol{v}_0 \to \boldsymbol{V}^\infty(p), \quad H_2 \to 0, \quad \text{when } r = (x^2 + y^2)^{1/2} \to \infty . \tag{7.76c}$$

In the above boundary condition (7.76b), $C_0(p)$ stands for the intersection of the relief $z = h(x, y)$ with the isobaric surface $p =$ const, according to $H = H_0(p)$, that is,

$$(x, y) \in C_0(p) \Leftrightarrow H_0(p) = h(x, y) . \tag{7.77}$$

On each isobaric surface which intersects the relief, there is a partial blocking phenomenon. The flow, being constrained to remain on its own isobaric surface, cannot ride over the topography, and must flow around it. For those isobaric surfaces which do not cross the relief, the situation is exactly the same as it was in Sect. 6.5. The difference from Sect. 6.5 is that the model generates a flow disturbed from (6.98) owing to the motion around relief, but only within the isobaric surfaces which intersect the relief. For those which lie entirely above it, the model is unable to generate perturbations. The same holds concerning the vertical vorticity. Let us stress the main differences, and for this purpose, instead of (6.99), we write

$$\boldsymbol{v}_0 = \boldsymbol{V}^\infty(p) + \boldsymbol{k} \wedge \boldsymbol{D}\psi_0 + \boldsymbol{D}\varphi_0 , \tag{7.78}$$

but the functions φ_0 and ψ_0 have to be computed (for isobaric surfaces which intersect the relief) in a different way. Namely, we have the following two problems: first,

$$\frac{\partial^2 \psi_0}{\partial x^2} + \frac{\partial^2 \psi_0}{\partial y^2} = \Omega_0, \quad \psi_0 = 0, \quad \text{on } C_0(p) , \tag{7.79a}$$

$$\psi_0 = \frac{\langle \Omega_0 \rangle}{2\pi} \log |\boldsymbol{x}| + o(1) \quad \text{when } |\boldsymbol{x}| \to \infty , \tag{7.79b}$$

and second,

$$\frac{\partial^2 \varphi_0}{\partial x^2} + \frac{\partial^2 \varphi_0}{\partial y^2} = 0, \quad \varphi_0 = o(1), \quad \text{when } |\boldsymbol{x}| \to \infty , \tag{7.80a}$$

$$\boldsymbol{D}\varphi_0 \cdot \boldsymbol{D}h + \boldsymbol{V}^\infty(p) \cdot \boldsymbol{D}h = 0 \quad \text{on } C_0(p) . \tag{7.80b}$$

We may write

$$\psi_0 = \frac{\langle \Omega_0 \rangle}{2\pi} \log |\boldsymbol{x}| - \frac{\boldsymbol{M}_0 \cdot \boldsymbol{x}}{2\pi |\boldsymbol{x}|^2} + \frac{\boldsymbol{x} \cdot \mathbf{N}_0 \cdot \boldsymbol{x}}{2\pi |\boldsymbol{x}|^4} + O\left(\frac{1}{|\boldsymbol{x}|^3}\right) , \tag{7.81}$$

$$\varphi_0 = -\frac{\boldsymbol{A} \cdot \boldsymbol{x}}{2\pi |\boldsymbol{x}|^2} + \frac{\boldsymbol{x} \cdot \mathbf{B} \cdot \boldsymbol{x}}{2\pi |\boldsymbol{x}|^4} + O\left(\frac{1}{|\boldsymbol{x}|^3}\right) , \tag{7.82}$$

but where

$$\langle \Omega_0 \rangle = \int\int_{\text{Ext. } C_0(\text{p})} \Omega_0(t, x', y', p)\, \mathrm{d}x'\mathrm{d}y', \quad \frac{\partial \langle \Omega_0 \rangle}{\partial t} = 0 \tag{7.83}$$

holds for those isochoric surfaces which intersect the relief. Here, we have the following for $\boldsymbol{v}_0$, according to (7.78):

$$\begin{aligned} \boldsymbol{v}_0 = \; & \boldsymbol{V}^\infty(p) + \frac{\langle \Omega_0 \rangle}{2\pi}(\boldsymbol{k} \wedge \boldsymbol{x})\frac{1}{|\boldsymbol{x}|^2} \\ & - \frac{1}{2\pi|\boldsymbol{x}|^2}\Big\{(\boldsymbol{k} \wedge \boldsymbol{M}_0) - 2\left[(\boldsymbol{M}_0 \,.\, \boldsymbol{x})(\boldsymbol{k} \wedge \boldsymbol{x})\right]\frac{1}{|\boldsymbol{x}|^2} \\ & + \boldsymbol{A} - \left[2(\boldsymbol{A} \,.\, \boldsymbol{x})\boldsymbol{x}\right]\frac{1}{|\boldsymbol{x}|^2}\Big\} + O\left(\frac{1}{|\boldsymbol{x}|^3}\right) . \end{aligned} \tag{7.84}$$

We may use (6.101e), provided that we define χ_0 by: $\boldsymbol{k} \wedge \boldsymbol{D}\psi_0 + \boldsymbol{D}\varphi_0 = \boldsymbol{D}\chi_0$, instead of (6.101f). We obtain

$$\begin{aligned} \chi_0 = \; & \frac{\langle \Omega_0 \rangle}{2\pi}\text{Arctg}\left(\frac{y}{x}\right) + \left[(\boldsymbol{k} \wedge \boldsymbol{M}_0) \,.\, \boldsymbol{x} - \boldsymbol{A} \,.\, \boldsymbol{x}\right]\frac{1}{2\pi|\boldsymbol{x}|^2} \\ & + \left[\boldsymbol{x} \,.\, \mathbf{N}_0 \,.\, (\boldsymbol{k} \wedge \boldsymbol{x}) + \boldsymbol{x} \,.\, \mathbf{B} \,.\, \boldsymbol{x}\right]\frac{1}{2\pi|\boldsymbol{x}|^4} + \dots , \end{aligned} \tag{7.85}$$

and also

$$\begin{aligned} H_2 = & -\gamma\left\{\left[\boldsymbol{k} \wedge \left(\frac{\partial \boldsymbol{M}_0}{\partial t} - \boldsymbol{V}^\infty_{(p)}\langle \Omega_0 \rangle\right) - \frac{\partial \boldsymbol{A}}{\partial t}\right] . \,\boldsymbol{x}\right\}\frac{1}{2\pi|\boldsymbol{x}|^2} \\ & - \frac{\gamma}{2}\left[\frac{\langle \Omega_0 \rangle}{2\pi}\right]^2\frac{1}{|\boldsymbol{x}|^2} - \frac{\gamma}{2\pi|\boldsymbol{x}|^2}\Big\{\boldsymbol{V}^\infty(p) \,.\, \left[(\boldsymbol{k} \wedge \boldsymbol{M}_0 - \boldsymbol{A}\right] \\ & - 2\left[(\boldsymbol{k} \wedge \boldsymbol{M}_0 - \boldsymbol{A}) \,.\, \boldsymbol{x}\right](\boldsymbol{V}^\infty(p) \,.\, \boldsymbol{x})\frac{1}{|\boldsymbol{x}|^2} \\ & + \left[\boldsymbol{x} \,.\, \frac{\partial \mathbf{N}_0}{\partial t} \,.\, (\boldsymbol{k} \wedge \boldsymbol{x}) + \boldsymbol{x} \,.\, \frac{\partial \boldsymbol{B}}{\partial t} \,.\, \boldsymbol{x}\right]\frac{1}{|\boldsymbol{x}|^2}\Big\} + O\left(\frac{1}{|\boldsymbol{x}|^3}\right) . \end{aligned} \tag{7.86}$$

Of course, for altitudes above the relief, a number of terms disappear by virtue of the relations

$$\frac{\partial \boldsymbol{M}_0}{\partial t} = \boldsymbol{V}^\infty(p)\langle \Omega_0 \rangle, \quad \mathbf{A} = 0 \quad \text{and } \mathbf{B} = 0 .$$

Now we see that, in order to compute $\boldsymbol{v}_2$ and then H_4, we need $\partial^2 H_2/\partial p^2$. However, unfortunately, from inspection of (7.86), we see that we cannot expect that H_2 to have as many derivatives as we need at the crossover altitude (when we consider the highest isobaric surface which intersects the relief). This is just an indication that the low-Mach-number expansion cannot be uniformly valid, including the vicinity of the crossover altitude. This phenomenon has been studied, in slightly different context, by Brighton (1978) and Hunt and Snyder (1980). We shall come back to this later. Before doing so, we consider the outer expansion, assuming that $h(x, y)$ vanishes in the vicinity of $|x| = \infty$. We may use without change (6.102a) to (6.103) but we must change (6.104a,b) to

$$\boldsymbol{v}_0 \approx \boldsymbol{V}^{\infty}(p) + M\frac{\langle \Omega_0 \rangle}{2\pi}(\boldsymbol{k} \wedge \boldsymbol{\xi})\frac{1}{|\boldsymbol{\xi}|^2}$$

$$-M^2\frac{1}{2\pi|\boldsymbol{\xi}|^2}\left\{(\boldsymbol{k} \wedge \boldsymbol{M}_0) + \boldsymbol{A} - 2[(\boldsymbol{M}_0 \,.\, \boldsymbol{x})(\boldsymbol{k} \wedge \boldsymbol{\xi}) + (\boldsymbol{A} \,.\, \boldsymbol{\xi})\boldsymbol{\xi}]\frac{1}{|\boldsymbol{\xi}|^2}\right\} + O(M^3)\,, \tag{7.87a}$$

$$H_0 + M^2 H_2 = H_0 + M^3\left\{-\gamma\left[\boldsymbol{k} \wedge \left(\frac{\partial \boldsymbol{M}_0}{\partial t} - \boldsymbol{V}^{\infty}(p)\langle \Omega_0 \rangle\right) - \frac{\partial \boldsymbol{A}}{\partial t}\right] .\, \boldsymbol{\xi}\right\}\frac{1}{2\pi|\boldsymbol{\xi}|^2}$$

$$+ M^4 O\left(\frac{1}{|\boldsymbol{\xi}|^2}\right) + \dots\,. \tag{7.87b}$$

Again we must choose $\beta = 1$ in (6.102c,d) for the far field, and (6.105) holds with (6.106); here, the correctness of (6.106) relies again on $\frac{\partial \langle \Omega_0 \rangle}{\partial t} = 0$. Of course, whenever no vertical vorticity exists beforehand, $\langle \Omega_0 \rangle$ is zero and the leading velocity field is zero. But now the solution (6.107) is no longer valid, and a comment is in order. Here, the position with respect to the crossover altitude does not matter. Once there is a relief, the solution in the far field, to the order just after the leading order, is no longer the continuation of the solution at order one relative to horizontal distance. This far-field solution mixes all the altitude levels and requires a full solution to (6.108), not one for which the last term (proportional to γ) in (6.108) vanishes identically. However we shall not attempt to write down this solution; instead, we shall stop this discussion here, because further research is needed concerning higher-order terms in the expansions. We observe only that here, when $|\boldsymbol{\xi}| \to 0$, we have the behaviour

$$\boldsymbol{v}_1^{\mathrm{ff}} \approx -\frac{1}{2\pi|\boldsymbol{\xi}|^2}\left\{\boldsymbol{k} \wedge \boldsymbol{M}_0 + \boldsymbol{A} - 2[(\boldsymbol{M}_0 \,.\, \boldsymbol{x})(\boldsymbol{k} \wedge \boldsymbol{\xi}) + (\boldsymbol{A} \,.\, \boldsymbol{\xi})\boldsymbol{\xi}]\frac{1}{|\boldsymbol{\xi}|^2}\right\} + \dots\,,$$

$$H_1^{\mathrm{ff}} \approx -\gamma\frac{1}{2\pi|\boldsymbol{\xi}|^2}\left[\boldsymbol{k} \wedge \left(\frac{\partial \boldsymbol{M}_0}{\partial t} - \boldsymbol{V}^{\infty}(p)\langle \Omega_0 \rangle\right) - \frac{\partial \boldsymbol{A}}{\partial t}\right] .\, \boldsymbol{\xi} + \dots\,,$$

The Problem of the Crossover Altitude

The first point to investigate concerns the behaviour of the leading-order solution near crossover. We assume that the relief may be expanded according to

$$h(x,y) = h_{\rm M} - h^{(2)}(x,y) - h^{(3)}(x,y) + \dots , \tag{7.88}$$

where $h^{(j)}$ is a polynomial of degree j, homogeneous with respect to $(x - x_{\rm M}, y - y_{\rm M})$, where $(x_{\rm M}, y_{\rm M})$ is the point of maximum altitude of the relief. Let $p_{\rm M}$ be the pressure level at the top of the relief, namely:

$$H_0(p_{\rm M}) = h_{\rm M} ; \tag{7.89}$$

we obtain an ellipse for the approximate form of $C_0(p)$ in (7.76b),

$$h^{(2)}(x,y) = \frac{1}{p_M} T_{0\rm M}(p - p_{\rm M}) . \tag{7.90}$$

We consider first the potential φ_0, and a moment's reflection shows that when $p \to p_{\rm M}$, we have

$$\varphi_0 = (p - p_{\rm M})^{1/2} \boldsymbol{F}(X,Y) \,.\, \boldsymbol{V}^\infty(p), \quad p > p_{\rm M} , \quad \text{but } \varphi_0 \equiv 0 \text{ if } p < p_{\rm M} , \tag{7.91a}$$

where

$$X = \frac{(x - x_{\rm M})}{(p - p_{\rm M})^{1/2}}, \quad Y = \frac{(y - y_{\rm M})}{(p - p_{\rm M})^{1/2}} . \tag{7.91b}$$

Now we must consider ψ_0. As a first approximation, we may write the following (we shall see later, however, that this is not fully satisfactory):

$$\psi_0^{(\rm a)}(t,x,y,p) = \frac{1}{2\pi} \int\!\!\int_{E_0(p)} \Omega_0(t,x',y',p) \log[(x-x')^2 + (y-y')^2]^{1/2} \, {\rm d}x' \, {\rm d}y' , \tag{7.92}$$

where $E_0(p)$ stands for the whole plane when $p < p_{\rm M}$ and for the exterior of $C_0(p)$ when $p > p_{\rm M}$. Starting from this approximation, we may look for ψ_0 as follows:

$$\psi_0 = \psi_0^{(\rm a)} + \psi_0^{(H)}, \quad \text{where } \frac{\partial^2 \psi_0^{(H)}}{\partial x^2} + \frac{\partial^2 \psi_0^{(H)}}{\partial y^2} = 0 , \tag{7.93a}$$

with the condition

$$\psi_0^{(\rm a)} + \psi_0^{(H)} = 0 \quad \text{on } C_0(p) . \tag{7.93b}$$

Again, a moment's reflection suggests that, to leading order,

$$\psi_0^{(H)} = (p - p_{\rm M})^{1/2} \boldsymbol{G}(X,Y) \,.\, \boldsymbol{D}\psi_0^{(\rm a)}(t, x_{\rm M}, y_{\rm M}, p_{\rm M}) , \tag{7.94}$$

provide we ignore the term $\psi_0^{(\rm a)}(t, x_{\rm M}, y_{\rm M}, p_{\rm M})$, which does not contribute to the velocity.

From (7.91a) and (7.94), we deduce that

$$\boldsymbol{v}_0 = A(X,Y)\boldsymbol{V}^\infty(p_{\rm M}) + B(X,Y)\, \boldsymbol{D}\psi_0^{(\rm a)}(t, x_{\rm M}, y_{\rm M}, p_{\rm M}) + \boldsymbol{v}_0^{\rm R}(t,x,y,p) , \tag{7.95}$$

and as a consequence we find that

$$H_2 = F_2(X,Y) + H_2^{\mathrm{R}}, \quad p > p_{\mathrm{M}}, \quad \text{but} \quad H_2 \equiv H_2^{\mathrm{R}}, \text{ if } p < p_{\mathrm{M}} \,. \tag{7.96}$$

We see also that $\frac{\partial H_2}{\partial p}$ is singular when $p \downarrow p_{\mathrm{M}}$. As a matter of fact, our rough analysis leading from (7.91) to (7.96) fails to consider the singular vorticity field due to convection by a singular velocity field. As a consequence, we should reconsider the analysis starting from (7.92), and we can conjecture that the proper correction to (7.95) would be $O[(p - p_{\mathrm{M}})^{1/2}]$ and that it would match with a higher-order term, $O(M)$ for $\boldsymbol{v}$, in the following expansion:

$$\boldsymbol{x} = \boldsymbol{x}_{\mathrm{M}} + M\boldsymbol{x}^*, \quad p = p_{\mathrm{M}} + M^2 p^*, \quad h = h_{\mathrm{M}} - M^2 h^{(2)}(x^*, y^*), \quad \boldsymbol{x}^* = (x^*, y^*), \tag{7.97a}$$

$$\boldsymbol{v} = \boldsymbol{v}^*, \quad \varpi = M\varpi*, \quad T = T^*, \quad H = h_{\mathrm{M}} + M^2 H^* \,, \tag{7.97b}$$

if we wished to derive a limiting process capable of retaining the singularity in $\partial H_2/\partial p$ when $p \downarrow p_{\mathrm{M}}$. With (7.97), from the primitive Kibel system of equations (6.69a–d), where $Ki = \infty$ (no Coriolis force), we derive the following set of leading-order equations in place of (7.76):

$$(\boldsymbol{v}^* \,.\, \boldsymbol{D}^*)\boldsymbol{v}^* + \varpi^* \frac{\partial \boldsymbol{v}^*}{\partial p^*} = 0, \quad \boldsymbol{D}^* \,.\, \boldsymbol{v}^* + \frac{\partial \varpi^*}{\partial p^*} = 0 \,, \tag{7.98a}$$

$$\boldsymbol{v}^* \,.\, \boldsymbol{D}^* T^* + \varpi^* \frac{\partial T^*}{\partial p^*} = 0, \quad T^* + p_{\mathrm{M}} \frac{\partial H^*}{\partial p^*} = 0 \,, \tag{7.98b}$$

with the boundary conditions *on the relief*

$$H_{\mathrm{g}}^* + h^{(2)} = 0, \quad \boldsymbol{v}_{\mathrm{g}}^* \,.\, \boldsymbol{D}^* P_{\mathrm{g}}^* = \varpi_{\mathrm{g}}^*, \quad p^* > 0 \,. \tag{7.98c}$$

Matching with the solution away from the crossover altitude requires, according to (7.96), that

$$H^* \approx \left(\frac{\partial H_0}{\partial p}\right)_{\mathrm{M}} p^* + H_2^{\mathrm{R}} + F_2(X^*, Y^*), \quad p^* \to +\infty \,, \tag{7.99a}$$

$$H^* \approx \left(\frac{\partial H_0}{\partial p}\right)_{\mathrm{M}} p^* + H_2^{\mathrm{R}}, \quad p^* \to -\infty \,, \tag{7.99b}$$

where $X^* = x^*/\sqrt{p^*}, Y^* = y^*/\sqrt{p^*}$, and $p^* \to +\infty$ is intended to mean

$$p^* \to +\infty \quad \text{with } X^* = \frac{x^*}{\sqrt{p^*}} \text{ and } Y^* = \frac{y^*}{\sqrt{p^*}} \text{ fixed} \,. \tag{7.100}$$

As a matter of fact, we cannot consider $p^* \to +\infty$ with $\boldsymbol{x}^*$ fixed, because this would carry us inside the relief. We can check easily that, when $p^* \to -\infty$,

$$\boldsymbol{v}^* \to \boldsymbol{v}_{0\mathrm{M}}, \quad H^* \to \left(\frac{\partial H_0}{\partial p}\right)_{\mathrm{M}} p^* + H_{2\mathrm{M}}, \quad T^* \to T_{0\mathrm{M}}, \quad \varpi \to 0 \tag{7.101}$$

is consistent with (7.98a,b). Here $\boldsymbol{v}_{0\mathrm{M}}$ is a constant vector, and $H_{2\mathrm{M}}$ and $T_{0\mathrm{M}}$ are constant values, which are expected to be the values of H_2 and T_0 right at the top of the relief, according to the solution of the problem (7.76). This is consistent with (7.99) if we assume that $H_2^R = H_{2M}$. Now we must examine the behaviour of the solution to equations (7.98a,b) under the condition (7.100). We assume

$$\varpi \to 0, \quad \boldsymbol{v}^* \to \boldsymbol{v}^+\left(t, \frac{\boldsymbol{x}^*}{\sqrt{p^*}}\right), \quad H \approx \left(\frac{\partial H_0}{\partial p}\right)_M p^* + H^+\left(t, \frac{\boldsymbol{x}^*}{\sqrt{p^*}}\right),$$
$$T^* \approx T_{0M} + O(p) \tag{7.102}$$

when we have used the notations:

$$\boldsymbol{X}^* = \frac{\boldsymbol{x}^*}{\sqrt{p^*}}, \quad \boldsymbol{D}^* \to \boldsymbol{D}^+\left(\frac{\partial}{\partial X^*}, \frac{\partial}{\partial Y^*}\right), \quad \boldsymbol{X}^* = (X^*, Y^*)\,, \tag{7.103}$$

and where we have taken care of the second equation of (7.98b).
In order to check (7.102), we need to check the following:

$$(\boldsymbol{v}^+ \,.\, \boldsymbol{D}^+)\boldsymbol{v}^+ + \boldsymbol{D}^+\left(\frac{H^+}{\gamma}\right) = 0, \quad \boldsymbol{D}^+ \,.\, \boldsymbol{v}^+ = 0\,. \tag{7.104}$$

We delay the verification of (7.104) in order to consider the boundary condition on the relief (7.98c). Using (7.102), we obtain (on $p^* = P_g^*$)

$$P_g^*\left[\left(\frac{\partial H_0}{\partial p}\right)_M + h^{(2)}\left(\frac{x^*}{\sqrt{P_g^*}}, \frac{y^*}{\sqrt{P_g^*}}\right)\right] + H^+\left(t, \frac{\boldsymbol{x}^*}{\sqrt{P_g^*}}\right) = 0\,, \tag{7.105a}$$

$$\boldsymbol{v}_g^+ \,.\, \boldsymbol{D}^+ P_g^* = 0\,, \tag{7.105b}$$

and H^+ in (7.105a) is negligible in the limit $p^* \to +\infty$. The function $P_g^*(t, \boldsymbol{x}^*)$ is obtained from $h^{(2)}(x^*/\sqrt{P_g^*}, y^*/\sqrt{P_g^*}) = -(\partial H_0/\partial p)_M$, and, from that the fact that the function $h^{(2)}$ is quadratic, we may extract P_g^* and obtain a closed-form formula for P_g^*,

$$P_g^* = -\left[\frac{1}{(\partial H_0/\partial p)_M}\right] h^{(2)}(\boldsymbol{x}^*)\,. \tag{7.106}$$

This has two consequences: the first is that P_g^* does not depend on t, and the second is that P_g^* is the same (for $p^* \to +\infty$) as that which would result from the process $M \to 0$ with $\boldsymbol{x}^*$ and p^* fixed being applied to $H(t, x, y, P_g(t, x, y)) = h(x, y)$. Now, in order that (7.105) holds, we need only the condition that $\boldsymbol{v}^+ \,.\, \boldsymbol{D}^+ h^{(2)} = 0$, on $h^{(2)} + (\partial H_0/\partial p)_M = 0$.

As a consequence, v^+ and H^+ are obtained by solving

$$(\boldsymbol{v}^+ \,.\, \boldsymbol{D}^+)\boldsymbol{v}^+ + \boldsymbol{D}^+\left(\frac{H^+}{\gamma}\right) = 0\,, \tag{7.107a}$$

$$\boldsymbol{v}^+ \,.\, \boldsymbol{D}^+ h^{(2)} = 0, \quad \text{on } h^{(2)}(\boldsymbol{X}) + \left(\frac{\partial H_0}{\partial p}\right)_{\mathrm{M}} = 0 \,, \tag{7.107b}$$

$$\boldsymbol{v}^+ \to \boldsymbol{v}_{0\mathrm{M}}, \quad |\boldsymbol{X}| \to \infty \,. \tag{7.107c}$$

In (7.107c), $\boldsymbol{v}_{0\mathrm{M}}$ means the value of the velocity field according to (7.76a–c) right at the top of the relief. We stress that the solution to (7.107a,b) with (7.107c), is not necessarily irrotational. More precisely, putting $\boldsymbol{k} \,.\, (\boldsymbol{D}^+ \wedge \boldsymbol{v}^+) = \Omega^+$ we have $(\boldsymbol{v}^+ \,.\, \boldsymbol{D}^+)\Omega^+ = 0$. However, the true vorticity is $\Omega^+/\sqrt{p}$, and this does not match with an $O(1)$ vorticity according to (7.76a–c)? Whether or not the solution to (7.107a,b) with (7.107c) is irrotational must be considered as an open question. The same is true concerning the possible occurrence of an $O(M)$ correction, rather than an $O(M^2)$ correction, to $\boldsymbol{v}$ due to vorticity effects near the top of the relief. We now leave this matter, which requires further research, and hope that the reader is convinced that a fully consistent theory of flow at low Mach number over relief will be obtained with some effort. We stress only two final points: the first concerns the separation which occurs almost inevitably on the lee side of a relief, and the second concerns the possible occurrence of an analogue of the Boussinesq states. We add that the non-uniformity that occurs at the top of a relief may also occur near a no flat region where it matches with a flat ground environment. More generally, some kind of non-uniformity may occur whenever $h(x, y)$ is not smooth.

7.3.2 Models Derived From the Euler Equations when $M \to 0$

First we consider the following Eulerian problem for the flow of a compressible, non-viscous adiabatic fluid, namely

$$\rho\left[St\frac{\partial \boldsymbol{u}}{\partial t} + (\boldsymbol{u} \,.\, \nabla)\boldsymbol{u}\right] + \frac{1}{\gamma M^2}(\nabla p + Bo\,\rho \boldsymbol{k}) = 0 \,,$$

$$S\frac{\partial \rho}{\partial t} + \boldsymbol{u} \,.\, \nabla \rho + \rho \nabla \,.\, \boldsymbol{u} = 0 \,, \tag{7.108a}$$

$$\gamma p\left(St\frac{\partial \rho}{\partial t} + \boldsymbol{u} \,.\, \nabla \rho\right) - \rho\left(St\frac{\partial p}{\partial t} + \boldsymbol{u} \,.\, \nabla p\right) = 0 \,,$$

and

$$\boldsymbol{u} \,.\, \nabla(h - z) = 0 \quad \text{on } z = h(x, y) \,,$$

$$x^2 + y^2 \to \infty: \quad \boldsymbol{u} \to \boldsymbol{U}_\infty(z), \quad \boldsymbol{U}_\infty(z) \,.\, \boldsymbol{k} = 0, \quad p \to p_\infty(z), \quad \rho \to \rho_\infty(z) \,. \tag{7.108b}$$

A Model Analogous to that of Drazin (1961)

Our basic assumption is $M \to 0$ with $\gamma, \boldsymbol{x} = (x, y, z)$ and t fixed and $|h(x, y)| = O(1)$, and we consider a low-Mach-number expansion:

$$(\boldsymbol{u}, p, \rho) \equiv U(x, t; M) = U_0(x, t) + M^2 U_2(x, t) + \dots \,. \tag{7.109}$$

As a result, we derive the following leading-order model problem (à la Drazin 1961):

$$w_0 = \boldsymbol{u}_0 \,.\, \boldsymbol{k} = 0, \quad p_0 = p_\infty(z), \quad \rho_0 = \rho_\infty(z), \quad \frac{\mathrm{d}p_\infty}{\mathrm{d}z} + Bo\,\rho_\infty = 0\,. \tag{7.110a}$$

For $\boldsymbol{v}_0 = (u_0, v_0)$, p_2 and ρ_2, we obtain the model problem

$$St\frac{\partial \boldsymbol{v}_0}{\partial t} + (\boldsymbol{v}_0 \,.\, \nabla)\boldsymbol{v}_0 + \frac{1}{\gamma}\boldsymbol{D}p_2 = 0\,, \tag{7.110b}$$

$$\boldsymbol{D} \,.\, \boldsymbol{v}_0 = 0\,, \tag{7.110c}$$

$$\frac{\partial p_2}{\partial z} + Bo\,\rho_2 = 0\,, \tag{7.110d}$$

where $\nabla = \boldsymbol{D} + (\partial/\partial z)\boldsymbol{k}$, and with the conditions

$$\boldsymbol{v}_0 \,.\, \boldsymbol{D}h = 0 \quad \text{on } z = h(x, y), \quad x^2 + y^2 \to \infty: \quad \boldsymbol{v}_0 \to \boldsymbol{U}_\infty(z), \quad p_2 \to 0, \quad \rho_2 \to 0\,. \tag{7.111}$$

The above model problem (7.110) and (7.111) is almost identical to the model problem considered by Drazin (1961), the only difference being in the conditions (7.111) for the second-order pressure and density. In both cases we have a two-dimensional flow along every plane $z =$ const, contouring without slip the intersection of the relief with this same plane $z =$ const. The solution is obviously steady, and Drazin (1961) and Brighton (1978) developed a closed-form solution for a irrotational flow. We observe again that, the vorticity cannot be generated by the above model and may persist only in some unsteady solution. The above two authors also considered higher-order approximations and Drazin (1961) recognized the necessity of using local expansions for $z - z_{\mathrm{M}} = O(\delta)$, where z_{M} is the altitude of the top of the relief, and also when $r = (x^2 + y^2)^{1/2} = O(1/\delta)$. The small parameter δ used by Drazin (1961) is such that the relief is given by the dimensionless equation $z = (1/\delta)\eta(\delta x, \delta y)$, and in the expansion (7.109) this small parameter δ is used in place of M^2. The situation near the top and the problem of the behaviour near infinity in the horizontal direction are both quite analogous to our earlier discussion in Sect. 7.3.1.

The Steady Two-Dimensional Eulerian Case

In dimensional form, the steady two-dimensional Euler equations can be written in the following form, for the velocity components u and w, the pressure p and the density ρ, all assumed depend on the coordinates x and z:

$$\rho\left[u\frac{\partial u}{\partial x}+w\frac{\partial u}{\partial z}\right]+\frac{\partial p}{\partial x}=0\,, \tag{7.112a}$$

$$\rho\left[u\frac{\partial w}{\partial x}+w\frac{\partial w}{\partial z}\right]+\frac{\partial p}{\partial z}+g\rho=0\,, \tag{7.112b}$$

$$\frac{\partial(\rho u)}{\partial x}+\frac{\partial(\rho w)}{\partial z}=0\,, \tag{7.112c}$$

$$\left(u\frac{\partial}{\partial x}+w\frac{\partial}{\partial z}\right)\left(\frac{p}{\rho^{\gamma}}\right)=0\,. \tag{7.112d}$$

We consider a mountain, and write the equation in the plane of the gravitational field as

$$z=h^0\eta\left(\frac{x}{l^0}\right),\quad -\frac{1}{2}\le\frac{x}{l^0}\le+\frac{l}{2}\,, \tag{7.113}$$

and the slip condition gives

$$w=u\frac{\mathrm{d}\eta}{\mathrm{d}x}\quad\text{on } z=h^0\eta\left(\frac{x}{l^0}\right)\,. \tag{7.114}$$

Again, our non-viscous, non-heat-conducting fluid is a perfect gas, and the equation of state is

$$T=\frac{p}{R\rho},\quad R=C_p-C_v\quad\text{and } \gamma=C_p/C_v\,. \tag{7.115}$$

First, from (7.112c), we introduce a stream function $\psi(x,z)$ such that

$$\rho u=-\frac{\partial\psi}{\partial z},\quad \rho w=+\frac{\partial\psi}{\partial x}\,. \tag{7.116}$$

Using (7.116), we derive, in place of (7.112d), the following first integral:

$$p=\rho^{\gamma}\Pi(\psi)\,, \tag{7.117}$$

where the function $\Pi(\psi)$ is arbitrary and conservative along each streamline. Then, from the two equations (7.112a,b), we derive a Bernoulli integral,

$$\frac{1}{2}(u^2+w^2)+\frac{\gamma}{\gamma-1}\rho^{\gamma-1}\Pi(\psi)+gz=I(\psi)\,, \tag{7.118}$$

where $I(\psi)$ is a second arbitrary function, conservative along each streamline. From the two equations (7.112a,b), we derive also a relation for the vorticity ω,

$$\omega \equiv \frac{\partial u}{\partial z} - \frac{\partial w}{\partial x} = -\rho \left[\frac{\mathrm{d}I}{\mathrm{d}\psi} - \frac{1}{\gamma - 1} \frac{p}{\rho} \frac{\mathrm{d} \log \Pi}{\mathrm{d}\psi} \right] . \tag{7.119}$$

The two arbitrary functions $I(\psi)$ and $\Pi(\psi)$ are determined from the behaviour conditions in the upstream unperturbed region, where $x \to -\infty$ and where z_∞ is the altitude of the unperturbed streamline. We assume that

$$\begin{aligned} &u = U_\infty(z_\infty), \quad w = 0, \quad p = p_\infty(z_\infty), \quad \rho = \rho_\infty(z_\infty) , \\ &T = T_\infty(z_\infty) \quad \text{at } x \to -\infty . \end{aligned} \tag{7.120}$$

With (7.120), we obtain the following relation:

$$\frac{\mathrm{d}I}{\mathrm{d}\psi} - \frac{1}{\gamma - 1} \frac{p}{\rho} \frac{\mathrm{d} \log \Pi}{\mathrm{d}\psi} \equiv -\frac{1}{\rho_\infty} \left(\frac{\mathrm{d}U_\infty}{\mathrm{d}z_\infty} - \frac{\gamma R}{\gamma - 1} \frac{\chi_\infty}{U_\infty} \right) , \tag{7.121a}$$

where

$$\chi_\infty \equiv N^2(z_\infty)(T - T_\infty) , \tag{7.121b}$$

and

$$N^2(z_\infty) \equiv \frac{1}{T_\infty} \left(\frac{\mathrm{d}T_\infty}{\mathrm{d}z_\infty} + g \frac{\gamma - 1}{\gamma R} \right) > 0 . \tag{7.121c}$$

Since the functions $I(\psi)$ and $\Pi(\psi)$ are both conservative along each streamline, we can determine the temperature T from the Bernoulli integral (7.118) in the following form:

$$T = T_\infty - \frac{\gamma - 1}{\gamma R} \left\{ \frac{1}{2} \left[u^2 + w^2 - U_\infty^2(z_\infty) \right] + g(z - z_\infty) \right\} . \tag{7.122a}$$

For the density, we obtain the relation

$$\rho = \rho_\infty \left(1 + \frac{T - T_\infty}{T_\infty} \right)^{1/\gamma - 1} . \tag{7.122b}$$

Finally, from the above results, we obtain an equation for $\psi(x, z)$ in an 'awkward' form (Zeytounian, 1979), but we shall not write this equation here. For our present purpose, we introduce (in place of ψ) the vertical deviation $\delta(x, z)$ of a streamline in the perturbed flow over a mountain relative of the unperturbed altitude at infinity upstream; namely, we write

$$z = z_\infty(\psi) + \delta(x, z) \tag{7.123a}$$

for the altitude of a perturbed streamline, and

$$\frac{\partial \psi}{\partial x} = -\rho_\infty U_\infty \frac{\partial \delta}{\partial x}, \quad \frac{\partial \psi}{\partial z} = \rho_\infty U_\infty \left[1 - \frac{\partial \delta}{\partial z} \right] . \tag{7.123b}$$

In place of the slip condition (7.113), we obtain

$$\delta\left(x, h^0 \eta \left(\frac{x}{l^0}\right)\right) = h^0 \eta \left(\frac{x}{l^0}\right) , \tag{7.124}$$

and we observe that at infinity upstream we have the relation

$$\psi = -\int_0^{z_\infty} \rho_\infty U_\infty \mathrm{d}z \equiv \psi(z_\infty) \Leftrightarrow z_\infty^{-1}(\psi) , \tag{7.125}$$

where $z_\infty^{-1}(\psi)$ is the inverse function of $\psi(z_\infty)$.

For the function $\delta(x,z)$, we derive the following second-order partial differential equation:

$$\begin{aligned}\frac{\partial^2\delta}{\partial x^2} &+ \frac{\partial^2\delta}{\partial z^2} + \left(\frac{\rho}{\rho_\infty}\right)^2 \frac{g}{U_\infty^2} N^2(z_\infty)\delta \\ &= -\frac{1}{2}\left(\frac{\rho}{\rho_\infty}\right)^2 \frac{\mathrm{d}}{\mathrm{d}z_\infty}\left\{\log\left[U_\infty^2 \exp\left(-\frac{S_\infty}{C_p}\right)\right]\right\} \\ &+ \frac{1}{2}\frac{\mathrm{d}}{\mathrm{d}z_\infty}\left\{\log\left[\rho_\infty^2 U_\infty^2 \exp\left(-\frac{S_\infty}{C_p}\right)\right]\right\}\left[\left(\frac{\partial\delta}{\partial x}\right)^2 + \left(\frac{\partial\delta}{\partial z}\right)^2\right. \\ &\left. +1 - 2\frac{\partial\delta}{\partial z}\right] + \left(\frac{\partial\log\rho}{\partial x}\right)\frac{\partial\delta}{\partial x} + \left(\frac{\partial\log\rho}{\partial z}\right)\left(\frac{\partial\delta}{\partial z} - 1\right) , \end{aligned} \tag{7.126}$$

$$\text{where} \quad S_\infty \equiv C_v \log\left(\frac{p_\infty}{\rho_\infty^\gamma}\right) .$$

For this equation (7.126), with (7.124), we have also the following three conditions:

$$\delta(x=-\infty, z_\infty) = 0, \quad \delta(x, z=H_\infty) = 0, \quad |\delta(x=+\infty, z)| < \infty , \tag{7.127}$$

where H_∞ is the altitude of the upper level (for instance, the tropopause, assumed to be a flat, horizontal plane), where the streamlines are undeflected.

The last condition of (7.127) is the only possible physical one, because of the lee-waves regime downstream of the mountain. However, in (7.126), we have also, as an unknown function, the density ρ, and consequently we must return to the two relations (7.122), which we transform to an equation for ρ in which δ is present:

$$\begin{aligned}\left(\frac{\rho}{\rho_\infty}\right)^{\gamma-1} &= \left\{1 + \frac{U_\infty^2}{2C_\mathrm{p}T_\infty}\left[\frac{\rho_\infty}{\rho}\right]^2\left[\left(\frac{\partial\delta}{\partial x}\right)^2 + \left(\frac{\partial\delta}{\partial z}\right)^2 - 2\frac{\partial\delta}{\partial z} + 1\right]\right. \\ &\left. + \frac{U_\infty^2}{2C_\mathrm{p}T_\infty}\left[\frac{2g}{U_\infty^2}\delta - 1\right]\right\} ,\end{aligned} \tag{7.128a}$$

and we have the following condition at infinity upstream:

$$\rho \to \rho_\infty(z_\infty) \quad \text{when } x \to -\infty \text{ and } \delta(x=-\infty, z_\infty) = 0 . \tag{7.128b}$$

The above problem (7.126)–(7.128) with (7.124), for the two functions δ and ρ is strongly nonlinear. Below, we consider a simplified case when:

$$U_\infty \equiv U^0_\infty = \text{const} \quad \text{and} \quad -\frac{\mathrm{d}T_\infty}{\mathrm{d}z_\infty} \equiv \Gamma^0_\infty = \text{const} \, , \tag{7.129a}$$

such that

$$T_\infty(z_\infty) = T_\infty(0)\left[1 - \frac{\Gamma^0_\infty}{T_\infty(0)} z_\infty\right] , \tag{7.129b}$$

and this linear disribution for $T_\infty(z_\infty)$ is very well justified in the usual meteorological situation in the troposphere when the lee-waves regime is considered, H_∞ being the height of the whole troposphere. The parameter

$$\mu^0_\infty = \frac{\Gamma^0_\infty H_\infty}{T_\infty(0)} \tag{7.129c}$$

is a reference parameter for the temperature profile at infinity upstream, and

$$\frac{T_\infty(z_\infty)}{T_\infty(0)} \equiv \Theta(\zeta_\infty) = 1 - \mu^0_\infty \zeta_\infty \, , \tag{7.129d}$$

with

$$\zeta_\infty \equiv z_\infty / H_\infty \, .$$

Now we introduce the non-dimensional density perturbation

$$\varpi = \frac{(\rho - \rho_\infty)}{\rho_\infty} \tag{7.130}$$

and the non-dimensional vertical displacement of the streamline $\Delta = \delta/h^0$, and since far ahead of the mountain there is a uniform flow with velocity components ($U^0_\infty = \text{const}, 0$), we write in place of (7.123a)

$$\zeta_\infty = Bo\left[\zeta - \frac{1}{\nu^0}\Delta\right] , \tag{7.131}$$

where z is reduced by the vertical length scale H_0 characterizing the lee waves process and z_∞ is reduced by $H_\infty \equiv RT_\infty(0)/g$. In this case we have two ratios,

$$\nu^0 = \frac{H_0}{h^0}, \quad Bo = \frac{H_0}{H_\infty} , \tag{7.132}$$

where ν^0 is the 'linearization' parameter, and in place of (7.124) we have the following dimensionless slip condition:

$$\Delta\left(\xi, \zeta = \frac{1}{\nu^0}\eta(\xi)\right) = \eta(\xi), \quad \text{where } \xi = \frac{x}{l^0} \, . \tag{7.133}$$

The second condition of (7.127) gives

$$\Delta(\xi, \zeta = 1/Bo) = 0 \, , \tag{7.134}$$

and the 'long-wave' approximation parameter is

$$\varepsilon = \frac{H_0}{l^0} \, . \tag{7.135}$$

For ϖ , defined by (7.130), we obtain the following relation from (7.128a):

$$(1+\varpi)^{\gamma-1} = 1 - \frac{1}{\Theta(\zeta,\Delta)}\left\{\frac{1}{2}(\gamma-1)\left(\frac{M_\infty^0}{\nu_0}\right)^2 \frac{1}{(1+\varpi)^2}\left[\varepsilon^2\left(\frac{\partial\Delta}{\partial\xi}\right)^2 + \left(\frac{\partial\Delta}{\partial\zeta}\right)^2 - 2\nu_0\frac{\partial\Delta}{\partial\zeta} + \nu_0^2\right] + \frac{\gamma-1}{\gamma}\frac{Bo}{\nu_0}\Delta - \frac{1}{2}(\gamma-1)(M_\infty^0)^2\right\} , \tag{7.136a}$$

where

$$\Theta(\zeta,\Delta) = 1 - Bo\,\mu_\infty^0\left[\zeta - \frac{1}{\nu_0}\Delta\right] . \tag{7.137}$$

Finally, for the function Δ, we obtain the following dimensionless equation from (7.126), when we take into account (7.129):

$$\Theta(\zeta,\Delta)\left\{\varepsilon^2\frac{\partial^2\Delta}{\partial\xi^2} + \frac{\partial^2\Delta}{\partial\zeta^2} - \frac{1}{1+\varpi}\left[\varepsilon^2\frac{\partial\Delta}{\partial\xi}\frac{\partial\varpi}{\partial\xi} + \frac{\partial\Delta}{\partial\zeta}\frac{\partial\varpi}{\partial\zeta} - \nu_0\frac{\partial\varpi}{\partial\zeta}\right]\right\} + \frac{Bo^2}{\gamma M_\infty^{0\,2}}S_0(1+\varpi)^2\Delta$$
$$= \frac{\nu_0}{2}Bo\,S_0\left\{\varpi(2+\varpi) - \left[\varepsilon^2\left(\frac{\partial\Delta}{\partial\xi}\right)^2 + \left(\frac{\partial\Delta}{\partial\zeta}\right)^2 - 2\nu_0\frac{\partial\Delta}{\partial\zeta}\right]\right\} , \tag{7.136b}$$

where the 'hydrostatic stability parameter'

$$S_0 = \frac{\gamma-1}{\gamma} - \mu_\infty^0 \tag{7.138}$$

characterizes the stratification of the unperturbed flow at infinity upstream. For the usual meteorogical values in the troposphere for dry air, we have

$$\gamma = 1.4, \quad T_\infty(0) = 288°\mathrm{C}, \quad \mu_\infty^0 = \frac{R}{g}\Gamma_\infty^0 \approx 0.19037, \quad \frac{\gamma-1}{\gamma} - \mu_\infty^0 \approx 0.09534 \, .$$

On the other hand, $[\gamma R T_\infty(0)]^{1/2} \approx 340.17\,\mathrm{m/s}$, and if $34\,\mathrm{m/s}, \geq U_\infty^0 \geq 10\,\mathrm{m/s}$, then we obtain $0.03 \leq M_\infty^0 \leq 0.1$. We observe that, for to the unknown function $\Delta(\xi,\zeta)$ solution of the equation (7.136b), we have the following boundary conditions for the lee waves problem over and downstream of the mountain:

$$\Delta\left(\xi, \frac{1}{\nu_0}\eta(\xi)\right) = \eta(\xi), \quad -\frac{1}{2} \leq \xi \leq +\frac{1}{2} \tag{7.139a}$$

$$\Delta(\xi=-\infty,\zeta_\infty) = 0, \quad \Delta\left(\xi,\zeta=\frac{1}{Bo}\right) = 0, \quad |\Delta(\xi=+\infty,\zeta)| < \infty \, . \tag{7.139b}$$

In (7.136b), the parameter

$$K_o^2 \equiv S_0\left(\frac{Bo^2}{\gamma M_\infty^{0\,2}}\right) = g\frac{\Gamma_A - \Gamma_\infty^0}{T_\infty(0)}\left(\frac{H_0}{U_\infty^0}\right)^2 , \tag{7.140}$$

where $\Gamma_A = (g/R)[(\gamma-1)/\gamma]$ is the dry adiabatic temperature gradient, and plays a fundamental role when $M_\infty^0 = U_\infty^0/[\gamma R T_\infty(0)]^{1/2} \ll 1$ but $K_o^2 = O(1)$.

Four Limiting Cases

We consider the following cases:

(i) The first case, which is linked to the *deep convection* considered in Sect. 6.4.3, is valid in the whole troposphere when $Bo = O(1)$, but with the following similarity rule between the low Mach number $M_\infty^0 \ll 1$ and $S_0 \gg 1$, i.e. assumed large:

$$S_0 = S^* M_\infty^{0\,2} \Leftrightarrow \Gamma_\infty^0 \approx \Gamma_A - \frac{g}{R}S^* M_\infty^{0\ 2}, S^* = O(1) . \tag{7.141a}$$

(ii) The second case is the *Boussinesq case* when the Boussinesq and Mach numbers are both small parameters, i.e. $Bo \ll 1$ and $M_\infty^0 \ll 1$, but when $S_0 = O(1)$, such that

$$\frac{Bo}{M_\infty^0} = B^* = O(1) \quad \Leftrightarrow H_0 \approx \frac{U_0}{g}\left[\frac{RT_\infty(0)}{\gamma}\right]^{1/2} . \tag{7.141b}$$

(iii) The third case is the *isochoric case* when $\gamma \gg 1$, i.e. it is assumed large, and the Mach number is low, i.e. $M_\infty^0 \ll 1$ such that $S_0 = O(1)$ and $Bo = O(1)$, but where

$$\gamma M_\infty^{0\,2} = M^* = O(1) \Leftrightarrow U_\infty^0 \approx [RT_\infty(0)]^{1/2} . \tag{7.141c}$$

(iv) The fourth case is related to a *very thin atmospheric layer*, when $S_0\, Bo \approx 1, Bo \ll 1$, and $\gamma M_\infty^{0\,2} \ll 1$, such that

$$\frac{Bo}{\gamma M_\infty^{0\,2}} \approx 1 \quad \Leftrightarrow H_0 \approx \frac{U_\infty^{0\,2}}{g} \quad \text{and } \Gamma_A - \Gamma_\infty^0 = \frac{U_\infty^{0\,2}}{gT_\infty(0)} . \tag{7.141d}$$

For each of the above cases, we can derive a low-Mach-number model problem from the full nonlinear problem (7.136b), (7.139) for the function $\Delta(\xi,\zeta)$, together with the relations (7.136a), (7.137) and (7.138) for ϖ, Θ and S_0, respectively. We observe that in case (i), if we assume also that $\nu_0 \gg 1$ or, according to the first relation of (7.132), that $h^0 \ll H_0$, such that

$$M_\infty^{0\,2}\nu_0 = O(1) , \tag{7.142}$$

we can derive from (7.136) a leading-order equation for $\Delta_0(\xi,\zeta) = \lim_{M_\infty^0 \to 0} \Delta$, which is very similar to an equation considered by Dorodnitsyn (1950). The reader can find in Zeytounian (1990, pp. 324–328) some results concerning the solution of this equation with linearized boundary conditions deduced from (7.134) and (7.139). Case (iv) is also considered in Zeytounian (1990, pp. 328–330). Case (ii) (the Boussinesq case) was considered in Sect. 6.2, and case (iii) (the isochoric case) was considered in Sect. 6.4.3. Concerning Case (i), without linearization, we obtain in the leading order a coupled system of two equations for $(\Delta_0(\xi,\zeta),\varpi_0(\xi,\zeta)) = \lim_{M^0_\infty \to 0}(\Delta,\varpi)$, namely

$$(1+\varpi_0)^{\gamma-1} = \frac{1 - Bo((\gamma-1)/\gamma)\zeta}{1 - Bo((\gamma-1)/\gamma)[\zeta - (1/\nu_0)\Delta_0]}\,, \tag{7.143a}$$

$$\begin{aligned}&\left\{1 - \frac{\gamma-1}{\gamma}Bo\left[\zeta - \frac{1}{\nu_0}\Delta_0\right]\right\}\left\{\varepsilon^2\frac{\partial^2\Delta_0}{\partial\xi^2} + \frac{\partial^2\Delta_0}{\partial\zeta^2}\right.\\ &\left. - \frac{1}{1+\varpi_0}\left[\varepsilon^2\frac{\partial\Delta_0}{\partial\xi}\frac{\partial\varpi_0}{\partial\xi} + \frac{\partial\Delta_0}{\partial\zeta}\frac{\partial\varpi_0}{\partial\zeta} - \nu_0\frac{\partial\varpi_0}{\partial\zeta}\right]\right\}\\ &+ \frac{Bo^2}{\gamma}S^*(1+\varpi_0)^2\Delta_0 = 0\,.\end{aligned} \tag{7.143b}$$

Just as in the linear case, with the assumption (7.142), we can derive from the above system (7.143) a single equation similar to the equation analysed by Dorodnitsyn (1950), Namely

$$\varepsilon^2\frac{\partial^2\sigma_0}{\partial\xi^2} + \frac{\partial^2\sigma_0}{\partial\zeta^2} + D_0(Bo\,\zeta)\sigma_0 = 0\,, \tag{7.144a}$$

where, in place of Δ_0, we have the new unknown function

$$\sigma_0 = \left[1 - Bo\frac{\gamma-1)}{\gamma}\zeta\right]^{1/2(\gamma-1)}\Delta_0\,, \tag{7.144b}$$

and the coefficient $D_0(Bo\,\zeta)$ is given by

$$D_0(Bo\,\zeta) = \frac{(Bo/\gamma)S^*}{[1 - Bo((\gamma-1)/\gamma)\zeta]} - \frac{(Bo/2\gamma)^2(2\gamma-1)}{[1 - Bo((\gamma-1)/\gamma)\zeta]^2}\,. \tag{7.144c}$$

7.4 The Howarth Initial-Stage Equation for $M \ll 1$ and $Re \gg 1$

In a numerical computation, when the motion of a body, starting from the rest, is considered in a compressible, heat-conducting viscous fluid medium, the problem is very stiff close to the initial time, mainly because of the emergence of short acoustic waves when the Mach number M is small. In the case of a weak viscosity in the problem, we have also a second small parameter, the inverse of a large Reynolds number, $1/Re$. Our purpose below is to derive, from the full NSF equations, a system of model equations that are consistent simultaneously close to $t = 0$ and near the wall $n = 0$ of a bounded, rigid, moving body, where n is the coordinate normal to the wall; this coordinate is a measure of the distance of a point in the external flow around the body from the wall Σ.

7.4.1 The Dominant NSF Equations Close to the Initial Time and Near the Wall

We work with dimensionless quantities and assume that the fluid is a thermally perfect gas. The movement of the wall Σ is defined by the vector $\boldsymbol{P}(t, \boldsymbol{s})$ and, in the vicinity of the wall $n = 0$, a point in the flow is represented by

$$\boldsymbol{M}(t, \boldsymbol{s}, n) = \boldsymbol{P}(t, \boldsymbol{s}) + n\boldsymbol{N}(t, \boldsymbol{s}) \, , \tag{7.145}$$

where $\boldsymbol{N}$ denotes the unit vector normal to Σ directed towards the fluid, and we write, for the velocity vector $\boldsymbol{u}$ of the exterior flow,

$$\boldsymbol{u} = \boldsymbol{u}_P + \boldsymbol{u}_\perp + w\boldsymbol{N}, \quad \boldsymbol{u}_\perp \,.\, \boldsymbol{N} = 0 \, , \tag{7.146}$$

where $\boldsymbol{u}_P = \partial P/\partial t$. The no-slip condition gives

$$\boldsymbol{u}_\perp = 0, w = 0 \quad \text{on } \Sigma \quad (n = 0) \, , \tag{7.147a}$$

On the wall $\Sigma(n = 0)$, we write the second condition, for the temperature T, as

$$\frac{\partial T}{\partial n} = 0 \quad \text{on } \Sigma \quad (n = 0) \, . \tag{7.147b}$$

The subscript '$\perp$' characterizes the projection of a vector on the tangent plane to $\Sigma(n = 0)$; $\boldsymbol{s} = (s_1, s_2)$, where (s_1, s_2) is a two-dimensional curvilinear system of coordinates on Σ. Close to the initial time $t = 0$ and near the wall, where $n = 0$, when $M \ll 1$ and $Re \gg 1$, it is necessary to introduce, in place of t and n, a new short time τ and a new coordinate η normal to Σ. First, we write a similarity rule between M and Re,

$$\frac{1}{Re} = M^a, \quad a > 0 \, , \tag{7.148}$$

and we assume that

$$\tau = \frac{t}{M^b} \quad \text{and } \eta = \frac{n}{M^c}, \quad b > 0, \quad c > 0 \, . \tag{7.149}$$

With (7.148) and (7.149), we derive the following dominant continuity equation, where ρ is the density, from the dimensionless NFS equations:

$$\frac{\partial \rho}{\partial \tau} + M^{b-c}\frac{\partial(\rho w)}{\partial \eta} + O(M^b) = 0 \, . \tag{7.150a}$$

For the horizontal velocity $\boldsymbol{u}_\perp$, we obtain the following dominant equation of motion:

$$\rho\left[\frac{\partial(\boldsymbol{u}_P+\boldsymbol{u}_\perp)}{\partial\tau}\right]_\perp + M^{b-c}\rho w\frac{\partial \boldsymbol{u}_\perp}{\partial\eta} - M^{b-2c+a}\frac{\partial}{\partial\eta}\left(\mu\frac{\partial \boldsymbol{u}_\perp}{\partial\eta}\right) + O(M^b) = 0\,. \tag{7.150b}$$

For w, we have

$$\rho\frac{\partial w}{\partial\tau} + M^{b-c}\rho w\frac{\partial w}{\partial\eta} + M^{b-c-2}\frac{\partial(p/\gamma)}{\partial\eta} - M^{b-2c+a}\frac{\partial}{\partial\eta}\left[(2\mu+\lambda)\frac{\partial w}{\partial\eta}\right] + O(M^b) = 0\,, \tag{7.150c}$$

where p is the pressure, such that $p=\rho T$. Finally, for the temperature T, we derive the dominant equation

$$\rho\frac{\partial T}{\partial\tau} + M^{b-c}\left[\rho w\frac{\partial T}{\partial\eta} + (\gamma-1)p\frac{\partial w}{\partial\eta}\right] - M^{b-2c+a}\frac{\gamma}{Pr}\frac{\partial}{\partial\eta}\left(k\frac{\partial T}{\partial\eta}\right) - M^{b-2c+a+2}\gamma(\gamma-1)\left[\mu\left|\frac{\partial \boldsymbol{u}_\perp}{\partial\eta}\right|^2 + (2\mu+\lambda)\left|\frac{\partial w}{\partial\eta}\right|^2\right] + O(M^b) = 0\,. \tag{7.150d}$$

We observe that the horizontal coordinates (s_1, s_2) do not play any role in obtaining the limiting model equations, and therefore we have not mentioned them. These coordinates play a role only as parameters in the condition (7.147b), via $\boldsymbol{u}_P$ and also via the initial conditions. Everything considered, if we want to rederive the approximate system of leading-order equations given by Howarth (1951, p. 160) from the above dominant equations (7.150), then it is necessary to assume also that the following expansions are valid for our unknown functions:

$$\boldsymbol{u}\perp = \boldsymbol{u}_{\perp\mathrm{H}} + \ldots,\ w = M[w_\mathrm{H} + \ldots],\ (p,\rho,T) = (1,1,1) + M^2(p_\mathrm{H},\rho_\mathrm{H},T_\mathrm{H}) + \ldots, \tag{7.151a}$$

$$(\mu,\lambda,k) = (1,1,1) + M^2(\mu_\mathrm{H},\lambda_\mathrm{H},k_\mathrm{H}) + \ldots\,. \tag{7.151b}$$

With the expansions (7.151), we obtain two relations for the positive scalars b and c from (7.150a–c):

$$b-c+1=2,\quad b-2c+a=0 \quad\Rightarrow\quad a=c-1\,. \tag{7.152a}$$

If we now want the expansions (7.151) to be relative to M^2 as small parameter, then it is necessary also that

$$b=4 \quad\Rightarrow\quad c=3 \quad \text{and } a=2\,. \tag{7.152b}$$

Finally, with an error of M^2, we obtain from (7.150a–d), when

$$M \to 0 \quad \text{with } \tau \text{ and } \eta \text{ fixed:} \tag{7.153}$$

the following linear model equations 'à la Howarth':

$$\frac{\partial \rho_{\rm H}}{\partial \tau} + \frac{\partial w_{\rm H}}{\partial \eta} = 0 \,, \tag{7.154a}$$

$$\left[\frac{\partial (\boldsymbol{u}_P + \boldsymbol{u}_{\perp \rm H})}{\partial \tau}\right]_\perp = \frac{\partial^2 \boldsymbol{u}_{\perp \rm H}}{\partial \eta^2} \,, \tag{7.154b}$$

$$\frac{\partial w_{\rm H}}{\partial \tau} + \frac{\partial (p_{\rm H}/\gamma)}{\partial \eta} = \frac{4}{3}\frac{\partial^2 w_{\rm H}}{\partial \eta^2} \,, \tag{7.154c}$$

$$\frac{\partial T_{\rm H}}{\partial \tau} - \frac{\gamma - 1}{\gamma}\frac{\partial p_{\rm H}}{\partial \tau} = \frac{1}{Pr}\frac{\partial^2 T_{\rm H}}{\partial \eta^2} + (\gamma - 1)\left|\frac{\partial \boldsymbol{u}_{\perp \rm H}}{\partial \eta}\right|^2 \,, \tag{7.154d}$$

$$p_{\rm H} = \rho_{\rm H} + T_{\rm H} \,. \tag{7.154e}$$

The above system of model equations (7.154) was first used by Howarth (1951) in his investigations related to the Rayleigh problem for a compressible fluid. In this case, the wall Σ is an infinite plane, immersed in viscous, heat-conducting, compressible fluid, and the motion is set up when this horizontal plane is set moving instantaneously from rest in a direction parallel to itself and thereafter maintened in uniform motion. As a consequence, we assume that

$$\boldsymbol{u}_P \,.\, \boldsymbol{N} = 0 \quad \text{and } \boldsymbol{u}_P = I(t)\boldsymbol{e} \,, \tag{7.155}$$

where $I(t)$ is the 'unit' Heaviside function, and $\boldsymbol{e}$ is a unit vector tangent to the infinite horizontal plane. In fact, if we do not consider the vicinity of a rounded nose of a bounded body (where another model works) then, because the derivatives of the horizontal coordinates (s_1, s_2) are absent, we can assume that in the system (τ, η), locally, the wall Σ is a flat plane normal to $\boldsymbol{N}$. Using $\boldsymbol{u}_P + \boldsymbol{u}_{\perp H} = u_H \boldsymbol{e}$, we obtain the following, for the horizontal component of the velocity, u_H:

$$\frac{\partial u_{\rm H}}{\partial \tau} = \frac{\partial^2 u_{\rm H}}{\partial \eta^2} \quad \Rightarrow u_{\rm H} = \frac{2}{\sqrt{\pi}}\int_{\eta/2\sqrt{\tau}}^{\infty} \exp(-\eta'^2)\, d\eta \,. \tag{7.156}$$

7.4.2 The Results of Howarth (1951) and Hanin (1960)

We obtain the following linear problem for $w_{\rm H}, p_{\rm H}, \rho_{\rm H}$ and $T_{\rm H}$:

$$\begin{gathered}\frac{\partial \rho_{\rm H}}{\partial \tau} + \frac{\partial w_{\rm H}}{\partial \eta} = 0, \\ \frac{\partial w_{\rm H}}{\partial \tau} + \frac{1}{\gamma}\frac{\partial p_{\rm H}}{\partial \eta} = \frac{4}{3}\frac{\partial^2 w_{\rm H}}{\partial \eta^2}, \\ \frac{\partial T_{\rm H}}{\partial \tau} - \frac{\gamma - 1}{\gamma}\frac{\partial p_{\rm H}}{\partial \tau} = \frac{1}{Pr}\frac{\partial^2 T_{\rm H}}{\partial \eta^2} + \frac{\gamma - 1}{\pi \tau}\exp\left(-\frac{\eta^2}{2\tau}\right),\end{gathered} \tag{7.157a}$$

with $p_{\mathrm{H}} = T_{\mathrm{H}} + \rho_{\mathrm{H}}$. When we take into account (7.156), the conditions are:

$$w_{\mathrm{H}} = 0 \text{ and } \frac{\partial T_{\mathrm{H}}}{\partial \eta} = 0 \quad \text{for } \eta = 0 ,$$
$$w_{\mathrm{H}} = 0, T_{\mathrm{H}} = 0, p_{\mathrm{H}} = 0 \text{ and } \rho_{\mathrm{H}} = 0, \quad \text{for } \tau = 0 \text{ and } \eta \uparrow \infty . \quad (7.157b)$$

This problem (7.157a), with (7.157b) is analysed in Howarth (1951). In particular, Howarth derives the following third-order equation for $p_{\mathrm{H}}(\tau, \zeta)$ when $Pr = 3/4$:

$$\frac{4\gamma}{3} \frac{\partial^3 p_{\mathrm{H}}}{\partial \eta^2 \, \partial \tau} + \frac{\partial^2 p_{\mathrm{H}}}{\partial \eta^2} - \frac{\partial^2 p_{\mathrm{H}}}{\partial \tau^2} = -\frac{\gamma(\gamma-1)}{\pi \tau} \frac{\partial}{\partial \tau} \left[\exp\left(-\frac{\eta^2}{2\tau} \right) \right] . \quad (7.158a)$$

The choice of $Pr = 3/4$ strongly simplifies the equation for p_{H}. For the hyperbolic equation (7.158a), we have the following homogeneous conditions:

$$p_{\mathrm{H}} = \frac{\partial p_{\mathrm{H}}}{\partial \tau} = 0 \quad \text{for } \tau = 0 \ (\eta > 0) \text{ and } \eta \uparrow \infty, \quad (7.158b)$$
$$\frac{\partial p_{\mathrm{H}}}{\partial \zeta} = 0 \quad \text{on } \eta = 0 \ (\tau > 0) . \quad (7.158c)$$

Via a Laplace transform, it is possible to obtain a solution of the above problem (7.158) for $p_{\mathrm{H}}(\tau, \eta)$. In the paper by Hanin (1960, Sect. 5), the reader can find an asymptotic solution of the equations (7.157a) with the conditions (7.158b,c), for large times. The outstanding property of the solution is the transition of the initial motion to the large-time flow. According to Hanin (1960, p. 197), in the *initial stage* the normal velocity and the variation of density are still small, and the dissipation in the main motion raises both the temperature and the pressure equally: they decrease from a constant value at the plate to zero in a narrow but expanding region. During the *transition period*, the character of the flow is changed by the appearance of a wave-like motion. The pressure variation is transformed into a progressing pulse, and a similar normal-velocity pulse builds up. A density disturbance is developed, consisting of a condensation wave moving forward and a rarefaction 'tail' ending at the plate. The temperature variation also exhibits the formation of a pulse ahead of the expanding boundary region. The *large-time flow* can be regarded as a superposition of two distinct flows. The *first flow* is related closely to the incompressible solution of Rayleigh's problem, but the density variation and the vertical velocity do not vanish, since the equation of state requires that the sum of the density and the temperature is equal to zero, from which the vertical velocity is then determined by continuity. The *second flow* describes, on the other hand, the propagation of an aperiodic, isentropic acoustic wave in a compressible fluid, modified by effects of viscosity and heat conduction (dissipation terms). Furthermore, these terms asymptotically satisfy the equations of motion with the dissipation terms omitted. Dissipation thus determines the wave indirectly by producing the generating disturbances at the plate, and this wave produces variations of all the relevant quantities. Finally, these variations form a single propagating pulse and at large times the peak of this pulse moves at a speed approaching the speed of sound; the peak amplitude decreases as $1/\tau^{1/4}$, and the thickness of the wave region increases as $1/\tau^{1/2}$.

7.5 Nearly Incompressible Hydrodynamics

The structure of the 'nearly incompressible' (NI) equations of motion for both magnetohydrodynamics and hydrodynamics, with either a polytropic or a full ideal-gas equation of state, has been described in three papers by Matthaeus and Brown (1988) and Zank and Matthaeus (1990, 1991); see also Ghosh and Matthaeus (1992). These equations generalize the equations adopted by Klainerman and Majda (1981, 1982). A common feature of these formalisms is that (according to Ghosh and Matthaeus 1992) the flow becomes nearly incompressible, in the sense that the leading-order solutions in an asymptotic hierarchy developed in powers of the reference Mach number M are solutions of the incompressible equations (but only if M tends to zero with the time–space variables fixed). Unfortunately, in combustion theory, for a time-dependent bounded domain, this is not the case. In all cases, pressure variations must scale as the Mach number squared, and specific orderings are imposed on the compressions and density variations in the initial data. Except in the case of an ideal gas with large temperature fluctuations, one finds that the leading-order density variations include important contributions that may be described as non-propagating, or 'pseudo-sound', density fluctuations. The latter are determined as a linear response to the incompressible pressure in the manner described by George, Beuther and Arnt (1984) for hydrodynamics. Authors working in the field of 'nearly incompressible hydrodynamics' use mainly the Navier–Stokes equations, inspired by the well-known book of Landau and Lifshitz (1959). In dimensionless form, the Navier–Stokes system is written as

$$\frac{\partial \rho}{\partial t} = -\nabla \cdot (\rho \boldsymbol{u}) , \tag{7.159a}$$

$$\begin{aligned}\frac{\partial \boldsymbol{u}}{\partial t} &= \boldsymbol{u} \wedge (\nabla \wedge \boldsymbol{u}) - \frac{1}{2}\nabla \left(\boldsymbol{u} \cdot \boldsymbol{u} + \frac{2}{M_{\mathrm{s}}^2(\gamma-1)}\rho^{\gamma-1}\right) \\ &\quad + \frac{\nu_{\mathrm{c}}}{\rho}\left[\nabla^2 \boldsymbol{u} + \left(\frac{1}{3} + \frac{\zeta_{\mathrm{c}}}{\nu_{\mathrm{c}}}\right)\nabla(\nabla \cdot \boldsymbol{u})\right] ,\end{aligned} \tag{7.159b}$$

and this system of two equations is closed by a dimensionless, polytropic specific relation $p = \rho^\gamma$, which eliminates the need for a separate energy or entropy equation. We note, again, that the above system (7.159), with $p = \rho^\gamma$, is physically 'admissible' only when, in (7.159b), the viscous coefficients ν_{c} and ζ_{c} are both zero (inviscid case). In numerical simulations, the

Navier–Stokes system (7.159) is initialized with constant density $\rho^0(\boldsymbol{x}) = 1$, and with random fluctuations in $\boldsymbol{u}(\boldsymbol{x})$ such that $\langle \boldsymbol{u} \,.\, \boldsymbol{u} \rangle = 1$, where the angle brackets denote a spatial average. Concerning the Mach number M_s in (7.159b), we note that the pressure scale is $p_0 = \rho_0 C_s^2 \gamma$, where C_s is the sound speed associated with the characteristic density ρ_0 and $M_s^2 = (U_0/C_s)^2$, where U_0 is a characteristic speed. Ghosh and Matthaeus (1992) define various Mach numbers, and for extremely low Mach numbers a simulation must run for several characteristic time units (and many sound-crossing times L_0/C_s, where the characteristic length scale L_0 is related to the characteristic timescale t_0 through $t_0 = L_0/U_0$) before the passage of one eddy turnover time. An important feature of the above system (7.159) is the computational robustness of these equations. In particular, the average mass $\langle \rho \rangle$ and the average total energy $\langle (1/2)\rho \boldsymbol{u} \,.\, \boldsymbol{u} \rangle + \langle [\rho^\gamma / M_s^2 \gamma(\gamma - 1)] \rangle$ are pseudo-spectral invariants of the ideal system. However, the average bulk momentum $\langle \rho \boldsymbol{u} \rangle$ is not a pseudo-spectral invariant of the ideal system. Nevertheless, the equations (7.159) are useful for the simulation of turbulent relaxation, since it can be shown that the momentum non-conservation is due to an aliasing contribution in the velocity field, which can be controlled with a sufficient amount of viscous dissipation provided that severe aliasing instabilities are not inherent in the numerical scheme itself.

We observe that various initial data are used. If the initial random velocity fluctuations are all solenoidal ($\nabla \,.\, \boldsymbol{u} = 0$) and the initial density incorporates a 'pseudo-sound' correction, namely: $\rho(\boldsymbol{x}) = 1 + \delta\rho_{\mathrm{PS}}(\boldsymbol{x})$ and $\delta\rho_{\mathrm{PS}}$ is associated with the non-propagating, incompressible pressure. Specifically, $\delta\rho_{\mathrm{PS}}$ is obtained by taking the divergence of the ideal form of (7.159b) under the assumption $\nabla \,.\, \boldsymbol{u} = 0$, solving for the incompressible pressure p_∞ through

$$\nabla^2 \mathrm{p}_\infty = -\gamma M_s^2 \nabla \,.\, (\boldsymbol{u} \,.\, \nabla \boldsymbol{u}) \,, \tag{7.160}$$

and assuming $\delta\rho_{\mathrm{PS}} = (1/\gamma) p_\infty$.

7.5.1 Hydrodynamic NI Turbulence

Obviously, the clarification of the relationship between the limiting forms of compressible turbulence and purely incompressible turbulence is an important step in advancing turbulence theory towards a full (numerical) treatment of compressible flows. In Ghosh and Matthaeus (1992), using guidelines based on advances in the mathematical structure of low-Mach-number flows, the authors of that paper described three perspectives on the nature of low-Mach-number compressible, polytropic hydrodynamics. Namely, the theory suggests that three distinct types of turbulence can occur in the low-Mach-number limit of polytropic flow: nearly incompressible flows dominated by vorticity, nearly pure acoustic turbulence dominated by compression, and flows characterizd by near-statistical equipartition of vorticity and compression. In Ghosh and Matthaeus (1992), the distinction between these kinds of turbulence was investigated by direct numerical simulation of 2D compressible hydrodynamic turbulence. More precisely, dynamical scalings of density fluctuations, examination of the ratio of transverse to longitudinal velocity fluctuations, and spectral decomposition of the fluctuations were employed to distinguish the nature of these low-Mach-number solutions.

It is important to note that a strong dependence on the initial data was observed, as well as a tendency for enhanced effects of compressibility at later times and at higher wavenumbers, as suggested by theories of nearly incompressible flows. Namely, flows with small initial density fluctuations remained nearly incompressible for the duration of the simulation, and, in agreement with NI turbulence theory, the density fluctuations remained scaled to the square of the sonic Mach number. Flows with larger initial density fluctuations became strongly compressive, and, in agreement with the theory of modally equipartitioned compressible (MEC) turbulence, the density fluctuations became linearly scaled to the Mach number (in agreement also with the model of Kraichnan 1955 for compressible hydrodynamic turbulence). This was no suggestion that the NI simulations ever evolved towards strongly compressive scalings.

For the Mach numbers considered in Ghosh and Matthaeus (1992), i.e. 0.1–0.5, and the Reynolds numbers used (initially, $Re = 250$, and for a Mach number of 0.4 we have $1/\sqrt{Re} \approx M^3$), all runs starting with a longitudinal velocity and a pseudo-sound density remained NI for the duration of the simulation, generally out to several tens of (or as many as 100) eddy turnover times. By this we mean that a global comparison of the energy decay in the compressible case showed that it remained extremely close to that for a perfectly incompressible computed solution. In addition, the scaling of the fluctuations of density, internal energy and longitudinal velocity remained in accordance with the asymptotic NI theory of Klainerman and Majda (1981, 1982). This indicates that the NI theory, originally derived for inviscid flows and shown to remain valid for a finite time, may remain an accurate description of these flows for much longer times when the turbulence decays. Presumably, this enhanced range of applicability is a consequence of the continual decrease of the effective Mach number in decaying turbulence.

The above conclusion, supported by computations, is not at all obvious, since the effect of decreasing Mach number must compete with the dynamical production of acoustic waves through higher-order effects in the asymptotic theory. On the other hand, it is not clear whether the above conclusion will obtain for higher Reynolds numbers, where smaller-scale acoustic effects may be even more important, and the decrease in Mach number may be more gradual. The results of computations described above are far from a complete (numerical) description of compressible-gas dynamics even at low Mach number and, in particular, the treatment of turbulence with the full ideal-gas equation of state at low Mach number requires a more elaborate theoretical framework (as in Zank and Matthaeus, 1990, 1991) and more demanding computations as well.

7.5.2 The Equations of NI Hydrodynamics (Non-Viscous Case with Conduction)

As is mentioned in Müller (1998), 'Employing multiple-time and space scale expansions, Zank and Matthaeus (1991) derive low-Mach number equations from the compressible Navier–Stokes equations'. Indeed, a detailed examination of the paper of Müller (1998) shows that his multiple-timescale, single-space-scale asymptotic analysis of the compressible Navier–Stokes equations has a resemblance to some equations derived in Zank and Matthaeus (1991) and also in Zank and Matthaeus (1990). Both approaches consider only equations, without any initial and boundary conditions, and it is not clear how the derived approximate NI equations can be used in the framework of a formulated fluid-dynamic problem. Now, we know, from various results obtained in Chaps. 3–6, that the theory of low-Mach-number flows is strongly dependent on the problem considered (such as external/internal aerodynamics, or atmospheric or else nonlinear acoustics), and also on the initial and boundary conditions.

Zank and Matthaeus (1990) find, by means of an asymptotic analysis, two distinct approaches to incompressibility for a low-Mach-number ideal (non-viscous) fluid, distinguished according to the relative magnitudes of the temperature, density, and pressure fluctuations. First, for heat-conduction-dominated fluids, the temperature and density fluctuations are predicted to be anticorrelated, and the classical passive scalar equation for the (fluctuations of the) temperature is recovered (see below), whereas a generalized 'pseudo-sound' relationship among the fluctuations is found for heat-conduction-modified fluids, together with a modified thermal equation (see below, also). Very pertinently, the Zank and Matthaeus (1990) write that: since the NI heat-conduction-dominated model and the pseudo-sound model lead to such dramatically different results with respect to the density and temperature correlations, it is clear that the choice of which model to apply to a given situation is most critical and requires detailed assumptions about which physical processes are dominant. Evidently, the interpretation of solar-wind data (as in George, Beuther, and Arnt, 1984) and interstellar data (Montgomery, Brown and Matthaeus 1987) favours the heat-conduction-modified, pseudo-sound picture. It is interesting to observe that, in contrast, a study of low-Mach-number gases or plasmas would reveal that the temperature and density fluctuations are anticorrelated and are associated with the heat-conduction-dominated limit. The approach of Zank and Matthaeus is to derive a modified system of fluid equations which retain the effects of compressibility (such as density fluctuations) weakly, yet contain the incompressible-fluid solutions as the leading-order, low-Mach-number solutions. Such an approach was initiated, for ideal polytropic compressible flows, by Klainerman and Majda (1981, 1982) and Majda (1984), who postulated a set of modified hydrodynamic equations and proved rigorously that their equations converge to the incompressible hydrodynamic equations with decreasing Mach number. These limiting modified (incompressible hydrodynamic) equations are called 'nearly incompressible' in Matthaeus and Brown (1988). With the usual normalizations, the following dynamical equations are considered:

$$\frac{\partial \rho}{\partial t} + \nabla \cdot (\rho \boldsymbol{u}) = 0, \quad \rho \frac{\partial \boldsymbol{u}}{\partial t} + \rho \boldsymbol{u} \cdot \nabla \boldsymbol{u} = -\frac{1}{\varepsilon^2} \nabla p, \quad \varepsilon^2 = \gamma M_s^2 \ll 1 \; , \quad (7.161)$$

where (for an ideal gas) $p = \rho T$.

Kreiss (1980) showed that if the solution is to vary on the scale of the slow time (t) alone, then it is necessary that several time derivatives (and, in particular, their values at $t = 0$) of the solution be bounded and of order 1. This procedure suppresses fast-scale variations (in particular, on the fast timescale $\tau = t/\varepsilon$) such as acoustic waves and allows the solution and its derivatives to be estimated independently of ε. The limit $\varepsilon \to 0$ can then be considered and asymptotic expansions derived. For $\partial \boldsymbol{u}/\partial t$ to be bounded independently of ε, it is necessary to choose the normalized pressure as $p = 1 + \varepsilon^2 \delta p$, and it can be shown, further, that $\partial^2 \boldsymbol{u}/\partial t^2$ is bounded if and only if $\nabla \cdot \boldsymbol{u} = 0$. Then, from the energy equation, it can be shown that the density ρ must be constant. Thus, application of Kreiss's principle (Kreiss, 1980) yields the equations of incompressible hydrodynamics directly, as constraints which eliminate all solutions which vary on fast timescales τ.

We denote by $\boldsymbol{u}^\infty$ and p^∞ the limiting solutions of (7.161) which vary on the scale of the slow time t only, i.e. solutions of the incompressible hydrodynamic equations

$$\frac{\partial \boldsymbol{u}^\infty}{\partial t} + \boldsymbol{u}^\infty \cdot \nabla \boldsymbol{u}^\infty = -\nabla p^\infty, \quad \nabla \cdot \boldsymbol{u}^\infty = 0 \; . \qquad (7.162)$$

Now, the equation of heat transfer is expressed in two forms by Zank and Matthaeus (1990):

$$\rho \left(\frac{\partial T}{\partial t} + \boldsymbol{u} \cdot \nabla T \right) - \left(\frac{\partial p}{\partial t} + \boldsymbol{u} \cdot \nabla p \right) = \frac{1}{Pr} \nabla^2 T \qquad (7.163a)$$

and

$$\rho \left(\frac{\partial T}{\partial t} + \boldsymbol{u} \cdot \nabla T \right) - \gamma p \, \nabla \cdot \boldsymbol{u} = \frac{\gamma}{Pr} \nabla^2 T \; . \qquad (7.163b)$$

Below, we consider solutions which are weakly perturbed from the slow-timescale solutions, i.e.

$$\boldsymbol{u} = \boldsymbol{u}^\infty + \varepsilon \delta \boldsymbol{u} \quad \text{and } p = 1 + \varepsilon^2 (p^\infty + p^*) \; , \qquad (7.164)$$

so that the fast-timescale modes vary only as $O(1/\varepsilon)$ at worst.

Heat-Conduction-Dominated Case

Some care should be exercised in choosing the scaling for the density and temperature fluctuations. The first choice is

$$T = T_0 + \varepsilon\,\delta T \quad \text{and } \rho = 1 + \varepsilon\,\delta\rho\,, \tag{7.165}$$

which corresponds to a fluid in which heat conduction dominates the dynamics. Together with the slow scale $\boldsymbol{x}$ (short-wavelength scale), we introduce a long-wavelength scale $\boldsymbol{\xi} = \varepsilon\boldsymbol{x}$, and consider (7.162) and (7.163a). In the absence of p^* (the contibution of the acoustic modes is neglected), we obtain

$$\frac{\partial\,\delta T}{\partial\tau} = 0 \quad \text{at } O(\varepsilon^0)\,, \tag{7.166a}$$

$$\frac{\partial\,\delta T}{\partial t} + \boldsymbol{u}^\infty\,.\,\nabla\delta T = \frac{1}{Pr}\nabla^2\delta T \quad \text{at } O(\varepsilon^1)\,, \tag{7.166b}$$

$$\boldsymbol{u}^\infty\,.\,\boldsymbol{\nabla}_\xi\delta T + \delta\boldsymbol{u}\,.\,\nabla\delta T + \delta\rho\frac{1}{Pr}\nabla^2\delta T - \frac{\partial p^\infty}{\partial t}$$

$$- \boldsymbol{u}^\infty\,.\,\nabla p^\infty = \frac{2}{Pr}\nabla\,.\,\nabla_\xi\,\delta T \quad \text{at } O(\varepsilon^2)\,, \tag{7.166c}$$

from which it can be seen that the temperature fluctuation δT, like the density, is a function of the slow time t only. Equation (7.166b) is the equation of heat transfer for nearly incompressible hydrodynamics and has been derived earlier by Zeytounian (1977) (see also Sect. 4.1.5 of the present book). In (7.166c) $\boldsymbol{\nabla}_\xi$ is the gradient operator relative to $\boldsymbol{\xi}$.

However, from (7.162) and (7.163b), we can derive also the following equation in place of (7.166b):

$$\frac{\partial\,\delta T}{\partial t} + \boldsymbol{u}^\infty\,.\,\nabla\delta T + \gamma\nabla\,.\,\delta\boldsymbol{u} = \frac{\gamma}{Pr}\nabla^2\delta T \quad \text{at } O(\varepsilon^1)\,. \tag{7.167}$$

Thus, for (7.166b) and (7.167) to be compatible, we require that the velocity fluctuations $\delta\boldsymbol{u}$ satisfy the non-solenoidal equation

$$\nabla\,.\,\delta\boldsymbol{u} = (\gamma - 1)\frac{1}{\gamma Pr}\nabla^2\delta T\,. \tag{7.168}$$

The final nearly incompressible equations come from the continuity and momentum equations (7.161). Namely, we obtain

$$\frac{\partial\,\delta\rho}{\partial t} + \boldsymbol{u}^\infty\,.\,\nabla\delta\rho + \nabla\,.\,\delta\boldsymbol{u} = 0, \quad \frac{\partial\,\delta\boldsymbol{u}}{\partial t} + \boldsymbol{u}^\infty\,.\,\nabla\delta\boldsymbol{u} + \delta\boldsymbol{u}.\nabla\boldsymbol{u}^\infty + \left(\frac{1}{\delta\rho}\right)\nabla p^\infty = 0\,. \tag{7.169}$$

Equations (7.166b)–(7.169) represent a new system of fluid equations (according to Zank and Matthaeus 1990) for NI heat-conduction-dominated hydrodynamics. On the other hand, a simple manipulation of (7.166b), (7.167) and the first equation of (7.169) yields at once the relation

$$\gamma\delta\rho = -(\gamma - 1)\delta T\,, \tag{7.170}$$

indicating that the two fluctuations are anticorrelated, which contrasts strongly with the pseudo-sound relation ($\delta p = C_s^2\delta\rho$).

Heat-Conduction-Modified Case

The implications of the analysis by Zank and Matthaeus (1990) are rather different for this case, where in place of (7.165), we have

$$T = T_0 + \varepsilon^2 \delta T \quad \text{and } \rho = 1 + \varepsilon^2 \delta\rho \,, \tag{7.171}$$

since, unlike the previous case, acoustic modifications are present in the NI equations. In this case, from (7.163), the two forms of the NI thermal-transport equation are found to be (rewriting in terms of the original variables)

$$\frac{\partial\,\delta T}{\partial t} + \boldsymbol{u}^\infty \,.\, \nabla\delta T - \frac{1}{Pr}\nabla^2\delta T - \frac{\partial p^*}{\partial t} - \boldsymbol{u}^\infty \,.\, \nabla p^* = \frac{\partial p^\infty}{\partial t} + \boldsymbol{u}^\infty \,.\, \nabla p^\infty \tag{7.172a}$$

and

$$\frac{\partial\,\delta T}{\partial t} + \boldsymbol{u}^\infty \,.\, \nabla\delta T + \frac{1}{\varepsilon}\nabla \,.\, \delta\boldsymbol{u} = \frac{\gamma}{Pr}\nabla^2\delta T \,, \tag{7.172b}$$

from which we obtain the compatibility condition

$$\frac{\partial p^*}{\partial t} - \boldsymbol{u}^\infty \,.\, \nabla p^* - (\gamma - 1)\frac{1}{Pr}\nabla^2\delta T + \frac{1}{\varepsilon}\nabla \,.\, \delta\boldsymbol{u} = -\frac{\partial p^\infty}{\partial t} - \boldsymbol{u}^\infty \,.\, \nabla p^\infty \,. \tag{7.173}$$

It can be shown that on the fast timescale and short-wavelength scale $(t, \boldsymbol{x})$, p^* satisfies an acoustic wave equation and may therefore be identified as the acoustic contribution to the total pressure. Furthermore, the compatibility condition (7.173) reveals that the incompressible-fluid fluctuations act as a source of acoustic waves (as in Lighthill, 1954). Here the model is closed by the modified momentum and continuity equations

$$\frac{\partial\,\delta\rho}{\partial t} + \boldsymbol{u}^\infty \,.\, \nabla\delta\rho + \frac{1}{\varepsilon}\nabla \,.\, \delta\boldsymbol{u} = 0 \,, \tag{7.174a}$$

$$\frac{\partial\,\delta\boldsymbol{u}}{\partial t} + \boldsymbol{u}^\infty \,.\, \nabla\delta\boldsymbol{u} + \delta\boldsymbol{u} \,.\, \nabla\boldsymbol{u}^\infty = -\frac{1}{\varepsilon}\nabla p^* \,. \tag{7.174b}$$

We observe that the equations of NI heat-conduction-modified hydrodynamics are linear about the incompressible flow, thus making them relatively tractable. It is easily seen that use of (7.172)–(7.174) yields

$$p^* + p^\infty - \delta T = -\frac{1}{\gamma}\delta T + \delta\rho \,. \tag{7.175}$$

7.5.3 The Equations of NI Hydrodynamics (Viscous case with Conduction)

In the second paper of Zank and Matthaeus (1991), a unified analysis delineating the conditions under which the equations of classical incompressible (Navier–Fourier) and compressible (NSF) Newtonian fluid dynamics are related in the absence of large-scale thermal, gravitational and field gradients is presented. Modified systems of fluid equations are derived in which the effects of compressibility are admitted only weakly in terms of the solutions of incompressible hydrodynamics (NI hydrodynamics). With the inclusion of heat conduction, it is found (again) that two distinct routes to incompressibility are possible, distinguished according to the relative magnitudes of the temperature, density and pressure fluctuations. This leads to two distinct models for thermally conducting, NI hydrodynamics: heat-fluctuation-dominated hydrodynamics (HFDH's) and heat-fluctuation-modified hydrodynamics (HFMH's).

Here we list the two systems of model equations, taken directly from Zank and Matthaeus (1991, Appendix B, p. 81).

Heat-Fluctuation-Dominated Hydrodynamic Equations

We assume

$$\boldsymbol{u} = \boldsymbol{u}^{\infty} + \varepsilon\, \delta \boldsymbol{u}, \quad p = 1 + \varepsilon^2 (p^{\infty} + p^*) \,, \tag{7.176a}$$

$$T = T_0 + \varepsilon\, \delta T \quad \text{and}\ \rho = 1 + \varepsilon\, \delta\rho \,, \tag{7.176b}$$

where $\boldsymbol{u}^{\infty}$ and p^{∞} again denote the solutions of the Navier incompressible, viscous system of two dynamical equations,

$$\frac{\partial \boldsymbol{u}^{\infty}}{\partial t} + \boldsymbol{u}^{\infty} \,.\, \nabla \boldsymbol{u}^{\infty} + \nabla p^{\infty} = \nabla^2 \boldsymbol{u}^{\infty}, \quad \nabla \,.\, \boldsymbol{u}^{\infty} = 0 \,. \tag{7.177}$$

With the above, we have the following model equations, rewriting in terms of the original variables:

$$\frac{\partial\, \delta\rho}{\partial t} + \boldsymbol{u}^{\infty} \,.\, \nabla \delta\rho + \nabla \,.\, \delta \boldsymbol{u} = 0 \,, \tag{7.178a}$$

$$\frac{\partial\, \delta \boldsymbol{u}}{\partial t} + \boldsymbol{u}^{\infty} \,.\, \nabla \delta \boldsymbol{u} + \delta \boldsymbol{u} \,.\, \nabla \boldsymbol{u}^{\infty} - \left(\frac{1}{\delta\rho}\right) \nabla p^{\infty} = \nabla^2 \delta \boldsymbol{u} + \left(\lambda + \frac{1}{3}\right) \nabla(\nabla \,.\, \delta \boldsymbol{u}) \,, \tag{7.178b}$$

$$\frac{\partial\, \delta T}{\partial t} + \boldsymbol{u}^{\infty} \,.\, \nabla \delta T = \frac{1}{Pr} \nabla^2 \delta T \,, \tag{7.178c}$$

$$\nabla \,.\, \delta \boldsymbol{u} = (\gamma - 1) \frac{1}{\gamma Pr} \nabla^2 \delta T \,. \tag{7.178d}$$

In (7.178b), the coefficient λ denotes the ratio of the bulk viscosity to the dynamic viscosity, both assumed to be constant. We observe also that

$$-\frac{\partial\, \delta\rho}{\partial t} - \boldsymbol{u}^{\infty} \,.\, \nabla \delta\rho = \frac{1}{\gamma}(\gamma - 1) \left(\frac{\partial\, \delta T}{\partial t} + \boldsymbol{u}^{\infty} \,.\, \nabla \delta T\right) \,, \tag{7.178e}$$

i.e. the density and temperature fluctuations are convected with the background turbulent flow field, and the solution for $\delta\rho$ is simply

$$-\,\delta\rho = \frac{\gamma-1}{\gamma}\delta T\,. \tag{7.178f}$$

It is obvious from (7.178f) that the passive scalar transport equation (7.178c) for the temperature fluctuations δT is equally valid for the density fluctuations $\delta\rho$, so that we may use instead

$$\frac{\partial\,\delta\rho}{\partial t} + \boldsymbol{u}^{\infty}\,.\,\nabla\delta\rho = \frac{1}{Pr}\nabla^2\delta\rho\,. \tag{7.178g}$$

The last result (7.178g) will have interesting consequences, both for investigations of turbulence in general and for studies of the solar wind, where density fluctuations are relatively easy to measure.

It should be noted that a more general system of NI equations can be derived if we include one more order in the expansion and retain the acoustic pressure p^*. In particular, such a system includes acoustic effects in a heat-fluctuation-dominated medium. We derive the following NI energy equation:

$$\frac{\partial p^*}{\partial t} + \boldsymbol{u}^{\infty}\,.\,\nabla p^* - \delta\boldsymbol{u}\,.\,\nabla\delta T - \frac{1}{\varepsilon}\left[\frac{\partial\delta T}{\partial t} + \boldsymbol{u}^{\infty}\,.\,\nabla\delta T - \frac{1}{Pr}\nabla^2\delta T\right]$$
$$= -\frac{\partial p^{\infty}}{\partial t} + \boldsymbol{u}^{\infty}\,.\,\nabla p^{\infty}\,. \tag{7.178h}$$

This equation is to be used instead of the classical thermal transport equation (7.178c).

Heat-Fluctuation-Modified Hydrodynamic Equations

In this second case, we assume:

$$\boldsymbol{u} = \boldsymbol{u}^{\infty} + \varepsilon\,\delta\boldsymbol{u}, \quad p = 1 + \varepsilon^2(p^{\infty} + p^*)\,, \tag{7.179a}$$

$$T = T_0 + \varepsilon^2\,\delta T \quad \text{and} \quad \rho = 1 + \varepsilon^2\,\delta\rho\,. \tag{7.179b}$$

With (7.179), we obtain the following system of equations:

$$\left[\frac{\partial}{\partial t} + \boldsymbol{u}^{\infty}\,.\,\nabla\right](\delta T - p^*) - \frac{1}{Pr}\nabla^2\,\delta T = \frac{\partial p^{\infty}}{\partial t} + \boldsymbol{u}^{\infty}\,.\,\nabla p^{\infty}$$
$$+\frac{1}{2}[D(\boldsymbol{u}^{\infty})]^2\,, \tag{7.180a}$$

$$\left[\frac{\partial}{\partial t} + \boldsymbol{u}^{\infty}\,.\,\nabla\right]\delta T + \frac{\gamma}{\varepsilon}\nabla\,.\,\delta\boldsymbol{u} - \frac{\gamma}{Pr}\nabla^2\,\delta T = \frac{\gamma}{2}[D(\boldsymbol{u}^{\infty})]^2\,, \tag{7.180b}$$

$$\frac{\partial p^*}{\partial t} + \boldsymbol{u}^{\infty}\,.\,\nabla p^* + \frac{\gamma}{\varepsilon}\nabla\,.\,\delta\boldsymbol{u} - \frac{\gamma-1}{Pr}\nabla^2\,\delta T$$
$$= -\frac{\partial p^{\infty}}{\partial t} + \boldsymbol{u}^{\infty}\,.\,\nabla p^{\infty} + \frac{\gamma-1}{2}[D(\boldsymbol{u}^{\infty})]^2\,, \tag{7.180c}$$

$$\frac{\partial\,\delta\rho}{\partial t} + \boldsymbol{u}^{\infty}\,.\,\nabla\,\delta\rho + \frac{1}{\varepsilon}\nabla\,.\,\delta\boldsymbol{u} = 0\ , \tag{7.180d}$$

$$\frac{\partial\,\delta\boldsymbol{u}}{\partial t} + \boldsymbol{u}^{\infty}\,.\,\nabla\,\delta\boldsymbol{u} + \delta\boldsymbol{u}\,.\,\nabla\boldsymbol{u}^{\infty}$$
$$= -\frac{1}{\varepsilon}\nabla p^* + \nabla^2\,\delta\boldsymbol{u} + \left(\lambda + \frac{1}{3}\right)\nabla(\nabla\,.\,\delta\boldsymbol{u})\,. \tag{7.180e}$$

We observe that the most important difference between the temperature fluctuations here and in the previous case is that δT is now a function of the fast time scale τ. Thus, associated with the acoustic fluctuations, there is a temperature perturbation on the same scale ($\partial\,\delta T/\partial\tau = \partial p^*/\partial\tau$). In equations (7.180a,b,c), we have $D(\boldsymbol{u}^{\infty}) = \partial u_i^{\infty}/\partial x_k + \partial u_k^{\infty}/\partial x_i$. The acoustic pressure satisfies a simple sound wave equation,

$$\frac{\partial^2 p^*}{\partial\tau^2} - \gamma\,\nabla^2 p^* = 0\,, \tag{7.181a}$$

and the high-frequency thermal fluctuations satisfy the convective–acoustic wave equation

$$\frac{\partial}{\partial\tau}\left[\frac{\partial^2\,\delta T}{\partial\tau^2} - \gamma\nabla^2\,\delta T\right] = 0\,, \tag{7.181b}$$

which indicates that high-frequency thermal fluctuations can convect with the flow (entropy and vorticity waves) as well as propagate at the speed of sound.

In Zank and Matthaeus (1991), it is anticipated that the applicability of the HFDH's and HFMH's model equations is likely to be far greater than their envisaged principal application to homogeneous turbulence and wave propagation in low-Mach-number flows.

References

C. Bailly, *Modélisation asymptotique et numérique de l'écoulement dû à des disques en rotation.* Doctoral Thesis, Université des Sciences et Technologies de Lille, order No. 1512, 1995.

H. Bénard. Rev. Génér. Sci. Pures Appl., **11** (1990), 1261.

P. M. W. Brighton. Q. J. Roy. Meteorol. Soc., **104** (1978), 289–307.

J. M. Burgers. Adv. Appl. Mech., **1** (1948), 171–199.

S. Chandrasekhar, *Hydrodynamic and Hydromagnetic Stability.* Clarendon Press, Oxford, 1981. See also Dover Publications, New York, 1981.

F. Coulouvrat. J. Acoustique, **5** (1992), 321–359.

D. G. Crighton, Acoustics as a branch of fluid mechanics. J. Fluid Mech., (1981), 261–298.

D. G. Crighton, Computational aeroacoustics for low Mach number flows. In: *Computational Aeroacoustics*, edited by J. C. Hardin and M. Y. Hussaini. Springer, New York, 1993, p. 340.

P. C. Dauby and G. Lebon. J. Fluid Mech., **329** (1996), 25–64.

A. A. Dorodnitsyn, Influence of the Earth's surface relief on airflow. Trudy Ts. Inst. Prognoz., **21**(48) (1950), 3–25.

P. G. Drazin. Tellus **13** (1961), 239–251.

P. G. Drazin and W. H. Reid, *Hydrodyamic Stability*. Cambridge University Press, Cambridge, 1981.

W. K. George, P. D. Beuther and R. E. A. Arnt, J. Fluid Mech., **148** (1984), 155.

S. Ghosh and W. H. Matthaeus. Phys. Fluids A, **4**(1) (1992), 148–164.

L. N. Gutman, *Introduction to the Nonlinear Theory of Mesoscale Meteorological Processes*. Israel Program for Scientific Translation, Jerusalem, 1972.

M. Hanin, On Rayleigh's problem for compressible fluids. Q. J. Mech. Appl. Math., **13** (1960), 184.

M. F. Hamilton and D. T. Blackstock (eds.), *Frontiers of Nonlinear Acoustics, Proceedings of the 12th International Symposium on Nonlinear Acoustics*. Elsevier Applied Science, London, 1990.

L. Howarth, Some aspects of Rayleigh's problem for a compressible fluid. Q. J. Mech. Appl. Math., **4** (1951), 157.

J. C. R. Hunt and W. H. Snyder. J. Fluid Mech., **96**(4) (1980), 671.

J. Kevorkian and J. D. Cole, *Perturbation Methods in Applied Mathematics*. Springer, New York, 1981.

S. Klainerman and A. Majda. Commun. Pure Appl. Math., **34** (1981), 481–524.

S. Klainerman and A. Majda, Compressible and incompressible fluids. Commun. Pure Appl. Math., **35** (1982), 629–651.

R. Kh. Kraichnan, On the statistical mechanics of an adiabatically compressible fluid. J. Acoust. Soc. Amer., **27** (1955), 438.

H.-O. Kreiss. Commun. Pure Appl. Math., **33** (1980), 399.

V. P. Kuznetsov. Sov. Phys. Acoust. **16** (1970), 467–470.

L. D. Landau and E. M. Lifshitz, *Fluid Mechanics*. Pergamon, Oxford, 1959.

M. J. Lighthill. Proc. Roy. Soc. Lond. A, **211** (1954), 564–587.

A. Majda, *Compressible Fluid Flow and Systems of Conservation laws in Several Space*. Springer, New York, 1984.

W. H. Matthaeus and M.R. Brown. Phys. Fluids, **31** (1988), 3634–3644.

M. B. Moffett and R. H. Mellen. J. Acoust. Soc. Amer., **61** (1977), 325–337.

D. Montgomery, M. R. Brown and W. H. Matthaeus. J. Geophys. Res., **92** (1987), 282.

B. Müller. J. Engg. Math., **34** (1998), 97–109.

D. J. Needham and J. H. Merkin, J. Fluid Mech., **184** (1987), 357–379.

S. Pavithran and L. G. Redeekopp, Stud. Appl. Math., **93** (1994), 209.

M. J. Takashima, J. Phys. Soc. Japan, **50**(8) (1981), 2745–2750. J. Phys. Soc. Japan, **50**(8) (1981), 2751–2756.

J. N. Tjøtta and S. Tjøtta, Nonlinear equation of acoustics. In: *Frontiers of Nonlinear Acoustics, Proceedings of the 12th International Symposium on Nonlinear Acoustics*. Elsevier Applied Science, London, 1990. edited by M. F. Hamilton and D. T. Blackstock. 1990, pp. 80–97.

P. J. Westervelt, Parametric acoustic array. J. Acoust. Soc. Amer., **35** (1963), 535–470.

E. A. Zabolotskaya and R. V. Khokhlov. Sov. Phys. Acoust., **15** (1969), 35–40.

G. P. Zank and W. H. Matthaeus, Nearly incompressible hydrodynamics and heat conduction. Phys. Rev. Lett., **64**(11) (1990), 1243–46.

G. P. Zank and W. H. Matthaeus. Phys. Fluids A, **3** (1991), 69–82.

G. P. Zank and W. H. Matthaeus. Phys. Fluids A, **5** (1993), 257–273.

R. Kh. Zeytounian, Criterion for filtering parasite solutions in numerical computations. Phys. Fluids, Suppl. II (1969), II-46–II-50.

R. Kh. Zeytounian, Analyse asymptotique des écoulements de fluides visqueux compressibles à faible nombre de Mach, I: Cas des fluides non pesants. USSR Comput. Math. Math. Phys. **17**(1) (1977), 175–182.

R. Kh. Zeytounian, Models for lee waves in a baroclinic compressible troposphere. Izv. Acad. Sci. USSR, Atmosph. Ocean. Phys., **15**(6) (1979), 498–507.

R. Kh. Zeytounian, Int. J. Engng. Sci. **27** (11) (1989) 1361–1366.

R. Kh. Zeytounian, *Asymptotic Modelling of Atmospheric Flows.* Springer, Heidelberg, 1990

R. Kh. Zeytounian, The Bénard–Marangoni thermocapillary instability problem: on the role of the buoyancy. Int. J. Engg. Sci., **35**(5) (1997), 455–466.

R. Kh. Zeytounian, *Asymptotic Modelling of Fluid Flow Phenomena.* Fluid Mechanics and Its Applications 64, Kluwer Academic, Dordrecht, 2002.

Epilogue

I hope that a reader who has read the preceding seven chapters will now have a good idea of why the development of a 'hyposonic flow theory' as a branch of Newtonian fluid mechanics is of interest. I think also that the curious reader will have well understood that this hyposonic theory is actually far from complete. On the one hand, many interesting problems are still open in hyposonic flows theory, and, on the other hand, some important technological, environmental and geophysical problems certainly need to be investigated in the framework of this theory. In fact, hyposonic flow theory has a fundamental, indispensable role in human terrestrial activities, and I am sure that in the next few years this theory will be a challenging branch of fluid mechanics.

As a conclusion, I wish touch on the problem of the unification of Newtonian fluid mechanics. It is well known that this is a fundamental problem in physics; this is very well discussed in Stephen Hawking's book *A Brief History of Time* (Bantam, New York, 1988). The unification of Newtonian fluid mechanics seemed to have been achieved long ago, mainly thanks to the efforts of Cauchy, Poisson, Navier, Stokes and Fourier in the 18th century, who derived the Navier–Stokes–Fourier (NSF) equations for a Newtonian fluid that is assumed to be a thermally perfect, viscous, heat conducting gas. However, the analytical theory of the Newtonian fluid dynamics is, in fact, a puzzle of very partial, approximate theories (equations), and we are actually far from having a good idea of the relation between the NSF equations and the partial, approximate equations. In particular, in the framework of university education, students have only a very 'shallow' knowledge about the NSF system of equations. Mainly, for students, fluid mechanics is a curious miscellany of various problems which do not seem to have any relation to each other. As a consequence, the 'unification' of Newtonian fluid mechanics is, in fact, a 'deconstruction' (from the French 'déconstruction', a word invented by Jacques Derrida; see *Charles Ramond, Le vocabulaire de Derrida*, Ellipses, Paris, 2001), which gives us the possibility to understand the intrinsic structure of the NSF equations and reveal the presence of a profound unity in the puzzle of the partial models of the fluid dynamics. In this context, the NSF system of equations is a closed, complete system; at least, for classical Newtonian fluids, it is an untouchable icon! A challenging approach would be to 'deconstruct' the NSF system in order to unify the puzzle of the various partial, approximate systems used in Newtonian fluid mechanics. Such a deconstruction approach would obviously be strongly linked to asymptotic modelling and, first of all, to theories related to high and low Reynolds numbers and low Mach numbers.

In this book, devoted to approximate models derived when the Mach number is a small parameter in the NSF equations, we have 'deconstructed' the complete NSF theory into a family of partial theories when the Mach number tends to zero in various space–time regions and for various flow configurations. It is well known that an analogous process is possible also for high and low Reynolds numbers (see, for instance, R. kh. Zeytounian, *Asymptotic Modelling of Fluid Flow Phenomena,* Kluwer Academic, Dordrecht, 2002).

Such an approach to Newtonian fluid mechanics would obviously also have a very important pedagogic impact on university education – we have the possibility, from the beginning, to start from the NSF full model and then discuss various partial models, which can be deduced in a rational way when the main parameters (the Mach, Reynolds, Prandtl, etc. numbers) in the NSF full model take values that correspond to various limiting processes. However, in this book, only a part of the puzzle has been reconstituted, in the framework of the NSF equations, and this reinforces the statuts of the NSF equations as an 'unified' system of equations of Newtonian fluid dynamics.

The various results described in this book were obtained during the last thirty years and have been published in detail for the first time here.

Index

Lecture Notes in Physics

For information about earlier volumes please contact your bookseller or Springer LNP Online archive: springerlink.com

Vol.627: S. G. Karshenboim, V. B. Smirnov (Eds.), Precision Physics of Simple Atomic Systems

Vol.628: R. Narayanan, D. Schwabe (Eds.), Interfacial Fluid Dynamics and Transport Processes

Vol.629: U.-G. Meißner, W. Plessas (Eds.), Lectures on Flavor Physics

Vol.630: T. Brandes, S. Kettemann (Eds.), Anderson Localization and Its Ramifications

Vol.631: D. J. W. Giulini, C. Kiefer, C. Lämmerzahl (Eds.), Quantum Gravity, From Theory to Experimental Search

Vol.632: A. M. Greco (Ed.), Direct and Inverse Methods in Nonlinear Evolution Equations

Vol.633: H.-T. Elze (Ed.), Decoherence and Entropy in Complex Systems, Based on Selected Lectures from DICE 2002

Vol.634: R. Haberlandt, D. Michel, A. Pöppl, R. Stannarius (Eds.), Molecules in Interaction with Surfaces and Interfaces

Vol.635: D. Alloin, W. Gieren (Eds.), Stellar Candles for the Extragalactic Distance Scale

Vol.636: R. Livi, A. Vulpiani (Eds.), The Kolmogorov Legacy in Physics, A Century of Turbulence and Complexity

Vol.637: I. Müller, P. Strehlow, Rubber and Rubber Balloons, Paradigms of Thermodynamics

Vol.638: Y. Kosmann-Schwarzbach, B. Grammaticos, K. M. Tamizhmani (Eds.), Integrability of Nonlinear Systems

Vol.639: G. Ripka, Dual Superconductor Models of Color Confinement

Vol.640: M. Karttunen, I. Vattulainen, A. Lukkarinen (Eds.), Novel Methods in Soft Matter Simulations

Vol.641: A. Lalazissis, P. Ring, D. Vretenar (Eds.), Extended Density Functionals in Nuclear Structure Physics

Vol.642: W. Hergert, A. Ernst, M. Däne (Eds.), Computational Materials Science

Vol.643: F. Strocchi, Symmetry Breaking

Vol.644: B. Grammaticos, Y. Kosmann-Schwarzbach, T. Tamizhmani (Eds.) Discrete Integrable Systems

Vol.645: U. Schollwöck, J. Richter, D. J. J. Farnell, R. F. Bishop (Eds.), Quantum Magnetism

Vol.646: N. Bretón, J. L. Cervantes-Cota, M. Salgado (Eds.), The Early Universe and Observational Cosmology

Vol.647: D. Blaschke, M. A. Ivanov, T. Mannel (Eds.), Heavy Quark Physics

Vol.648: S. G. Karshenboim, E. Peik (Eds.), Astrophysics, Clocks and Fundamental Constants

Vol.649: M. Paris, J. Rehacek (Eds.), Quantum State Estimation

Vol.650: E. Ben-Naim, H. Frauenfelder, Z. Toroczkai (Eds.), Complex Networks

Vol.651: J. S. Al-Khalili, E. Roeckl (Eds.), The Euroschool Lectures of Physics with Exotic Beams, Vol.I

Vol.652: J. Arias, M. Lozano (Eds.), Exotic Nuclear Physics

Vol.653: E. Papantonoupoulos (Ed.), The Physics of the Early Universe

Vol.654: G. Cassinelli, A. Levrero, E. de Vito, P. J. Lahti (Eds.), Theory and Appplication to the Galileo Group

Vol.655: M. Shillor, M. Sofonea, J. J. Telega, Models and Analysis of Quasistatic Contact

Vol.656: K. Scherer, H. Fichtner, B. Heber, U. Mall (Eds.), Space Weather

Vol.657: J. Gemmer, M. Michel, G. Mahler (Eds.), Quantum Thermodynamics

Vol.658: K. Busch, A. Powell, C. Röthig, G. Schön, J. Weissmüller (Eds.), Functional Nanostructures

Vol.659: E. Bick, F. D. Steffen (Eds.), Topology and Geometry in Physics

Vol.660: A. N. Gorban, I. V. Karlin, Invariant Manifolds for Physical and Chemical Kinetics

Vol.661: N. Akhmediev, A. Ankiewicz (Eds.) Dissipative Solitons

Vol.662: U. Carow-Watamura, Y. Maeda, S. Watamura (Eds.), Quantum Field Theory and Noncommutative Geometry

Vol.663: A. Kalloniatis, D. Leinweber, A. Williams (Eds.), Lattice Hadron Physics

Vol.664: R. Wielebinski, R. Beck (Eds.), Cosmic Magnetic Fields

Vol.665: V. Martinez (Ed.), Data Analysis in Cosmology

Vol.666: D. Britz, Digital Simulation in Electrochemistry

Vol.667: W. D. Heiss (Ed.), Quantum Dots: a Doorway to Nanoscale Physics

Vol.668: H. Ocampo, S. Paycha, A. Vargas (Eds.), Geometric and Topological Methods for Quantum Field Theory

Vol.669: G. Amelino-Camelia, J. Kowalski-Glikman (Eds.), Planck Scale Effects in Astrophysics and Cosmology

Vol.670: A. Dinklage, G. Marx, T. Klinger, L. Schweikhard (Eds.), Plasma Physics

Vol.671: J.-R. Chazottes, B. Fernandez (Eds.), Dynamics of Coupled Map Lattices and of Related Spatially Extended Systems

Vol.672: R. Kh. Zeytounian, Topics in Hyposonic Flow Theory

Printed in the United States
143494LV00005B/20/P